PRECISION INDUSTRIAL LUBRICATION

A TECHNICAL FRAMEWORK
FOR INCREASED EQUIPMENT RELIABILITY

MICHAEL D. HOLLOWAY

Industrial Press

Industrial Press, Inc.

1 Chestnut Street
South Norwalk, Connecticut 06854
Phone: 203-956-5593
Toll-Free in USA: 888-528-7852
Email: info@industrialpress.com

Author: Michael D. Holloway
Title: Precision Industrial Lubrication: A Technical Framework for Increased Equipment Reliability
Library of Congress Control Number: 2025944927

© by Industrial Press, Inc.
All rights reserved. Published in 2026.
Printed in the United States of America.

ISBN (print):   978-0-8311-3701-4
ISBN (ePUB):   978-0-8311-9677-6
ISBN (eMOBI): 978-0-8311-9678-3
ISBN (ePDF):   978-0-8311-9676-9

Publisher/Editorial Director: Judy Bass
Copy Editor: Judy Duguid
Compositor: Patricia Wallenburg, TypeWriting
Proofreader: David Johnstone
Indexer: WordCo. Indexing Services
Cover Designer: Jeff Weeks

books.industrialpress.com
ebooks.industrialpress.com
1 2 3 4 5 6 7 8 9 10

CONTENTS

PREFACE

In engineering, friction is a force we quantify. In chemistry, it's a source of heat, wear, and reaction. In philosophy, at least in the Buddhist tradition, friction is the tension between intention and action, self and ego, awareness and impulse.

I've spent my career analyzing failures not just in machines, but in how people interact with them. Bearings that seize. Pumps that cavitate. Gearboxes that die young. When you trace these failures to their roots, the technical causes might be lubrication starvation, contamination, or fatigue. But the true origin often lies in something less tangible: Someone was in a hurry. A step was skipped. A signal was ignored. Maintenance was performed out of habit, not out of presence.

That's what I mean by Buddhist friction.

It's not mystical. It's practical. It's the kind of friction that arises when we let intention slip and autopilot take over. When a technician adds grease because "that's what we always do," without checking condition or load trends. When a shaft is aligned "well enough" because the dial indicators are close enough to lunchtime. These moments introduce wear not just on components, but also on trust, discipline, and reliability culture.

Buddhism teaches that suffering stems from clinging, aversion, and delusion. Engineering failures follow a similar arc. We cling to shortcuts. We avoid discipline. We delude ourselves into thinking "goodenough" is good engineering. But machines don't lie. They simply respond.

Mindful maintenance, like mindful living, is rooted in awareness. It's aligning a coupling with full attention, not because the spec says so, but because we understand that misalignment means vibration, heat, inefficiency, and eventually, failure. It's understanding that even routine tasks carry consequences. Torque, cleanliness, fit; they all matter.

Over the years, I've seen brilliant engineers fail because they operated without presence. And I've seen maintenance techs with grease on their knuckles carry out work with the kind of intentionality that rivals that of any laboratory chemist or design engineer.

Reducing friction in machines starts with reducing friction in how we think about them. When we bring mindfulness into maintenance, when we pay attention, we prevent not just mechanical failure but the cultural fatigue that comes with always reacting, always fixing, always wondering why the same failures keep coming back.

The machines remember. So should we.

The Seven-Fold Path of Principles of Mindful Maintenance

➤ **Be Present at the Point of Contact**
Whether it's a torque wrench, grease gun, or sensors, know what you're doing, why you're doing it, and what might happen if you don't.

➤ **Don't Trust Habit, Trust Condition**
Replace interval-based habits with evidence-based actions. Use data to decide, not routine.

➤ **Treat Every Component Like It Has a Memory**
Because it does. Misalignment, contamination, and poor lubrication leave physical traces and future consequences.

➤ **Clean Before You Measure**
Surface condition is signal quality. Dirt is noise. Don't compromise your inputs.

➤ **Respect the Small Things**
Thread engagement, washer orientation, torque sequence; these are the granular details where failure often begins.

➤ **Finish What You Start**
Don't leave deferred fixes, skipped calibrations, or mystery alarms for the next shift. Reliability is a baton, not a solo act.

➤ **If You Can't Explain It, You Can't Improve It**
Every technician should be able to describe what they did, what changed, and why it matters. Clarity creates culture.

INTRODUCTION

*P*recision Industrial Lubrication provides a technical framework for implementing lubrication reliability programs in industrial environments. The content is structured to provide baseline lubrication knowledge to support practitioners who are responsible for system uptime, performance, and life-cycle optimization. The following are considered:

LUBRICATION IS AN ENGINEERING FUNCTION

The first thing to know is that lubrication is not a secondary maintenance task. It is a core engineering function tied directly to mechanical reliability, asset integrity, and operational efficiency. Mismanaged lubrication is a primary contributor to equipment failure. Proper application is strategic, not reactive.

FOCUS AREAS

This manual focuses on several major areas:

> **Lubricant selection.** This area addresses the alignment of lubricant characteristics with equipment duty cycles, environmental conditions, and load profiles. This includes base oil selection, additive package evaluation, formulation for electrified powertrains, and compliance with the original equipment manufacturer's and environmental

standards. Data-informed decision-making replaces spec-sheet guessing. Performance-based selection tools and condition-based monitoring support dynamic optimization.

➤ **Lubrication equipment.** Here the focus is on hardware and system selection for precise lubricant delivery. Topics include trade-offs between manual and automated methods, details of centralized systems, and IIoT-enabled solutions for flow control and delivery verification.

➤ **Contamination control.** This area details the specification and deployment of filters, desiccant breathers, and off-line conditioning systems. It highlights the cost impact of particle and moisture ingress, and it enforces the link between lubricant cleanliness and component service life.

➤ **Supplier and service management.** Standards for engaging third-party lubricant suppliers and service contractors are defined, and criteria for service-level agreements, enforceable key performance indicators, and audit procedures are discussed. In addition, the importance of treating external resources as extensions of internal reliability programs, not transactional vendors, is stressed.

➤ **Preventive maintenance and work order integration.** The focus here is on providing structure for incorporating lubrication into preventive maintenance. This area outlines how lubrication tasks integrate into computerized maintenance management systems, emphasizing frequency control, routing logic, and work verification. As well, the focus remains on functional reliability, not calendar compliance.

➤ **Procedures, staffing, and training.** Documentation standards are specified for lubrication tasks. Manpower planning, staffing ratios, and technician qualification requirements are discussed in detail. In addition, certification paths and continuous training requirements are outlined, and operating procedures are standardized across equipment classes and environments.

➤ **Lubrication information systems.** This area covers architecture and integration of lubrication data into plantwide systems. It emphasizes system compatibility with computerized maintenance management systems and enterprise resource planning platforms. It defines metrics for data collection, analysis, and decision-making. Focus areas include:

- Oil analysis coordination
- Failure mode and effects analysis and root cause failure analysis application

- ➤ Development of key performance indicators for lubrication performance
- ➤ Exception reporting and escalation protocols
- ➤ **Supporting topics.** Additional content addresses:
  - New equipment specifications
  - Lubricant storage, handling, and dispensing
  - Consumption tracking and conservation initiatives
  - Regulatory and warranty compliance
- ➤ **Appendixes.** You can find the Appendixes online at Precisionindustriallubrication.com. The Appendixes include lubrication and failure analysis basics, common failure mechanisms, and reference tables for lubricant applications by equipment class and operating condition.

This manual is engineered for practitioners responsible for mechanical reliability. The focus is on execution, performance tracking, and program sustainability. The idea behind this book is to have a manual that will have to be replaced because it will be used so often the binding will fail!

I hope it becomes an indispensable part of your toolbox and provides excellent guidance on the long road to superior equipment reliability.

MDH

ACKNOWLEDGMENTS

want to thank my past employers and exes for helping me truly understand the influence of friction, wear, and contamination—and for providing clarity on what it means to have proper separation at all times.

I want to thank my partner for teaching me that no one ever regretted good lubrication.

And last, I want to acknowledge and thank Judy Bass and Patricia Wallenburg for their belief, encouragement, and hard work. Without them, this work would not become definitive.

CHAPTER 1

SELECTION OF LUBRICANTS

THE SCIENCE OF FIT: MATCHING LUBRICANTS TO OPERATIONAL REALITIES

Lubricant selection has traditionally focused on base viscosity and OEM (original equipment manufacturer) recommendations, but modern operational contexts demand a much more nuanced approach. The "science of fit" involves matching lubricant performance characteristics to the actual operating realities of the machinery, including ambient conditions, load variations, contamination risks, and duty cycles.

Fit-for-purpose lubrication starts with a clear operational profile. For example, high-speed bearings in clean, climate-controlled facilities require low-viscosity synthetic oils with exceptional oxidative stability, while heavily loaded gearboxes in dusty environments may call for high-viscosity mineral oils fortified with solid or semisolid additives. Beyond mechanical factors, one must consider organizational variables like maintenance intervals, operator skill levels, and availability of sampling programs.

Data collection is critical in this science. Thermography, vibration analysis, and oil condition monitoring are used not only for doing diagnostics but also for refining the lubricant selection process. Historical failure data is also useful, as it reveals systemic weaknesses that may require changes in lubricant type or formulation. The objective is a tailored, predictive, and adaptive approach that evolves alongside the operational life cycle of the asset.

Beyond Viscosity and OEM Tables

Lubricant selection has long defaulted to viscosity grades and OEM specifications. In many plants, that's still the starting point, and unfortunately also the end point. The problem with this method is that it doesn't consider the true operating environment or mechanical realities of the equipment. A pump operating under intermittent thermal load in a coastal facility will not thrive on the same oil that's appropriate for a continuous-duty motor in a climate-controlled lab. OEM tables assume textbook conditions. Field conditions are rarely textbook.

A fit-for-purpose lubricant is not selected from a chart. It's engineered into the process based on measurable, contextual data. This includes not just equipment type and OEM tolerances, but the actual use case, ambient exposure, failure history, and the reliability maturity of the maintenance organization.

Defining the Operational Profile

Start with a machinery operating context map. Four variables form the baseline: speed, temperature, load, and duty cycle.

High-speed machinery, such as electric motors or high-speed centrifuges, places emphasis on shear stability and low internal friction. For these, low-viscosity synthetic base stocks with high oxidative resistance are required. Low-speed, high-load systems, like conveyors or extruders, require heavier viscosities with pressure-resistant additive packages, such as EP (extreme pressure) or solid boundary additives.

Environmental conditions are often overlooked. Dust and grit in mining or cement applications require lubricants with contamination resistance and separation stability. Moisture in washdown areas demands water-separating or hydrophobic formulations. Temperature fluctuations affect both viscosity stability and additive depletion rates. Chemical vapors in process plants can catalyze degradation or accelerate additive consumption, especially for ester-based synthetics or Group III oils without robust antioxidant packages.

Operational stability must be analyzed. Constant operation suggests thermal equilibrium, which allows lubricant temperatures to plateau. Cyclic operations introduce stress spikes and fatigue that can degrade lubricants through oxidation, cavitation, or foaming. Transient duty cycles, such as in robotic or servo systems, need formulations that stabilize rapidly without compressibility issues or temperature-induced viscosity collapse.

Mechanical Realities and Lubricant Demands

Lubrication performance is dictated by component design and loading regime.

High-speed, low-load equipment requires shear-stable, thermally resistant synthetics. ISO (International Organization for Standardization) VG 22 or 32 oils with high viscosity index (VI) are common. These must resist thinning at elevated temperatures while maintaining film strength at start-up.

Low-speed, high-torque systems, such as gearboxes, extruders, and open gears, require thick-film-forming oils or greases, often fortified with solid lubricants like MoS_2 or graphite. These slow-moving parts operate under boundary or mixed lubrication regimes for much of the cycle. Base oil viscosity and additive film strength become critical. EP additives must be selected with awareness of copper compatibility, depending on metallurgy.

Clean-room environments necessitate low-volatility, nontoxic lubricants with minimal residue and zero contamination potential. PFPEs (perfluoropolyethers), PAOs (polyalphaolefins), or NSF H1 synthetics dominate here. By contrast, heavy industry, mining, pulp and paper, and petrochemicals prioritize load-carrying capacity, resistance to abrasive ingress, and long drain intervals. Oils here need enhanced detergent/dispersant systems, oxidation stability, and often tackiness to remain in place.

Application-Specific Examples

- ➤ **Bearings.** The lubricant must accommodate the dN factor, housing geometry, and relube interval. The grease thickener type (e.g., lithium complex, polyurea) must match thermal and chemical exposures.
- ➤ **Gears.** The API GL classification isn't sufficient. Spur, helical, and worm gears each place different demands on the lubricant film, shear rate, and EP performance.
- ➤ **Hydraulics.** Anti-wear (AW) hydraulics fluids (ISO 32–68) need cleanliness, oxidation resistance, and controlled air release. Water tolerance and thermal stability are critical in high-cycling actuators.
- ➤ **Compressors.** Air, ammonia, or CO_2 compressors require low volatility, resistance to chemical attack, and clean burn-off characteristics. Oil carryover into downstream systems must also be considered.

Base oil selection drives core behavior. Mineral oils in Groups I–III have variable oxidative and thermal stability; synthetics (Groups IV–V) offer superior thermal range and longer life. The viscosity index indicates temperature-dependent behavior; high VI oils resist thinning at high temp and thickening at low temp. Additive systems, including anti-wear, antioxidants, detergents, dispersants, demulsifiers, and tackifiers, must be tuned to match the load, speed, and contaminant ingress profile.

The following are initial internal audit questions designed to assess whether lubrication practices align with the mechanical realities and application-specific demands described in your content. These questions are grouped by core principle or equipment type and aim to identify gaps in lubricant selection, application, and maintenance strategy.

General Mechanical and Lubricant Compatibility

- ➤ How is the lubricant selected to match the specific load and speed conditions of each component?
- ➤ What criteria are used to determine viscosity grade (e.g., ISO VG 22 versus 320) for each application?
- ➤ Is there documentation showing that lubricant selection accounts for boundary, mixed, or hydrodynamic regimes?
- ➤ How is compatibility with metallurgy (e.g., copper components) verified for EP additives?
- ➤ Are lubricant properties (VI, volatility, oxidative stability) reviewed regularly as operating conditions evolve?

High-Speed, Low-Load Applications

- ➤ Are synthetics used in high-speed applications shear-stable and thermally resistant?
- ➤ What is the minimum acceptable viscosity index for oils used in high-speed spindles or turbines?
- ➤ Are there procedures in place to confirm that film strength is sufficient at both start-up and peak operating temperatures?

Low-Speed, High-Load Systems

- ➤ Are solid lubricants (e.g., MoS_2, graphite) used appropriately in slow-moving, high-torque applications?
- ➤ How is base oil viscosity matched to the torque and loading characteristics of the gearbox or extruder?
- ➤ Are additives selected based on compatibility with the metallurgy of the system?

Cleanroom and Contamination-Sensitive Applications

➤ Are all lubricants used in cleanrooms verified to be low volatility and NSF H1 or equivalent?

➤ What controls are in place to prevent contamination from lubricants in sensitive manufacturing environments?

➤ Are PFPEs or PAOs evaluated against operational temperature and load requirements?

Heavy Industrial Applications (Mining, Petrochem, Pulp and Paper)

➤ How is lubricant tackiness evaluated and verified to prevent sling-off or washout?

➤ Are drain intervals extended based on lubricant performance testing or OEM recommendations?

Application-Specific Audits

Bearings

➤ How is the dN factor used to determine appropriate lubricant viscosity and type?

➤ Is grease thickener chemistry matched to the thermal and chemical environment of the bearing?

➤ How is the relubrication interval calculated and validated?

Gears

➤ Does the lubricant selection go beyond API GL classification to account for gear type (spur, helical, worm)?

➤ How is shear stability tested or verified in gear oil applications?

➤ Are EP and anti-scuff additives optimized for the specific gear geometry and contact pressure?

Hydraulics

➤ Are anti-wear hydraulic fluids selected based on actuator cycling rates and system cleanliness requirements?

➤ Is fluid water tolerance verified through periodic Karl Fischer or demulsibility testing?

➤ Are thermal stability and air-release properties monitored regularly?

Compressors

➤ Is oil carryover into downstream systems measured or mitigated?

➤ Are lubricant volatility and clean burn-off characteristics documented for each gas type (air, ammonia, CO_2)?

➤ What protections are in place against chemical degradation of the lubricant from process gas?

Base Oil and Additive Audit

➤ Is the base oil (Groups I–V) documented for all lubricants in use?

➤ Are base oil thermal and oxidative stabilities matched to the duty cycle and environment?

➤ How are additive packages (e.g., antioxidants, detergents, tackifiers) verified for appropriateness?

➤ Is VI tracked as part of lubricant performance evaluation in variable-temperature environments?

ORGANIZATIONAL AND HUMAN FACTORS

Lubricant performance isn't just chemistry; it's also execution. The best product will fail in a plant with poor maintenance culture or inconsistent practices.

Skill level dictates lubricant selection. In facilities with limited training or high turnover, complex synthetic blends or lubricants with tight contamination tolerances pose a risk. Simpler, robust formulations with broad operational windows are more appropriate. Centralized lubrication systems or color-coded manual greasing programs reduce operator-induced variability, but only when rigorously implemented.

Oil change intervals often reflect budget pressure rather than engineering requirements. When drain intervals are extended to reduce cost without data to justify the decision, lubricant degradation and varnish formation become systemic. In these environments, high-oxidation-resistance fluids with effective dispersants can mitigate the damage, but this is a patch, not a solution.

Condition monitoring availability changes everything. Plants with no oil analysis capability must overlubricate and overmaintain to stay safe. Plants with reliable sampling protocols can move toward true condition-based intervals and precision lubricant matching. Spotty, inconsistent, or improperly interpreted data leads to worse outcomes than no data at all.

Training in lubricant handling, especially storage, transfer, and contamination control, is often insufficient. Grease guns are reused across incompatible chemistries. Oil drums are stored outdoors or left open. These small acts compromise the lubricant long before it reaches the machine. Product labeling, standard operating procedures (SOPs) for handling lubricants, and cross-contamination prevention are low-cost fixes that deliver immediate reliability gains.

DATA-INFORMED LUBRICANT MATCHING

Data closes the gap between assumed performance and actual requirements.

Thermography identifies hotspots, indicating poor heat transfer, fluid starvation, or friction buildup. This data helps correlate load zones with oil film behavior, guiding viscosity selection or additive fortification.

Vibration analysis detects changes in bearing condition, load alignment, and resonance, all of which can point to lubrication breakdown or film collapse. Recurrent high-frequency signals in the ultrasonic band often indicate microwelds or asperity contact from lubricant failure.

Tribometry provides coefficient of friction, film strength, and wear scar dimensions under controlled test conditions. While lab-based, these values can model field behavior when matched with operating temperature and load inputs. Tribological data informs boundary versus hydrodynamic regimes, guiding additive selection.

Wear particle analysis reveals the real-time effectiveness of lubrication. A rising trend in ferrous wear, despite constant loads and speeds, signals lubricant starvation, contamination, or inadequate EP performance. The shape of a particle matters, such as cutting wear versus fatigue versus corrosion, all point to distinct root causes which can be addressed.

Real-world duty cycles and load profiles, pulled from SCADA (supervisory control and data acquisition), PLCs (programmable logic controllers), or historical data, allow engineers to overlay actual operating behavior against assumed norms. Many "intermittent" systems turn out to run near-continuously under partial load, an environment that stresses thermal stability more than originally spec'd.

Historical failure data is the engineer's most reliable ally. Tracking repeat failures on similar components shows patterns of lubricant underperformance or incompatibility. Whether it's sludge in hydraulic reservoirs, varnish on servo valves, or accelerated gearbox pitting, the failure modes tell the truth. Lubricant changes should not be reactive; they should be guided by this empirical evidence.

CASE STUDIES IN MISFIT AND CORRECTION

Example 1: A food processing facility used H1 white mineral oil in a high-load gearbox operating near ovens. Failures occurred every 9 months. Thermography revealed thermal spikes beyond 110°C. Switching to a high-VI synthetic ester with an EP additive extended life to 3 years. The root cause was thermal break-down and lubricant thinning.

Example 2: A wind farm reported generator bearing wear despite using synthetic grease. Vibration data indicated fluctuating load spikes. The grease, although high performance, was too soft for the dynamic loading. Switching to a stiffer, polyurea-based grease with tackifier additives stabilized the vibration profile and reduced iron levels by 60%.

Example 3: A cement plant experienced premature pump failures. Oil analysis showed water ingress and high ISO particle counts. Investigation revealed poor oil storage and transfer methods, not lubricant quality. Implementing sealed containers and desiccant breathers eliminated the problem. Same oil, different handling, radically different result.

Example 4: A steel mill using ISO 460 gear oil in a reversing mill gear reducer reported foam and thermal excursions. Foaming caused pressure surges and air entrainment. Switching to an antifoam fortified blend with better air-release properties solved the issue. The misfit wasn't in viscosity but in formulation stability under cycling shear.

Each correction was quantifiable—reduced unplanned downtime, lower lubricant consumption, and fewer maintenance interventions. More importantly, these changes aligned lubricant performance with machine reality rather than legacy assumptions.

ADAPTIVE STRATEGIES FOR DYNAMIC OPERATIONS

Aging assets introduce shifting baselines. Internal clearances widen, surface finishes degrade, and thermal conductivity falls. These changes alter the lubrication regime. A gearbox that once operated in hydrodynamic mode may now function under mixed or boundary conditions. Lubricants must evolve accordingly. Heavier base oils, friction modifiers, or solid EP additives may become necessary. Aging pumps with shaft wear or misalignment may require viscosity adjustments to maintain film integrity.

Production variability adds further complexity. In facilities with frequent start/stop cycles, seasonal spikes, or on-demand batch runs, lubricants must tolerate thermal cycling, low-temperature start-up, and inconsistent loads. For example, hydraulic oils used in outdoor forklifts require different cold-flow properties than those used indoors under constant conditions. Summer-to-winter viscosity shifts can push borderline fluids into failure. In high-latitude or desert climates, seasonal requalification of lubricants may be necessary to maintain reliability.

Custom blending offers targeted performance where COTS (commercial off-the-shelf) products fall short. Specialty formulations may include tackifiers for adhesion in high-vibration zones, ashless AW packages for systems with sensitive valves, or ester-based synthetics for high thermal resistance in tight spaces. Customization becomes cost-effective when applied to high-value assets, extended intervals, or systems with known reliability constraints. This is particularly relevant in industries like wind power, aviation, and semiconductor manufacturing.

Engineered lubricants, formulated in collaboration with suppliers, allow adjustment of pour point, volatility, base oil polarity, and additive balance to fit specific wear mechanisms, contaminant risks, or application demands. These are not shelf solutions; they are engineered into reliability programs based on measured field behavior.

THE ROLE OF DIGITALIZATION AND AI

Smart sensors embedded in rotating equipment can monitor temperature, pressure, flow rate, dielectric constant, ferrous debris, and water contamination in real time. This data enables dynamic lubricant evaluation rather than fixed-interval analysis. Algorithms compare live data against defined thresholds, flagging conditions that warrant a formulation change or contamination control intervention.

Machine learning models trained on historical performance data can detect subtle patterns that indicate misfit. These models incorporate factors like ambient temperature, load spectrum, uptime percentage, oil degradation curves, and wear rates. Output includes predictive lubricant suitability scores, risk forecasts, and proactive formulation suggestions.

Digital twins replicate a machine's mechanical and thermal behavior, simulating how different lubricants perform under modeled duty cycles. These simulations accelerate decision-making, reducing the trial-and-error period during lubricant transitions. High-fidelity models also predict how changes in operating conditions, like increased cycle time or torque fluctuation, affect lubricant stability.

Integrating AI into lubrication management shifts the process from calendar-driven to condition-driven, from generic selection to precision engineering. The lubricant becomes an engineered control variable, no different than a bearing preload or a seal specification.

From Reactive Selection to Predictive Fit

Lubricant selection isn't static. It's a fluid decision process embedded within the operational life cycle of the asset. What works at commissioning may no longer apply after 10,000 hours of load cycling, thermal drift, and minor misalignments. Fit must be reassessed continually.

Precision lubrication drives measurable returns: fewer failures, longer component life, reduced lubricant consumption, tighter control of contamination, shorter MTTR (mean time to repair), and better asset availability. Each of these metrics links directly to bottom-line outcomes, OEE (overall equipment effectiveness), energy efficiency, and cost per unit of production.

Embedding the science of fit requires cultural integration. Lubrication becomes an engineering parameter, not a consumable. Teams must move past OEM tables and embrace a data-driven, asset-specific approach. Fit-for-purpose lubrication isn't a best practice. It's a baseline expectation in any operation seeking reliability and performance.

Table 1.1 is a structured "how-to" guide for method development and implementation, tailored for lubrication engineering. The steps in the table follow a scientific and systems-oriented approach while being generalizable across industries.

HOLISTIC CRITERIA FOR LUBRICANT SELECTION

Viscosity, while fundamental, is only the starting point in lubricant selection. Holistic evaluation includes base oil composition, additive package chemistry, lubricant compatibility with seals and materials, volatility, biodegradability, and performance under thermal stress.

Modern selection practices involve detailed specification sheets and test data, including viscosity index, pour point, flash point, and oxidation stability. ISO, ASTM, and SAE standards serve as benchmarks but are not sufficient on their own. For example, two lubricants with the same ISO viscosity grade may behave very differently under high-shear conditions.

Load-bearing capacity (measured through FZG or Timken tests), anti-wear and extreme pressure properties, demulsibility, and corrosion protection are all essential parameters. Selection must also account for the system's operating temperature range, exposure to water or chemicals, and load variability.

TABLE 1.1 Guide for Method Development and Implementation

STEP	DESCRIPTION
1. Define the objective	Determine the goal of the lubricant method (e.g., extend the drain interval; verify viscosity under load; detect contamination). Establish performance metrics and failure thresholds.
2. Gather background information	Research equipment specifications, OEM lubricant requirements, and operating conditions (speed, load, temperature, environment). Review ASTM/ISO methods for reference.
3. Define inputs, variables, and constraints	Identify variables such as viscosity, TAN (total acid number), oxidation, cleanliness (ISO 4406), additive depletion, and contamination. List necessary instrumentation (e.g., viscometer, FTIR [Fourier transform infrared spectroscopy], particle counter).
4. Design the method (procedure development)	Develop detailed steps for sampling, analysis, interpretation, and lubricant changeout. Define specific acceptance/rejection criteria (e.g., viscosity ±10%, ISO 4406 code limits).
5. Validate the method	Run pilot tests on equipment with a known lubricant condition. Use repeat trials to evaluate reproducibility, sensitivity, and alignment with failure history or OEM specs.
6. Document the method (SOP or protocol)	Write a complete SOP with objectives, method steps, safety considerations, sample handling, data sheets, and pass/fail limits. Ensure compatibility with ISO 9001 or ISO 17025 systems.
7. Train personnel and perform initial rollout	Train technicians on SOP execution. Observe real-world implementation and gather usability feedback. Adjust SOP for clarity or ergonomics.
8. Monitor and evaluate	Establish periodic reviews using KPIs (key performance indicators)—for example, percent of in-spec results, oil life extension. Use trend analysis to detect early signs of failure or method drift.
9. Practice continuous improvement	Update the method based on field results, new lubricant formulations, or equipment design changes. Maintain version control and stakeholder sign-off.

Additionally, compatibility with legacy lubricants and the potential for additive clash must be assessed, especially when switching products. A true holistic evaluation also includes life-cycle considerations, such as drain intervals, reusability, and end-of-life disposal impacts.

The Limits of Viscosity Alone

Viscosity is the most widely recognized property in lubricant selection and serves as the foundational starting point. It defines a lubricant's flow resistance, with kinematic viscosity measuring how a fluid flows under gravity and dynamic viscosity reflecting its resistance under force. ISO and SAE classifications group fluids based on their viscosity at defined temperatures. These categories provide a baseline for identifying whether a fluid can form and maintain an adequate lubricating film under expected operating conditions. However, this parameter alone offers no insight into thermal degradation, oxidation stability, additive chemistry, or material compatibility.

Failures tied to viscosity-only selection are common in field operations. Bearings operating under dynamic or shock loads may experience film collapse even when the lubricant's viscosity is technically correct. In hydraulic systems, varnish accumulation and oxidation by-products often emerge in fluids that meet ISO viscosity grades but lack sufficient thermal and oxidative resilience. Seal degradation can occur when the base oil or additive package is chemically incompatible, regardless of viscosity classification. These failures reveal the limits of relying on viscosity as a proxy for overall performance.

Modern reliability programs require a shift toward performance-based lubricant engineering. This approach incorporates field data, failure history, and operating conditions into the selection process. Empirical testing, such as oxidation stability, wear resistance, and compatibility testing, becomes essential. The goal is to match lubricant behavior with system demands, accounting for temperature variation, load cycling, contaminant ingress, and maintenance capabilities. Viscosity remains a critical attribute, but it must be part of a broader, evidence-driven evaluation framework.

Base Oil Chemistry and Its Influence

Lubricant base oils form the bulk of any formulation and determine a wide range of performance characteristics. API Groups I and II base oils are solvent-refined mineral oils with moderate oxidation resistance and relatively high volatility. Group III base oils, although mineral-derived, are hydrocracked to improve thermal stability and viscosity index. Group IV base oils are PAOs, synthetic hydrocarbons known for their low volatility, high oxidative stability, and excellent performance in extreme temperatures. Group V includes all other base stocks, such as esters, polyalkylene glycols (PAGs), and silicone fluids, each with unique solvency, polarity, and temperature characteristics.

Polarity directly influences a base oil's solvency power and additive compatibility. Polar base oils, such as esters, have excellent detergent and dispersant capacity but may cause seal swelling or degradation. Nonpolar oils, like PAOs, offer low reactivity and thermal resilience but often require cosolvents or solubilizing agents to support additive packages. Solvency affects a lubricant's ability to suspend contaminants, dissolve degradation by-products, and prevent deposit formation in high-temperature applications.

Oxidative stability is another critical factor tied to base oil chemistry. Group I base oils degrade rapidly under cyclic thermal loading, leading to sludge and varnish. Synthetic

Groups IV and V base oils maintain chemical integrity under stress, extending oil life and protecting components in severe service environments. Deposit formation risks can often be traced back to base oil structure and reactivity with oxygen, heat, or metallic surfaces.

Application-specific requirements dictate the optimal base oil selection. High-temperature bearings benefit from synthetic esters or PAOs that resist thermal breakdown. Sealed-for-life systems demand base oils with low volatility and high oxidative stability to reduce maintenance and extend service intervals. Cold-weather applications require high-VI oils with low pour points to ensure start-up flow and pressure stability.

Additive Package Functionality

While the base oil provides the foundation, additive packages determine a lubricant's functional performance. Additives are blended to modify friction, protect against wear, prevent corrosion, control oxidation, manage contaminants, and stabilize fluid behavior under stress. Common additive types include anti-wear agents like zinc dialkyldithiophosphate (ZDDP), EP additives such as sulfur-phosphorus compounds, detergents for neutralizing acids, dispersants for maintaining insolubles in suspension, antioxidants for extending fluid life, tackifiers for adhesion, and demulsifiers for water separation.

Effective formulations require synergy among additives. Overuse of one type can suppress the function of another. High levels of EP additives, for example, may reduce AW performance or contribute to filter plugging. Some friction modifiers destabilize boundary films in systems requiring consistent anti-wear protection. Additive solubility within the base oil is also critical; polar additives may fall out of solution in nonpolar oils, especially under temperature extremes.

Incompatibility between additive packages and system materials can lead to operational failures. Additive residues can cause sludge formation, valve sticking, and filter fouling. Some additives degrade elastomers, leading to premature seal failure, oil leakage, and equipment downtime. Even filters can suffer, with swelling or degradation of filter media from aggressive chemistries.

Additive selection must be guided by machine function and duty profile. Gearboxes subjected to high sliding loads require robust EP packages. Precision hydraulic systems need clean, low-ash fluids with AW additives and antifoam agents. Turbines operate best with oxidation-resistant formulations that minimize deposit formation. Older engines with nonhardened valve trains may require low-detergent oils to avoid stripping protective films from soft metals. Matching the additive profile to operational demands is central to achieving system reliability and oil longevity.

Compatibility Considerations

Chemical compatibility between lubricants and component materials affects system integrity and reliability. Elastomers used in seals and O-rings respond differently to various base stocks and additive chemistries. Ester-based synthetics can soften or swell Viton and polyurethane seals, while nitrile rubber (NBR) may harden or crack depending on additive polarity and aging conditions. Failures here lead to leakage, pressure loss, and contamination ingress.

Beyond seals, lubricants can degrade paint films, gaskets, and adhesives. Aggressive additives or base oils with high solvency may lift coatings or dissolve binders, compromising equipment protection and cleanliness. Filter media can lose structural integrity when exposed to incompatible fluids, increasing bypass frequency or causing filter collapse under load.

Switching lubricants without evaluating additive compatibility introduces risk. Additive clash may result in soap precipitation, sludge formation, or copper corrosion. This is especially problematic when transitioning between high-SAPS (sulfated ash, phosphorus, sulfur) and low-SAPS formulations or when mixing Group I oils with synthetic Group IV or V fluids.

Before transitioning lubricants, conduct compatibility trials. Blend tests under controlled conditions reveal reaction tendencies. Small-scale system trials, supported by oil analysis and filter inspections, provide real-world data. Use complete flush procedures when incompatibility is confirmed, and condition new filters during the transition phase to mitigate risk.

Physical Property Benchmarks

The viscosity index indicates how viscosity changes with temperature. A high VI ensures that the lubricant resists thinning at high temperatures and thickening at low temperatures. This property is crucial in systems subject to wide thermal cycling, such as mobile hydraulics or outdoor gearboxes.

The pour point measures the lowest temperature at which oil remains flowable. Low pour points are essential in cold-start applications, especially in regions with subzero ambient conditions. Pour point depressants are added to improve flow in such environments, reducing start-up wear.

The flash point reflects the lowest temperature at which a lubricant emits flammable vapors. Lubricants with low flash points are more volatile and prone to evaporation or fire hazards. Compressor and turbine oils must meet or exceed minimum flash points to avoid operational risk.

Oxidation and thermal stability determine how long a lubricant can operate before degradation begins. Tests like RPVOT (rotating pressure vessel oxidation test) and PDSC (pressure differential scanning calorimetry) provide accelerated life estimates. Lubricants with high oxidative stability form fewer acids and deposits and less varnish, especially in systems with tight clearances and high temperatures.

Foam formation reduces film strength and increases the risk of cavitation. Foam inhibitors mitigate bubble formation and improve air release. In high-speed gear systems and hydraulic circuits, air entrainment can cause pressure fluctuations and increase wear. Measured air release time reflects the lubricant's ability to release entrapped gases and maintain consistent performance.

Performance Testing Criteria

Mechanical stress tests validate whether a lubricant can withstand real-world loading conditions. FZG testing determines a lubricant's ability to prevent gear scuffing under increasing load stages. Higher load stages indicate better EP additive effectiveness. Timken OK Load quantifies the load at which a rotating bearing fails under boundary lubrication; higher values reflect stronger protective films.

The four-ball wear and weld tests evaluate anti-wear and extreme pressure characteristics. The wear test measures the average scar diameter left on ball bearings under defined conditions. Smaller scars indicate superior protection. The weld point measures the load at which the balls fuse together, reflecting EP performance limits.

Demulsibility is a measure of how well a lubricant separates from water. In environments with high moisture ingress, such as food plants, paper mills, and marine applications, fast separation reduces emulsified wear and protects metal surfaces from corrosion. Demulsibility is typically measured in minutes until full separation.

Corrosion protection is evaluated using standardized methods. ASTM D665 assesses rust inhibition in the presence of distilled or salt water. ASTM D130 uses copper strips to measure corrosive attack by sulfur or phosphorus additives. Passing results in these tests indicate a lubricant's ability to protect ferrous and yellow metals from aggressive environments.

Application-Specific Considerations

Lubricants must be selected to match the specific operating context of the machine. Thermal range is a major factor. Systems that cycle between ambient start-up and elevated operating temperatures need fluids with a high VI to maintain consistent viscosity and pressure generation. Sudden thermal spikes require oxidative resilience and low volatility to prevent breakdown.

Contamination is a leading cause of lubricant degradation. Water ingress accelerates additive depletion, causes rust, and creates emulsions that reduce film integrity. Dust contributes to abrasive wear and filter overloading. Chemical contamination from process fluids demands fluids with strong resistance to acid or solvent attack.

Load patterns affect film formation. Steady-state systems favor hydrodynamic lubrication. Variable loads and shock conditions require lubricants with superior boundary protection and film persistence. Gearboxes and hydraulics operating under fluctuating torque require shear-stable viscosity to prevent collapse during high-stress transitions.

Drain intervals are dictated by sump size, contaminant ingress, and thermal load. Systems with small reservoirs and high cycling rates deplete additives faster and require more frequent changes. Relubrication schedules must reflect field conditions, not calendar intervals.

Shear stability impacts performance in high-speed versus low-speed systems. Gear oils with polymeric thickeners must resist permanent viscosity loss under mechanical stress. High-shear environments, such as hydraulic pumps, turbo compressors, or high-RPM spindles, require fluids that retain film thickness over time.

Life-Cycle and Sustainability Factors

Lubricants must now meet both performance and environmental criteria. Biodegradability is governed by standards like OECD 301B and ISO 15380. Ester- and PAG-based synthetics meet these standards and are preferred in environmentally sensitive areas such as forestry, agriculture, and marine operations.

Reusability through in-service reconditioning can extend fluid life and reduce consumption. Vacuum dehydration, fine filtration, and centrifugal separation remove contaminants and restore performance. Some fluids allow additive replenishment, depending on formulation and supplier support.

Oil aging and additive depletion affect disposal classification. Fluids oxidized or contaminated with heavy metals or solvents may qualify as hazardous waste. Disposal costs increase with contamination level and volume. End-of-life planning should consider waste minimization and compliance with the Resource Conservation Recovery Act (RCRA) and local regulations.

Total cost of ownership includes more than price per gallon. It accounts for drain interval, labor, downtime, parts replacement, and disposal costs. Higher-upfront-cost synthetics may reduce long-term operating expenses by extending component life and reducing unplanned outages.

Standards and Specification Sheets: Tools, Not Final Answers

Standards provide a framework for comparing lubricants, but they do not capture field performance. ISO VG defines viscosity range. SAE J300 applies to engine oils. ASTM and DIN protocols govern testing for wear, oxidation, and corrosion. API and ACEA categories group oils by base type and additive load.

Specification sheets include properties such as viscosity, VI, flash point, and pour point. These are helpful but limited. Sheet values reflect fresh oil under controlled test conditions, not degraded oil under operational stress. Understanding test method designations (e.g., ASTM D445 for kinematic viscosity) is necessary for interpreting values accurately.

Two lubricants with identical ISO VG ratings may perform differently due to additive package, base oil group, or oxidation stability. For example, ISO 68 gear oil and ISO 68 turbine oil serve different functions and contain different chemistries.

Vendor transparency matters. Test reproducibility, batch consistency, and post-sale support influence lubricant reliability. Confirm critical properties through third-party testing or in-house lab validation where operational risk is high. Do not rely solely on marketing claims or data sheet summaries.

CASE STUDIES IN HOLISTIC SELECTION

In practice, the limitations of viscosity-only lubricant selection are evident across industries. A manufacturing facility operating a fleet of mobile hydraulic equipment reported frequent pump cavitation during cold starts. The selected ISO 46 fluid met viscosity requirements but had a high pour point and poor low-temperature flow. Upon switching to a synthetic ISO 46 with a lower pour point and higher VI, start-up pressure stabilized, wear metals in oil analysis decreased, and downtime dropped by 22%.

A textile plant experienced varnish formation in its high-speed bearings despite using a lubricant with the correct viscosity. Investigation revealed that the Group I mineral base oil lacked oxidative stability, resulting in thermal degradation and insoluble formation. Upgrading to a Group IV PAO-based oil with enhanced antioxidants eliminated varnish accumulation. Bearing temperatures dropped, and intervals between maintenance interventions extended from 4 months to over a year.

Additive incompatibility was the root cause of a critical failure in a food processing facility. After switching to an H1-approved synthetic ester-based gear oil, operators reported widespread O-ring swelling and gearbox leakage. The previous lubricant was a PAO with neutral polarity. The ester's higher polarity reacted with the plant's Viton seals, causing them to soften and deform. A return to a PAO-based formulation resolved the issue, and seal life normalized. A post-incident review revealed that compatibility testing had been skipped during procurement.

In a mining operation, frequent gearbox failure in conveyor systems was attributed to lubricant shear-down and metal-to-metal contact. The existing mineral oil had no EP package and was prone to film collapse under heavy shock loads. Transitioning to a high-performance synthetic with shear-stable VI improvers and a robust EP additive system doubled gearbox life. Gear pitting was significantly reduced, and the life of the oil was increased by three-fold, which was verified by oil analysis. The cost of the new lubricant was 3.5 times higher per liter, but total annual savings from reduced failures and labor exceeded $180,000.

A wind turbine fleet applied a predictive lubricant selection model to optimize the oil change schedule. By incorporating thermographic, spectrometric, and ferrographic data, engineers identified early signs of oxidative stress and filter loading. The analysis led to adoption of a custom-blended synthetic tailored to the site's temperature swings and load profiles. Filter life extended by 40%, lubricant consumption dropped by 18%, and gearbox reliability metrics improved across the fleet. The ROI of the formulation switch paid back in less than 6 months.

Holistic lubricant selection consistently delivers operational and economic benefits. When viscosity is just one variable among many, and when formulations are tailored to the machine, environment, and process, uptime increases and failure rates decline. These case studies reflect the practical consequences of aligning lubricants to real-world operating conditions rather than generic specifications.

Table 1.2 summarizes holistic lubricant selection and the how-to steps, pitfalls, and remedies that should be considered.

TABLE 1.2 Holistic Lubricant Selection—How-to Steps, Pitfalls, and Remedies/Best Practices

SELECTION CRITERIA	HOW-TO STEPS	POTENTIAL PITFALL	REMEDY/BEST PRACTICE
Viscosity and VI	Start by defining operating temperature and speed. Select ISO VG grade. Confirm high VI for thermal stability across temperature cycles.	Choosing correct viscosity grade but with poor VI, causing thinning under heat or thickening in cold start.	Select lubricants with high VI (>140 for multiclimate systems). Use VI improvers only when shear stability is confirmed.
Base oil chemistry	Identify the base oil group (I–V) based on thermal load, volatility, and oxidative demands. Match polarity to desired solvency and additive retention.	Selecting a base oil with low oxidative stability or wrong polarity, leading to deposit formation or additive fallout.	Use synthetic base oils (Groups III–V) for severe conditions. Match polarity to additive solubility and material exposure.
Additive package	Match the additive function to the machine role (e.g., AW for hydraulics, EP for gearboxes) Ensure synergy between additives. Cross-reference with OEM specs.	Overloading one additive (e.g., EP) causing antagonism with AW or filter clogging.	Balance additive package with OEM guidance and real-world duty cycles. Avoid mixing incompatible formulations.
Compatibility	Check elastomer, paint, and metal compatibility with both base oil and additive chemistry. Perform blend tests before switching products.	Seal swell, paint stripping, or additive clash when switching lubricants without proper compatibility testing.	Run lab compatibility and field pilot trials. Fully flush when changing base oil type or additive class.
Physical property benchmarks	Confirm VI, pour point, flash point, oxidation stability, and foam resistance through lab data. Use correct ASTM/DIN methods for each property.	Relying on specs without understanding limits of lab testing under ideal conditions.	Use field testing (e.g., FTIR, RULER [Remaining Useful Life Evaluation Routine]) to supplement spec data. Review method numbers on sheets for test context.
Performance testing	Conduct FZG, Timken, four-ball wear, and demulsibility tests for real-world performance. Compare against in-service field data when available.	Using lubricants that pass bench tests but fail under shock, shear, or contaminant conditions.	Verify results with both lab and in-situ trials. Use historical wear data to correlate lab results with reliability.
Application-specific needs	Analyze the thermal profile, load variability, contamination risk, and sump size. Choose a lubricant with proper shear stability and oxidation resistance.	Failing to account for actual load cycling, water ingress, or real-world duty profile.	Use a matrix of environmental, operational, and maintenance inputs. Select based on worst-case exposure, not averages.
Life cycle and sustainability	Select biodegradable base oils for sensitive environments. Assess reusability through reconditioning methods. Plan for disposal costs and fluid life.	Ignoring end-of-life disposal, resulting in hazardous waste or regulatory violations.	Choose products with verified biodegradability or extended service life. Implement condition monitoring for in-service extension.
Specification sheets and standards	Use spec sheets as a baseline. Validate critical data with third-party or in-house tests. Understand method codes and test limitations.	Overtrusting vendor data without validation, leading to field failures despite "matching" specs.	Request test reports and validation data. Audit supplier labs or use independent verification before full deployment.

BASE OILS AND ADDITIVES

Base oils form the backbone of any lubricant, with additive packages acting as performance modifiers. Compatibility between these components is critical to ensuring performance, reliability, and longevity.

API Groups I–V base stocks each offer different properties in solvency, volatility, oxidative stability, and temperature behavior. For instance, Group IV PAOs provide superior oxidative stability but poor additive solubility, often requiring cosolvents or esters for balance.

Additive packages vary widely based on application: detergents, dispersants, antioxidants, anti-wear agents, EP additives, VI improvers, and friction modifiers are combined based on need. Compatibility matrices are developed to ensure that additive packages perform as expected in the chosen base oil and application. This matrix-driven approach is critical in preventing additive dropout, varnish formation, or unexpected deposit behavior.

When formulating or selecting lubricants for multipurpose or extreme applications, attention must be given to additive synergy or antagonism. For example, certain friction modifiers can degrade seal materials or interfere with anti-wear chemistries. Similarly, ZDDPs, widely used for anti-wear properties, can harm catalytic converters in engine oils or interfere with certain yellow metals in hydraulic systems.

Base Oils and Additives—A Strategic Compatibility Matrix

Base oils carry the structural load of the lubricant. Additives deliver the function. The relationship between the two determines whether the lubricant performs or fails. Base oil polarity dictates additive solubility. Thermal load defines oxidative demand. Mechanical stress and environmental exposure drive chemical stability requirements. Poor pairing between the base oil and additive package leads to fallout, varnish, or deposit formation. Matrix-based formulation methods are used to ensure chemical alignment and functional synergy across operational extremes.

Group I base oils are solvent-refined mineral oils with high aromatic content and low saturation. They exhibit good solvency but oxidize rapidly and perform poorly in high-temperature environments. Group II oils are hydrotreated, offering improved oxidative stability and lower volatility compared with Group I, but they still carry limited performance under extreme thermal cycling. Group III oils are hydrocracked and often marketed as synthetics. They have high VI, better oxidative resistance, and lower sulfur, making them suitable for more demanding environments than Groups I or II.

Group IV base stocks are PAOs, built from synthetic alpha-olefin monomers. They offer low volatility, excellent oxidative resistance, and consistent viscosity across temperature ranges. Their primary limitation is poor solvency. Additives do not dissolve well in PAOs without help. Cosolvents, usually esters or alkylated aromatics, are often required. Group V includes all other synthetics: esters, polyalkylene glycols, silicone fluids, and other specialty chemistries. These are used when conventional base oils can't meet application-specific challenges like biodegradability, extreme temperature range, or high-polarity needs.

Base oil solvency determines how well the lubricant suspends additives, contaminants, and degradation products. Esters and PAGs, being polar, are strong solvents and support stable additive dispersal even under oxidative stress. PAOs, being nonpolar, have poor solvency and risk additive fallout without polarity compensation. Group I oils, while solvent-rich, degrade quickly and generate sludge and varnish under heat and oxygen.

Volatility affects oil loss and flash risk. Groups I and II base oils exhibit higher evaporation and contribute to fluid consumption in high-temp applications. Groups IV and V oils hold volatility to a minimum, preserving volume and extending service intervals. Flash point and volatility also affect flammability, particularly in compressors, turbines, and other systems near ignition sources.

Oxidative and thermal stability determine whether the lubricant remains chemically active or breaks down into acids, sludge, and varnish. PAOs and esters resist oxidation and maintain functional integrity in high-load, high-heat systems. Group I oils fail rapidly under these conditions, forming varnish that gums valves, sticks rings, and clogs fine filters. Groups II and III oils provide middle-ground performance and are commonly used in extended-drain mineral oil applications when supported by antioxidant packages.

Cold flow and pour point performance are critical for equipment starting in subzero temperatures. PAOs offer superior low-temp fluidity and low pour points, enabling rapid lubrication during start-up. Groups I and II oils often require pour point depressants and may still thicken beyond acceptable limits in cold environments. Esters and some PAGs maintain good cold-start properties but can bring compatibility issues if improperly matched with seals or legacy systems.

Each base oil group brings specific limitations. PAOs must be chemically balanced with cobase stocks to hold additives in solution and prevent filter plugging under thermal stress. Esters may react with certain elastomers, causing swelling, softening, or embrittlement in Viton, NBR, and polyurethane seals. Group I base oils break down rapidly under cyclic heat load, and their use should be restricted to low-stress, legacy, or controlled-condition systems where thermal oxidation is minimal.

No base oil is universally optimal. Selection requires alignment with additive chemistry, material compatibility, and mechanical duty. The matrix model, mapping base oil characteristics against additive functionality, prevents unintended interactions and ensures reliability under load, heat, and time. Field performance is driven by chemical balance, not just viscosity or label designation.

Table 1.3 summarizes selection criteria for the matrix model along with how-to steps, pitfalls, and remedies.

Additive Package Categories and Their Functions

Additives modify the base oil's behavior and deliver targeted performance in wear protection, oxidation resistance, contaminant management, and viscosity stability. Each additive class serves a functional role, and the final formulation is defined by the sum of their contributions.

Detergents clean internal surfaces by neutralizing acids and removing deposits. Common metallic detergents include calcium sulfonates, magnesium phenates, and overbased sodium compounds. Dispersants suspend insolubles such as soot, varnish precursors, and sludge to prevent agglomeration and deposit formation. Polyisobutylene succinimide (PIB-SI) is a widely used ashless dispersant.

Antioxidants delay fluid degradation by scavenging free radicals and decomposing peroxides. Phenolic antioxidants operate at lower temperatures and delay oxidation onset. Aminic antioxidants function at higher temperatures and extend oxidative resistance further into the service life.

Anti-wear additives form sacrificial films under boundary lubrication conditions. ZDDP and TCP (tricresyl phosphate) are two of the most effective. EP additives activate under high load and temperature, creating a protective tribochemical layer. Sulfur and phosphorus compounds dominate this space in gear and heavy-duty applications.

Viscosity index improvers are polymers that expand with temperature to counteract viscosity loss. They are critical in multigrade engine oils and hydraulic fluids that must operate in broad ambient conditions. These modifiers shear down over time, and stability depends on polymer architecture and system dynamics.

Friction modifiers reduce surface contact resistance, improving energy efficiency and fuel economy. Organic molybdenum compounds and esters are common in engine oils and some gear lubricants. Foam inhibitors prevent air entrapment and collapse microbubbles that form during agitation. Silicone-based agents and polyacrylates are often used. Demulsifiers promote separation between oil and water, critical for systems exposed to washdown or ingress.

Application determines additive architecture. Engine oils demand a high load of detergents and dispersants to manage combustion by-products, plus robust AW and anti-

TABLE 1.3 Selection Criteria for the Matrix Model plus How-to Steps, Pitfalls, and Remedies/Best Practices

SELECTION CRITERIA	HOW-TO STEPS	POTENTIAL PITFALL	REMEDY/BEST PRACTICE
Base oil solvency	Assess polarity of the base oil to determine solvency. Use esters or PAGs for high-solvency needs; PAOs need cosolvents.	Using low-polarity oils (e.g., PAOs) without cosolvents leads to additive fallout and varnish formation.	Blend PAOs with esters or use Group III+ stocks to increase polarity. Conduct solubility trials.
Additive compatibility	Identify additive types and check for known antagonisms (e.g., ZDDP with yellow metals or friction modifiers versus AW agents).	Friction modifiers may clash with AW additives or cause elastomer degradation. ZDDPs harm catalytic surfaces.	Use additive compatibility matrices to avoid antagonisms. Confirm with lab or OEM testing.
Base oil thermal stability	Match base oil group to temperature exposure. Use Group IV or V for sustained high-temp systems. Use Group I for low-stress only.	Group I or II oils degrade rapidly under thermal stress, forming sludge and clogging filters.	Use Groups III–V oils with proven oxidative stability. Reserve Group I for legacy equipment only.
Additive-base oil solubility	Ensure the additive system dissolves fully in the base oil. Use polarity-matched cosolvents to prevent dropout in PAOs.	Incompatible solubility leads to filter plugging, deposit formation, or additive dropout at operating temp.	Balance with polarity-adjusted cobase stocks. Validate solubility under thermal cycling conditions.
Volatility management	Use Groups IV and V oils to reduce evaporation in high-heat systems. Confirm the flash point meets or exceeds the application threshold.	High volatility results in oil loss, increased consumption, and fire hazard in compressors and turbines.	Select base stocks with flash points > 200°C for high-temp uses. Verify via NOACK volatility data.
Oxidation resistance	Select base oils with inherent oxidative resistance (PAOs, esters). Supplement with antioxidant packages as needed.	Poor oxidative stability shortens lubricant life and leads to varnish, acid formation, and wear.	Prioritize Groups IV and V oils for oxidative resilience. Use RPVOT or PDSC testing for validation.
Low-temperature fluidity	Use PAOs or esters for extreme cold. Confirm the pour point via ASTM D97. Avoid Group I oils in subzero climates.	Poor cold-flow characteristics delay lubrication during start-up, increasing wear and failure risk.	Use high-VI synthetics in mobile or arctic conditions. Apply pour point depressants if necessary.
Seal and material compatibility	Check seal material specs against selected base oil and additives. Conduct compatibility trials when using esters or unknowns.	Ester-induced swelling or hardening of seals causes leaks, pressure loss, or seal failure.	Use Viton or FKM seals for ester exposure. Perform immersion testing for chemical resistance.
Matrix-based formulation approach	Map additive chemistry against base oil polarity, solvency, and application needs. Use small-batch testing to confirm synergy.	Unmapped chemistry causes instability under load, additive precipitation, and accelerated degradation.	Establish a formulation matrix aligned to system demands. Pilot-test all new chemistries in live systems.

oxidant protection. Gear oils prioritize EP strength and thermal stability, often sacrificing oxidation life for higher film pressure tolerance. Hydraulic fluids require cleanliness, thermal stability, and anti-wear properties with minimal deposit potential. Compressors need oxidation resistance, low volatility, and foam control.

Regulatory frameworks constrain formulation options. Low-SAPS engine oils protect exhaust aftertreatment devices. H1 food-grade lubricants exclude toxic additives, relying on inert thickeners and benign AW agents. Ashless fluids are required in systems with low-tolerance valves or sensitive emission control systems. Compliance adds complexity to the additive matrix and must be integrated early in the selection process.

Table 1.4 is an additive package selection guide.

Compatibility Matrix: Building the Strategic Fit

Compatibility must be engineered. The matrix begins with additive solubility mapping. Additives must remain dissolved and stable in the chosen base oil throughout the full thermal and mechanical operating range. Polar additives require polar base oils or cosolvents. Nonpolar oils like PAOs resist solvency and need polarity-balancing cobase stocks to prevent fallout.

Temperature and load affect chemical interaction. High load accelerates additive film activation. Elevated temperatures can drive additive decomposition, volatility loss, or deposit formation. The matrix links these conditions to additive type, reactivity window, and stability curve.

Additive synergy enhances performance when components cooperate. Dispersants and detergents work best together. Antioxidants delay breakdown when paired. Antagonism must be avoided. ZDDP can suppress friction modifiers. Overbased detergents may neutralize acid scavengers. The matrix flags known conflict zones.

Formulation must be validated beyond the lab. Bench tests simulate thermal aging, load cycles, and contaminant ingress. But field validation confirms additive performance in real operating environments. Compatibility issues often emerge only under live operating stresses, gear load reversals, extended drains, or variable duty cycles.

Additive fallout is a known risk in PAO systems. Without solubility support, AW or EP agents settle out, causing varnish and plugging filters. Friction modifiers may interfere with ZDDP, reducing wear protection. Excess polarity from esters can attack seals, leading to leakage and degradation. Sulfur and phosphorus EP additives can corrode yellow metals like copper and brass in hydraulic and compressor systems.

PAO blends the benefit from cosolvents or esters to stabilize additive dispersal. These should be selected for their thermal limits and seal compatibility. Additive dosing

TABLE 1.4 Additive Package Selection Guide

ADDITIVE CATEGORY	HOW-TO STEPS	POTENTIAL PITFALL	REMEDY/BEST PRACTICE
Detergents	Select metal detergents based on system cleanliness needs and base oil compatibility. Use calcium sulfonates for neutralization and magnesium for cleaning	Overbased detergents may increase ash content, leading to filter loading or deposit formation	Use detergent levels matched to deposit loading. Monitor ash levels in oil analysis
Dispersants	Use ashless PIB-SI for soot-heavy or deposit-prone systems. Validate effectiveness in sludge-prone or high-temp environments	Dispersants may lose effectiveness if the base oil lacks solvency or the system runs at extreme temperatures	Ensure dispersant stability in the selected base oil. Supplement with antioxidants to extend effectiveness
Antioxidants	Combine phenolic and aminic antioxidants to span the operating temperature range. Validate via RPVOT or PDSC testing	Single antioxidant systems degrade quickly under cyclic thermal stress, reducing fluid life	Use synergistic antioxidant blends. Monitor depletion through oil condition analysis
AW agents	Use ZDDP or TCP in systems with boundary lubrication. Match concentration to surface metallurgy and load cycles	ZDDP may poison catalysts or damage yellow metals. Excess can increase sludge formation	Balance AW levels to metallurgy. Consider ZDDP-free options for emissions-sensitive systems
EP additives	Apply sulfur- and phosphorus-based EP additives for gears and high-load contacts. Test for copper compatibility	Excessive EP additives may corrode soft metals or foul filters	Limit EP concentration based on metallurgy and system material compatibility testing
VI improvers	Select VI improvers based on shear stability and viscosity span. Confirm retention over the expected service interval	Low shear stability VI improvers degrade early, causing viscosity loss and wear risk	Use high-stability VI improvers, and validate viscosity profile with ASTM D445 over service life
Friction modifiers	Choose organic moly or esters for improved energy efficiency. Ensure compatibility with AW/EP chemistries	Incompatibility with other additives may destabilize boundary films or affect seal materials	Blend carefully with AW/EP systems. Validate with lab friction tests and elastomer compatibility trials
Foam inhibitors	Apply silicone or polyacrylate agents in high-agitation environments. Test foam suppression and air release	Foam inhibitors may not prevent entrained air under rapid cycling or high return flows	Evaluate foam in both static and dynamic states. Confirm with ASTM D892 testing
Demulsifiers	Use demulsifiers in systems with high water ingress risk. Confirm the separation rate under service conditions	Demulsifiers may strip too aggressively, destabilizing emulsions in critical systems	Balance the demulsifier load to application needs. Use visual separation and Karl Fischer tests
Application-specific architecture	Tailor the additive balance to the machine type: AW + antioxidant for hydraulics; EP + dispersants for gear oils	Generic formulations miss key performance needs, causing premature failure or varnish	Design with load, thermal range, and contamination in mind. Validate with application-specific testing
Regulatory compliance constraints	Screen additive options for compliance with SAPS limits or food-grade requirements. Select inert or benign alternatives where needed	Formulation may violate environmental or industry-specific limits (e.g., food grade, low SAPS)	Integrate compliance early. Use NSF-listed or OEM-approved chemistries to avoid reformulation

must be controlled. Too much sulfur can lead to gear corrosion; too much dispersant can lead to foaming or filter loading. Fine-tuning the balance among AW, EP, and FM (friction modifier) additives avoids internal competition.

Performance testing under harsh environmental conditions is mandatory. Simulations should include high thermal cycling, water ingress, and contaminant load. ASTM D2893, D943, and D892 provide controlled platforms, but field trials under worst-case conditions remain the standard for confirming strategic compatibility.

CASE EXAMPLES OF COMPATIBILITY-DRIVEN SUCCESS AND FAILURE

A compressor station using a PAO-based lubricant began to experience premature valve fouling and filter plugging. Analysis revealed additive dropout caused by poor solubility in the nonpolar base stock. The EP and AW agents had settled out under thermal cycling and degraded the internal cleanliness. Switching to a PAO/ester blend restored additive stability and eliminated fouling within two maintenance intervals.

A food processing facility deployed synthetic grease rated H1 for incidental contact. After one lubrication cycle, O-rings swelled, and gearhead seals began to weep. The ester-based base oil had reacted with the facility's common use of NBR seals. Reformulation with a PAO-thickened alternative stabilized the elastomers and restored equipment containment without sacrificing food-grade compliance.

In a municipal mixed-fleet operation, high-SAPS engine oils containing ZDDP were used across vehicles with and without modern emission systems. Catalyst poisoning was identified during emissions testing. The phosphorus in ZDDP had migrated to the surface of the catalytic converters, degrading conversion efficiency. The operation transitioned to low-SAPS formulations for aftertreatment-equipped vehicles and maintained ZDDP only in legacy engines.

A power plant experienced varnish in steam turbine bearings during summer peak load. The mineral-based turbine oil exhibited rapid oxidation under thermal stress. Engineers deployed a custom-blended Group III/ester formulation with enhanced antioxidants and varnish inhibitors. Oil analysis showed no insoluble increase over the next 18 months, and bearing operating temperatures dropped 10°C. Downtime and varnish remediation costs were eliminated for two seasons.

Guidelines for Lubricant Selection and Formulation

➤ Determine whether base oil or additive performance dominates the system needs. High-speed, high-temperature systems often demand base oils with low volatility and oxidative stability. Low-speed, high-load or contaminated systems may depend more on additive strength.

➤ Select base oils that complement the additive system. PAOs require polar cobase stocks to hold additives in solution. Esters improve solvency but may impact elastomer compatibility. Groups I and II stocks provide natural solvency but sacrifice oxidation resistance.

➤ Evaluate long-term chemical and thermal stability. Regulatory compliance, biodegradability, and food safety restrictions must be factored in at the formulation stage. Consider drain intervals, exposure to contaminants, and temperature extremes.

➤ Use supplier technical data, but confirm through internal validation. Conduct compatibility testing with seals, metals, and filters. Use field trials, thermal simulation, and trend analysis to confirm that the formulation supports reliability goals under operational conditions.

SELECTING THE RIGHT GREASE FOR THE RIGHT APPLICATION

Grease is often underestimated in reliability programs despite its importance in sustaining equipment longevity and uptime. The correct selection of grease formulation: base oil, thickener, and additive system must match environmental, mechanical, and operational demands. Failures due to misapplication are avoidable with a grounded understanding of grease chemistry, field-tested case data, and equipment-specific requirements.

Soap-Based Greases

Soap-based greases continue to dominate industrial use because of their versatility. Simple lithium soap greases are the most common general-purpose greases. These are typically formulated with lithium stearate and offer good mechanical stability and pumpability. However, their upper temperature limits are modest, and water resistance is average. They are suited for low- to medium-duty applications such as automotive chassis fittings and electric motors.

Lithium complex greases, formulated using lithium 12-hydroxystearate and a complexing agent, have been applied extensively in general manufacturing, agriculture, and fleet operations. For example, in a fleet maintenance program studied by the U.S. Department of Transportation, vehicles lubricated with lithium complex grease demonstrated a 22% increase in bearing service life compared to vehicles using older calcium soap-based grease. This is attributed to the lithium grease's better mechanical stability and wider operating temperature range. However, in high-speed or high-thermal applications, lithium soap may soften, bleed oil, or oxidize prematurely.

Aluminum complex greases offer high dropping points, excellent water resistance, and strong adhesive properties. They are often used in food-grade applications due to their compatibility with H1 food-contact standards. In a dairy processing facility in Wisconsin, aluminum complex grease reduced the frequency of bearing failure on washdown-exposed agitators by 40% when compared to standard lithium grease. However, aluminum complex greases can exhibit lower oxidative stability in high-load, high-thermal applications.

Barium complex greases possess excellent load-carrying capability, water resistance, and mechanical stability. Their superior adhesion and resistance to corrosion make them ideal for extreme pressure applications in marine and heavy-duty equipment. Despite their exceptional performance, their use has diminished due to the toxicological concerns surrounding barium compounds. Some formulations have been restricted or discontinued in certain regions due to safety regulations.

Non-Soap Thickened Greases

Calcium sulfonate greases are non-soap thickeners that are gaining popularity in heavy equipment sectors due to their superior water washout resistance, inherent corrosion protection, and high-temperature stability. A mining operation in Northern Ontario found a reduction in pin and bushing wear rates by 35% after switching to calcium sulfonate complex grease in loaders exposed to wet slurry and freeze-thaw cycles. Calcium sulfonate greases offer remarkable performance in environments with water ingress or mechanical shock.

Polyurea greases have seen widespread use in electric motors and sealed-for-life applications. In a long-term assessment by NEMA, polyurea-thickened greases extended motor bearing life by 40% in HVAC applications, especially when compared to lithium-thickened equivalents. However, polyurea can exhibit compatibility issues with other grease types, and inconsistency in production can yield unpredictable results if not well controlled.

Clay-based greases, formulated using bentonite or hectorite, are structurally stable and non-melting, making them suitable in oven conveyor chains or kiln bearings. Yet, they often exhibit oil bleed and poor water resistance. In a thermal processing facility in Texas, a switch to synthetic clay-based grease eliminated drip and smoke issues at 230°C but required weekly relubrication due to lubricant loss during shutdown cycles.

PTFE (polytetrafluoroethylene) as an additive offers low friction and anti-wear benefits. A laboratory test conducted by the German Federal Institute for Materials Research showed that PTFE-enhanced grease reduced friction coefficient by 30% in oscillating needle bearings. These greases are suited for high-speed and plastic-compatible interfaces but offer limited support under high loads or severe boundary lubrication.

Specialty Synthetic-Based Greases

Synthetic-based greases, especially those using Esters (diesters or polyol esters), demonstrate excellent cold-temperature behavior and oxidative resistance. These are often specified in aerospace, cryogenic, and food-processing applications. A NASA materials compatibility trial (NASA TM-2013-217984) found that ester-based greases-maintained lubricity down to −70°C and remained stable after thermal cycling between −60°C and 100°C. However, esters may hydrolyze in the presence of moisture or reactive metals such as zinc or copper.

Silicone greases, while chemically inert and thermally stable, provide poor extreme pressure performance. In dielectric and sealing applications such as rubber O-rings and connectors, silicone greases excel due to their non-reactivity and thermal consistency. A case study by Dow Corning showed that electrical insulation failures dropped by 50% in solar inverter connectors after silicone grease was applied to minimize oxidation and moisture ingress.

See Tables 1.5 and 1.6 for more information.

TABLE 1.5   Grease Chemistry Properties

GREASE TYPE	TEMP RANGE (°C)	LOAD CAPACITY	WATER RESISTANCE	OXIDATION STABILITY	COMMON APPLICATIONS
Simple lithium	–20 to 120	Moderate	Fair	Fair	Chassis, small motors
Lithium complex	–40 to 150	Moderate	Moderate	Good	Bearings, motors, ag equipment
Calcium sulfonate	–30 to 160	High	Excellent	Excellent	Wet environments, construction
Aluminum complex	–20 to 150	Moderate	Excellent	Moderate	Food processing, washdown environments
Barium complex	–20 to 150	Very high	Excellent	Very Good	Marine, heavy equipment
Polyurea	–30 to 250	Low-moderate	Good	Excellent	Motors, sealed bearings
Clay-based	–20 to 250	Low	Poor	Fair	Ovens, kilns, furnaces
PTFE-enhanced	–50 to 180	Low	Good	Good	High-speed, plastic components
Ester-based	–60 to 150	Moderate	Good	Very Good	Aerospace, food processing
Silicone-based	–60 to 250	Low	Excellent	Excellent	Electrical insulation, connectors

TABLE 1.6   Decision Matrix for Grease Selection

APPLICATION CONDITION	RECOMMENDED GREASE TYPE
General purpose	Simple lithium
High load + moisture	Calcium sulfonate complex
High temp + long life	Polyurea or lithium complex
Plastic compatibility	PTFE-enhanced or silicone-based
Low temp performance	Ester-based
Electrical insulation	Silicone-based
Extreme temp (non-melting)	Clay-based
Washdown environment	Aluminum complex
Marine heavy-duty service	Barium complex (if permitted by policy)

The decision to use specific grease chemistry must not rest solely on lab data or marketing literature. Field validation, equipment design constraints, and relubrication logistics form the foundation of proper grease selection. When aligned properly, grease is not just a consumable; it becomes an enabler of reliability.

CHAPTER 2

LUBRICATION PROCEDURES

WRITING LUBRICATION SOPS FOR CONSISTENCY AND COMPLIANCE

Standard operating procedures for lubrication are essential tools for achieving consistency, reliability, and regulatory compliance in industrial maintenance programs. A well-crafted SOP ensures that tasks are performed uniformly regardless of personnel, shift, or site location. This reduces error, extends asset life, and improves audit readiness.

Effective lubrication SOPs include:

➤ Clear objectives and scope of the procedure
➤ Equipment-specific details (ID, location, lube point references)
➤ Required tools and materials
➤ Detailed step-by-step instructions with quantities and specifications
➤ Precautionary measures, PPE requirements, and lockout/tagout (LOTO)

The SOP should reference OEM guidelines but also be adapted to field realities and observed conditions. SOPs must be controlled documents that are updated regularly and distributed through centralized systems to ensure currency.

Lubrication SOPs standardize task execution and eliminate procedural drift. Each SOP must align with equipment condition, site constraints, and lubricant properties.

Objectives and scope define the task boundaries and identify the asset, the lube point, and the expected outcome. General statements such as "lubricate as needed" are unacceptable. Specify exact volumes, intervals, and lubricant types.

Include equipment identifiers, system tags, and location codes. Match these with CMMS (computerized maintenance management system) tags and physical nameplates. Cross-reference OEM recommendations but override with field-validated data when operational conditions differ, such as ambient heat, high dust loading, or duty cycle variances.

List tools explicitly. Include grease guns, torque wrenches, filter carts, drain pans, and fittings. Lubricants must be designated by ISO viscosity grade, application type (H1, R&O, EP), and specific product code. Storage requirements and handling precautions for each product must be integrated.

Step-by-step task sequences begin with safety controls. Include PPE, lockout/tagout references, and verification of de-energized status. Define inspection criteria for seals, breathers, and vent paths. Describe surface cleaning requirements before fitting access. Specify how to dispense—pressure, duration, number of strokes, or mass (e.g., grams, cc).

Document purge and venting instructions, especially in closed systems. Address grease bleed, potential cross-contamination, and lubricant displacement during reapplication. Include end-of-task checks such as sight glass verification, reservoir level, and system pressure.

SOPs must be revision-controlled. Change logs and approval stamps are required. Host documents within the centralized document management system, linked directly to CMMS task records. Ensure that access permissions are role-specific.

SOP validation is mandatory before deployment. Field techs should perform dry runs under observation. Deviations, ambiguities, or equipment access constraints must be resolved and documented. Training sessions and periodic reviews reinforce adherence. Nonconformance should trigger SOP audits and corrective updates.

PROCEDURE DRIFT: HOW INFORMAL PRACTICE UNDERMINES POLICY

Procedure drift occurs when informal or undocumented practices diverge from written SOPs. Over time, this erosion leads to inconsistent task execution, missed steps, and elevated risk of failure or noncompliance.

Causes of drift include:

➤ Perceived inefficiency or complexity in the written SOP

➤ Inadequate training or supervision
➤ Equipment modifications not reflected in the procedure
➤ Informal knowledge transfer without documentation

Mitigating drift requires periodic observation of task execution, feedback loops for SOP improvement, and an organizational culture that treats SOPs as dynamic, actionable tools rather than static directives.

Technician engagement in SOP creation and revision fosters ownership and adherence. SOP compliance metrics and root cause analysis of deviations help identify high-risk procedures and retraining needs.

Procedure drift occurs when actual field practices deviate from documented lubrication SOPs. This divergence, often gradual, leads to variability in lubricant type, volume, frequency, and application method. Unchecked, drift introduces the risk of bearing damage, lubricant incompatibility, seal degradation, and audit noncompliance.

Common drivers of drift include perceived inefficiency of the documented procedure, especially when SOPs require multiple steps or specialized tools. Technicians may bypass steps under time pressure or in the belief that certain actions are unnecessary. Informal training, undocumented tribal knowledge, or reliance on verbal instruction accelerates procedural erosion.

Physical equipment changes, such as new fittings, modified access panels, or upgraded reservoirs, often go unrecorded. If the SOP is not immediately revised, personnel rely on memory or improvisation, further decoupling the task from its validated baseline.

Mitigation requires direct observation of live task execution, not just review of work orders. Auditors must verify lubricant type, point identification, volume dispensed, and adherence to safety controls. Deviation logs should be maintained in the CMMS and reviewed monthly.

Involve frontline technicians in SOP creation and revision. Their field input ensures feasibility, eliminates ambiguity, and enhances buy-in. Use deviation data to trigger retraining or corrective action. Where drift is persistent, deploy visual aids at the point of use: lube maps, tagged fittings, QR-coded task guides.

SOPs must be dynamic. Update cycles should align with equipment modifications, lubricant changes, and failure analyses. Lock revisions behind version control, and distribute only through centralized platforms. Assign compliance metrics, and tie them to individual and departmental KPIs. Where SOP drift causes failure, document the event, and initiate formal root cause analysis.

CREATING VISUAL WORK INSTRUCTIONS FOR LUBE TECHNICIANS

Visual work instructions enhance clarity, reduce misinterpretation, and support multi-lingual or cross-functional teams. For lubrication, this includes annotated diagrams of equipment, color-coded lubrication points, photos of access locations, and icons denoting personal protective equipment (PPE) or safety steps.

Best practices include:

➤ Using real images of the specific asset, not generic drawings
➤ Overlaying lubricant type, volume, and method directly onto the image
➤ Providing QR codes linking to digital SOPs or training videos
➤ Employing standardized symbols for grease, oil, filters, etc.

These instructions should be laminated and posted near the asset or embedded within digital route management systems. Interactive formats on tablets allow for technician inputs, digital sign-offs, and time-stamped task records.

Visual instructions reduce onboarding time for new hires and reinforce procedural discipline for all skill levels. Visual work instructions reduce ambiguity and improve compliance across varying skill levels and language barriers. For lubrication tasks, visual tools streamline execution by showing exact lube points, correct methods, and required safety steps directly on images of the equipment.

Use actual photographs of the target asset, not stock diagrams. Annotate each lubrication point with clear callouts indicating lubricant type (e.g., ISO VG 220 gear oil, NLGI 1 grease), volume, and application method (e.g., hand pump, zerk fitting, drip feed). Color-code lube points to differentiate oil, grease, or special lubricants. Indicate inspection windows, drain ports, and fill locations.

Integrate icons for PPE (gloves, goggles), LOTO points, and hot surface warnings. QR codes can link to digital SOPs, mobile instructional videos, or OEM manuals. Add visual indicators of normal versus abnormal conditions; examples include sight glass level images and photos of clean versus contaminated filters.

Standardize symbols across all visual work aids: droplet icon for oil, grease gun symbol for fittings, triangle for caution. Avoid clutter by limiting each visual to a specific procedure or lubrication zone. Layering details across multiple images works better than crowding one sheet.

Mount laminated versions near the equipment or embed images into tablet-based route platforms. Digital tools allow for real-time sign-offs, time stamps, and exception logging. Integration with CMMS ensures traceability and automatic updates when changes occur.

Visual instructions are essential for cross-training, backfilling roles, and onboarding. They act as reference tools for seasoned technicians and training scaffolds for new hires. Use them to reinforce procedural discipline, minimize technician variability, and reduce error from memory-based work.

JOB SAFETY AND ENVIRONMENTAL CONTROLS IN LUBRICATION TASKS

Lubrication activities can present a variety of safety and environmental risks, particularly when dealing with pressurized systems, high temperatures, elevated work, or hazardous fluids. SOPs must embed safety and environmental controls as an integral part of each task. Key controls include:

- PPE specifications based on lubricant SDS. The MSDS (Material Safety Data Sheet) is an outdated acronym superseded by Safety Data Sheet (SDS)
- Lockout/tagout requirements
- Spill containment and cleanup protocols
- Handling and storage instructions for hazardous substances
- Proper disposal or recycling of used lubricants and filters

Job hazard analysis (JHA) should be performed for lubrication procedures with elevated risk profiles. Environmental controls may also include secondary containment for mobile carts, absorbent mats at fill stations, and labeling systems for fluid compatibility. Compliance with OSHA (Occupational Safety and Health Administration), EPA (Environmental Protection Agency), and industry-specific regulations (such as NSF or ISO 21469) should be verified and documented within the procedure framework.

Lubrication tasks intersect with safety-critical operations and regulated environmental practices. Procedures must embed controls that mitigate risks from pressure, temperature, chemical exposure, and workspace hazards.

PPE requirements must be defined per lubricant SDSs, including gloves, goggles, face shields, and flame-resistant clothing where applicable. For high-pressure systems or elevated applications, include additional PPE such as fall protection or arm guards.

Lockout/tagout must be enforced before accessing rotating equipment, pressurized lines, or internal components. LOTO points should be clearly labeled and included in task instructions. Technicians must verify energy isolation before proceeding.

Spill control protocols include secondary containment for mobile carts, spill kits stationed at lube areas, and absorbent mats at reservoirs and fill ports. Use dripless quick-connect fittings and closed-system transfer containers to prevent environmental release.

Handling instructions for hazardous lubricants, such as those containing PAOs, esters, or solvents, must specify closed-transfer techniques, compatible containers, and segregation from reactive materials. Label all fluid containers with product name, compatibility codes, and hazard markings.

Used lubricant, rags, and filters must be collected in designated receptacles for recycling or hazardous waste disposal per EPA guidelines. Label disposal drums with waste profiles and maintain documentation for waste manifests.

Environmental safeguards extend to airborne and waterborne contamination. Use filtered breathers, sealed fill ports, and enclosed service procedures to prevent vapor release or fluid entry into storm drains. In food, pharma, or cleanroom environments, apply NSF H1 or ISO 21469-compliant practices, including hygiene zones, color-coded tools, and food-grade certifications.

JHA must precede nonroutine lubrication tasks, such as large-volume transfer, confined space greasing, or elevated refill operations. Identify ignition sources, slip hazards, pinch points, and ergonomic strains.

Audit procedures against OSHA 29 CFR 1910, EPA SPCC plans, and local environmental health and safety (EHS) standards. Record training, inspections, and compliance metrics in CMMS or EHS management systems to ensure continuous safety performance in lubrication workflows.

PROCEDURE HARMONIZATION ACROSS MULTISITE OPERATIONS

For organizations operating multiple facilities, harmonizing lubrication procedures ensures standardization, improves performance benchmarking, and simplifies training. However, differences in local regulations, equipment models, or climate can necessitate regional customization.

Harmonization strategy includes:

➤ Creating master procedures with site-specific appendices
➤ Identifying core procedural elements that are nonnegotiable
➤ Developing a centralized SOP management system
➤ Conducting inter-site audits to identify deviations and share best practices

Shared templates, common terminology, and uniform metrics create a foundation for alignment. Site autonomy can still be preserved where justified by local operating realities, but within a governed structure. Standardizing lubrication procedures across multisite operations reduces variability, streamlines technician training, and enables uniform reliability metrics. Master SOPs must define core lubrication principles, such as correct lubricant selection, application methods, and contamination control practices, that apply across all facilities.

Develop a master procedure framework that includes nonnegotiable elements: task sequences, safety requirements, lubricant specifications by asset class, and compliance references. Use appendices or site-specific modules to adapt for local equipment variations, climate conditions, or regulatory mandates.

Establish a centralized SOP management system with controlled versioning, access permissions, and audit trails. Integration with enterprise CMMS platforms ensures synchronization across departments and locations. SOPs must be accessible at the point of use through mobile devices or printed visual aids.

Deploy inter-site audits to verify alignment, capture undocumented deviations, and share proven practices across facilities. Use audit findings to revise master procedures or update site-specific appendices with justification.

Terminology, icons, and measurement units must be standardized. Define lubrication task metrics such as PMC (preventive maintenance compliance), MTBLRF (mean time between lubrication-related failures), and lubricant consumption rate in consistent formats across all locations.

Provide cross-training using harmonized procedures. Multisite technicians can be redeployed with minimal retraining, improving workforce flexibility and response time.

Preserve local autonomy only when justified by technical, legal, or environmental constraints. Require documentation and approval for site-specific deviations to ensure governance and traceability.

AUDITING LUBRICATION PROCEDURES FOR REGULATORY AND TECHNICAL ACCURACY

Routine auditing of lubrication procedures ensures they remain compliant with evolving regulations and reflect current technical realities. An effective audit process evaluates the document's completeness, accuracy, and field usability.

Audit elements include:

➤ Verification against OEM and industry standards
➤ Alignment with current lubricant types and volumes
➤ Safety and environmental control compliance

➤ Actual field execution observed and compared with SOP
➤ Document control: version history, revision dates, authoring accountability

Audits should be conducted at scheduled intervals or triggered by equipment failures, safety incidents, or regulatory changes. Audit results should be used not only to revise SOPs but to improve training, sourcing, and procedural discipline. Lubrication procedures must undergo routine audits to ensure alignment with operational requirements, safety mandates, and technical specifications. Audits identify gaps between written procedures and field execution, preventing procedural drift and ensuring regulatory readiness.

Verify each SOP against current OEM documentation, industry best practices, and site-specific equipment modifications. Confirm that lubricant types, volumes, application methods, and intervals reflect actual machine requirements and storage conditions. Adjustments in lubricant formulation or supplier changes must trigger an SOP review.

Inspect embedded safety and environmental protocols. Confirm that PPE requirements, spill response, disposal guidelines, and LOTO procedures align with the latest SDS, OSHA standards, and facility-specific policies. Environmental controls, such as fluid labeling, containment, and storage, must meet EPA and ISO 14001 requirements.

Conduct live task observations. Compare actual technician performance with SOP instructions. Document discrepancies, omissions, or work-arounds. Evaluate whether the procedure supports consistent, efficient execution under field conditions.

Review document control integrity. Confirm the presence of version history, revision justification, approval signatures, and distribution logs. All active procedures must be accessible in their most current form at point of use, whether printed or digital.

Audits should be scheduled annually, semiannually for critical assets, or immediately following failures, near misses, or compliance incidents. Findings must be tracked in a corrective action system with closure timelines, assigned ownership, and verification of resolution.

Use audit data to refine SOP templates, standardize formats across facilities, and update technician training materials. Audits provide measurable input for continuous improvement in lubrication management systems.

THE ROLE OF TRIBAL KNOWLEDGE IN PROCEDURE FAILURES

Tribal knowledge refers to informal, experience-based practices shared among workers but undocumented in formal systems. While it can be valuable, over-reliance on tribal knowledge often leads to procedural inconsistency, risk exposure, and decline of reliability.

Procedure failures resulting from tribal knowledge include:

➤ Using outdated methods passed down without validation
➤ Skipping steps perceived as unnecessary
➤ Improvising lubrication volumes or intervals

Organizations must actively capture tribal knowledge through structured interviews, technician workshops, and post-task debriefs. Validated practices can then be formalized into SOPs.

A healthy balance is necessary: recognizing the insights of experienced personnel while ensuring that all critical procedures are documented, reviewed, and continuously improved. Tribal knowledge introduces variability into lubrication tasks by replacing documented procedures with informal, experience-based shortcuts. These practices often go unverified and unrecorded, leading to inconsistent execution and degraded equipment performance.

Common failures include reliance on inherited methods that no longer reflect current equipment configurations or lubricant formulations. Technicians may skip steps they view as nonessential or redundant, such as purging old grease or cleaning fittings, assuming prior success guarantees reliability. Improvised lubrication schedules, based on "feel" or historic frequency rather than data, create the risk of overlubrication or underlubrication.

Failures rooted in tribal knowledge are difficult to trace because they bypass formal systems. When equipment fails, root cause analysis often reveals undocumented variations in task execution. These gaps compromise compliance, obscure accountability, and invalidate warranty protections.

Capturing tribal knowledge requires structured field engagement. Use observation sessions, debrief interviews, and cross-functional workshops to uncover undocumented practices. Evaluate these practices for validity through data analysis and expert review. Validated knowledge should be standardized into formal SOPs with clear authorship, version control, and training alignment.

Effective programs balance experiential insight with procedural discipline. Veteran input should inform SOP evolution, not override it. All lubrication tasks must converge toward documented, repeatable, auditable standards, reinforced by CMMS integration and continuous improvement cycles.

CHAPTER 3

REGULATORY OVERLAYS AND ENVIRONMENTAL CONSTRAINTS IN SELECTION

Environmental regulations have an increasingly powerful influence on lubricant selection, especially in sensitive or high-regulation zones such as marine, food processing, mining, and agriculture.

Lubricants must comply with a range of international, national, and industry-specific regulations, including REACH (EU), EPA VGP (USA), NSF H1/H2 (food grade), and ISO 21469 (hygiene). Biodegradability, toxicity, and bioaccumulation potential are key evaluation factors. Synthetic esters, polyalkylene glycols, and vegetable-based oils are often considered in these scenarios for their lower environmental impact.

Product labeling and Safety Data Sheet transparency are mandatory for cross-border movement of lubricants, and manufacturers must provide detailed disclosures of chemical constituents. Import/export limitations and disposal requirements also shape lubricant sourcing decisions.

Where regulations dictate limitations on certain additives (such as chlorine, phosphorus, or sulfur), performance trade-offs must be carefully engineered. In such cases, advanced additive technology and molecular tailoring of base oils become tools for meeting both performance and compliance requirements.

THE EXPANDING REGULATORY LANDSCAPE

Environmental and safety regulations now drive lubricant formulation decisions in multiple industries. Historically, lubricant design prioritized performance and cost. Today, it must also conform to an expanding matrix of legal mandates and certification regimes. Regulatory pressure has shifted from advisory guidelines to enforced limitations, compelling manufacturers and end users to adapt or face penalties, supply disruptions, or operational shutdowns.

In marine applications, the U.S. EPA Vessel General Permit (VGP) mandates the use of environmentally acceptable lubricants (EALs) in all oil-to-water interface points. These formulations must meet stringent biodegradability, toxicity, and bioaccumulation criteria. Conventional mineral oils are disqualified unless proved compliant under EAL standards.

Food processing facilities are governed by NSF H1 and ISO 21469 certifications. Lubricants must be physiologically safe for incidental contact and manufactured under hygienic conditions. Additive packages are limited, ruling out high-performance chemistries that might otherwise increase wear protection or oxidative stability. This constraint requires balancing lubricity with food safety and GMP standards.

Mining operations face regional and national environmental regulations targeting toxicity, groundwater contamination, and disposal. Synthetic esters and polyalkylene glycols are increasingly favored over traditional Group I and Group II oils for their low toxicity and potential for biodegradability. In some jurisdictions, lubricant selection must be documented as part of an environmental impact statement or compliance audit.

Agricultural and construction sectors must also consider biodegradability and toxicity, especially where fluid leaks can contact soil or crops. EU directives and local ordinances may impose restrictions on chlorine- or zinc-containing additives. Machine manufacturers in these sectors are integrating biodegradable fluid compatibility into equipment design, forcing alignment between OEM and fluid supplier.

The modern reliability engineer must be fluent in regulatory overlays, understanding how they reshape lubricant selection and life-cycle decisions. Compliance is no longer optional; it's structurally embedded in fluid performance specifications and material compatibility constraints.

KEY REGULATORY FRAMEWORKS AND STANDARDS

International and National Mandates

> ➤ REACH (Registration, Evaluation, Authorisation and Restriction of Chemicals) regulates the production and use of chemical substances in the EU. Lubricant formulations must undergo registration and safety evaluation, with full disclosure of all ingredients above threshold concentrations. Failure to comply results in restricted market access and product bans.
>
> ➤ TSCA (Toxic Substances Control Act) in the U.S. governs chemical inventories and new substance approvals. Manufacturers must report, test, and sometimes reformulate products based on EPA risk assessments. This directly impacts base oil and additive suppliers bringing new technologies to market.
>
> ➤ GHS (Globally Harmonized System) standardizes hazard classification and labeling worldwide. Lubricants must be labeled with consistent pictograms, signal words, and hazard statements. Safety Data Sheets must follow the GHS 16-section format. Accurate labeling is essential for compliance in international logistics and user safety.

Industry-Specific Directives

> ➤ The EPA's Vessel General Permit mandates the use of EALs in all equipment operating over water. Fluids must demonstrate biodegradability (per OECD 301B or ISO 10708), low toxicity (per OECD 201/202/203), and minimal bioaccumulation. Conventional mineral oils are typically noncompliant.
>
> ➤ NSF H1/H2 and ISO 21469 set hygienic and compositional standards for food-grade lubricants. NSF H1 oils are permitted for incidental food contact; ISO 21469 goes further, requiring GMP-certified production processes. Additive selection is restricted to FDA-approved substances under 21 CFR 178.3570.
>
> ➤ OECD 301 and 310 series tests evaluate inherent and ultimate biodegradability of lubricant constituents. These serve as international benchmarks for ecotoxicity screening, especially in sectors with environmental discharge exposure.

➤ Ecolabel certifications such as EU Ecolabel, Blue Angel (Germany), and Nordic Swan (Scandinavia) incorporate biodegradability, renewable content, packaging, and emissions performance into their criteria. These certifications are increasingly used by OEMs and fleet operators as procurement filters for compliant lubricants.

Table 3.1 summarizes the key regulatory frameworks and standards.

TABLE 3.1 Summary of Key Regulatory Frameworks and Standards

REGULATION/ STANDARD	SCOPE/FOCUS	KEY IMPACT ON LUBRICANTS
REACH (EU)	Chemical registration, safety evaluation, ingredient disclosure	Requires disclosure and registration of lubricant ingredients above thresholds; noncompliance restricts market access
TSCA (USA)	Chemical inventory control, risk assessment, reformulation mandates	Impacts base oil/additive introduction; may require product reformulation per EPA findings
GHS (Global)	Hazard classification, labeling, Safety Data Sheet format	Mandates uniform safety labeling and SDS; affects international shipping and compliance
EPA Vessel General Permit	Use of biodegradable, low-toxicity lubricants near/in water	Mandates environmentally acceptable lubricants for marine use; excludes conventional mineral oils
NSF H1/ISO 21469	Food-grade lubricant composition and hygienic manufacturing	Restricts additives to food-safe lists; ISO 21469 includes production GMP compliance
OECD 301/310	Biodegradability testing of lubricant components	Used for ecotoxicity screening; essential for lubricants in discharge-sensitive sectors
Ecolabel certifications	Environmental performance (biodegradability, emissions, packaging)	Used by OEMs/fleets to prequalify lubricants based on sustainability and eco-profile

ENVIRONMENTAL PERFORMANCE CRITERIA

Biodegradability

Biodegradability defines how quickly and completely a lubricant breaks down in the environment. "Readily biodegradable" fluids achieve ≥ 60% degradation within 28 days, meeting criteria set by standardized tests like OECD 301B and ISO 9439. Inherently biodegradable products degrade slower or incompletely, often leaving persistent by-products. Formulations relying on synthetic esters or saturated vegetable oils tend to score higher, while traditional mineral oils and poorly refined base stocks rarely meet these benchmarks.

Aquatic Toxicity

Lubricants must not endanger aquatic organisms in either short- or long-term exposure. Acute toxicity is measured via LC50 (lethal concentration for 50% of the test population) or EC50 (effective concentration for 50% of the test population), typically on fish, Daphnia, or algae. Chronic toxicity thresholds demand even lower concentration tolerances. Compliance requires avoidance of substances known to bioaccumulate or cause sublethal harm. Zinc, chlorine, organophosphates, and certain phenols are frequently excluded or limited due to their documented aquatic toxicity.

Bioaccumulation Potential

Lubricants and their constituent chemicals are evaluated for potential to accumulate in organisms. Calculated using the logarithmic partition coefficient, lipid solubility values above 3 generally signal higher bioaccumulation potential. Lubricants with persistent, bioaccumulative, and toxic characteristics, such as polybutylene terephthalate (PBT), are disqualified under many environmental standards. Base oil structure plays a central role; branched-chain hydrocarbons and stable aromatic compounds are more resistant to metabolic breakdown, increasing their regulatory risk profile.

Table 3.2 summarizes the environmental performance criteria.

TABLE 3.2 Summary of Environmental Performance Criteria

CRITERIA	DEFINITION	STANDARD TEST METHODS	BENCHMARK LIMITS	HIGH-RISK COMPONENTS	BEST PRACTICE FORMULATIONS
Biodegradability	Measures how rapidly and completely a lubricant breaks down in the environment	OECD 301B, ISO 9439, OECD 310	Readily biodegradable: ≥ 60% degradation within 28 days	Poorly refined mineral oils, aromatic hydrocarbons	Synthetic esters, saturated vegetable oils, biodegradable polyalkylene glycols
Aquatic toxicity	Assesses the toxicity of lubricants to aquatic organisms under short- and long-term exposure	OECD 201 (algae), 202 (Daphnia), 203 (fish)	Acute toxicity LC50/EC50 >100 mg/L preferred; chronic values significantly lower	Zinc compounds, chlorine, organophosphates, phenols	Formulations free of heavy metals and persistent toxins
Bioaccumulation potential	Evaluates potential of lubricant components to accumulate in living organisms	Log Kow (octanol–water partition coefficient), OECD 305	Log Kow < 3 preferred to minimize bioaccumulation risk	Branched-chain hydrocarbons, stable aromatics	Use low-persistence base stocks and avoid PBT-designated substances

46

FORMULATION CONSTRAINTS AND COMPLIANCE STRATEGIES

Additive Restrictions

Environmental directives often limit the use of chlorine, sulfur, and phosphorus, measured collectively as ash content, due to their potential environmental and catalytic impacts. Alternatives include boron-based compounds, molybdenum derivatives, and organic friction modifiers designed to maintain performance without breaching ecotoxic thresholds. Ashless anti-wear agents and dispersants are also gaining traction in environmentally sensitive formulations.

Base Oil Selection for Eco-Compliance

Eco-compliant lubricants often rely on nontoxic, biodegradable base stocks such as synthetic esters (HEES), polyalkylene glycols (HEPG), and saturated vegetable oils (HEPR). These offer favorable biodegradability and toxicity profiles. Compared with Groups I–IV mineral or synthetic base stocks, these fluids perform better in OECD and ISO environmental evaluations but may face challenges in oxidative stability or seal compatibility. Selection must balance regulatory compliance with performance metrics under load, temperature, and contamination exposure.

Life-Cycle Impact and Disposal Management

End-of-life lubricant management is tightly regulated. Spent lubricants must be classified for disposal under frameworks like RCRA in the U.S. or the Waste Framework Directive in the EU. Many regions enforce take-back programs or mandate container recovery. Fluid reconditioning or on-site recycling (e.g., centrifugation, dehydration, or polishing) can extend lubricant life and reduce disposal volumes. Documentation and labeling must reflect all hazardous constituents and disposal recommendations to comply with local transport and waste codes.

DOCUMENTATION AND TRANSPARENCY REQUIREMENTS

Safety Data Sheets and Labeling

Lubricants must be accompanied by Safety Data Sheets that meet the minimum disclosure requirements under GHS and REACH. Critical data includes chemical composition, hazard classifications, handling precautions, and emergency procedures. Import/export scenarios require harmonized labeling, including signal words, pictograms, and hazard statements. Variations in national implementation of GHS necessitate review of regional compliance specifics, especially when shipping across borders.

Technical Data Sheets and Certification

Technical Data Sheets must present validated performance data, material compatibility, and safety parameters. Performance testing should conform to recognized standards (e.g., ASTM, ISO, DIN) to ensure repeatability. Certifications for environmental or food-grade compliance, such as NSF H1 or EU Ecolabel, require either third-party audit or robust self-declaration. Accuracy in documentation is essential for regulatory inspections, customer audits, and warranty support, especially when dealing with equipment operated in regulated zones.

Table 3.3 summarizes the documentation and transparency requirements.

ENGINEERING TRADE-OFFS IN REGULATED APPLICATIONS

Managing Performance Without Prohibited Additives

In applications where substances like ZDDP or chlorinated paraffins are restricted, formulators must substitute with borate esters, ashless phosphorus compounds, or organomolybdenum additives. These alternatives often provide lower anti-wear performance under boundary lubrication regimes. The absence of chlorine and sulfur-bearing compounds also reduces oxidative stability and load-carrying capacity, requiring compensatory base oil selection and higher additive treat rates.

TABLE 3.3 Summary of Documentation and Transparency Requirements

REQUIREMENT TYPE	PURPOSE	REGULATORY STANDARDS	KEY REQUIREMENTS	COMPLIANCE NOTES
Safety Data Sheets	Disclose chemical composition, hazards, and safety measures per GHS/REACH standards	GHS, REACH, OSHA HCS (Hazard Communication Standard), 29 CFR 1910.1200	Include composition, hazard classification, handling instructions, PPE, and first aid measures in 16-section format	Must be reviewed for each country of distribution due to regional GHS implementation differences
GHS-compliant labeling	Ensure consistent hazard communication with pictograms, signal words, and hazard statements	GHS (UN Purple Book), regional adaptations (EU CLP, OSHA GHS)	Apply correct pictograms, signal words (e.g., "Danger," "Warning"), and H-statements relevant to product risks	Label accuracy is critical for customs clearance and downstream hazard management
Technical Data Sheets	Present validated lubricant performance, compatibility, and usage limits in standard formats.	ASTM, ISO, DIN for test data validation	Report viscosity, VI, flash point, pour point, corrosion test results, and compatibility ratings based on recognized methods	All data should be derived from standardized test procedures for reproducibility and customer assurance
Certifications and declarations	Demonstrate compliance with environmental, food-grade, or sustainability certifications	NSF H1, ISO 21469, EU Ecolabel, Blue Angel, Nordic Swan	Provide an audit trail or self-declaration with method references; maintain records for inspection or customer review	Third-party certifications increase market credibility; misrepresentation may result in fines or delisting

Real-World Performance Adjustments

In the field, operators may reduce oil change intervals to maintain performance when using environmentally compliant fluids. This offsets diminished oxidative resistance and thermal stability. Condition monitoring tools, such as particle count analysis, FTIR for oxidation, and RULER tests for antioxidant depletion, become essential. These programs support safe extension of drain intervals and prevent unplanned failures in systems constrained by regulatory lubricant requirements.

SECTOR-SPECIFIC IMPLEMENTATION

Marine Applications

Environmentally acceptable lubricants are required in U.S. waters under EPA's VGP regulation. These are used in stern tube bearings, thrusters, and deck machinery. Lubricants must resist water washout while remaining biodegradable and nontoxic. Synthetic esters and PAGs are common, often formulated with minimal ash and nontoxic additives to meet dual-performance criteria.

Food and Beverage Industry

Lubricants must meet NSF H1 (incidental contact), H2 (noncontact), or 3H (direct food contact) classifications. Formulations exclude toxic metals, prioritize white oils or food-grade synthetics, and use FDA-approved thickeners and additives. Incidental contact safety is validated through migration testing and toxicological assessments.

Agriculture and Forestry

Biodegradable hydraulic fluids are mandated in areas with potential soil or water exposure. ISO 15380 provides test and classification protocols for HEES, HEPG, HEPR, and HETG fluid types. Compatibility with standard elastomers, thermal stability, and field service life are required for adoption.

Mining and Construction

In regions with strict environmental controls, hydraulic and gear oils must meet local toxicity thresholds and demonstrate minimal groundwater impact. Case studies include the use of vegetable-based fluids in open-pit mining and biodegradable synthetics in forestry harvesters operating near protected watersheds.

Future Regulatory Trends

Regulatory pressure is increasing toward the development of zero-toxicity lubricant classes. This involves eliminating compounds with chronic ecotoxicity, endocrine disruption potential, or long-term bioaccumulation behavior. Authorities are focusing on full life-cycle environmental impacts rather than just end-of-pipe controls. Global harmonization efforts are intensifying across REACH (EU), EPA (USA), and the GHS (UN) to create consistent classification, labeling, and chemical disclosure practices. This will impact formulation strategies and material registration processes worldwide. Additionally, sustainability metrics such as carbon intensity, renewability of base stocks, and recyclability are emerging as specification criteria in tenders and product approvals, especially in public and large-scale infrastructure projects.

CHAPTER 4

FUNCTIONAL SUBSTITUTION: WHEN ONE LUBRICANT MUST DO THE JOB OF TWO

In complex industrial environments, operational simplification can be as valuable as performance optimization. Functional substitution is the process of selecting or engineering a lubricant capable of meeting the requirements of multiple applications, thereby reducing the SKU count, the inventory cost, and the risk of cross-contamination.

To achieve successful substitution, lubricants must demonstrate overlapping performance characteristics across domains such as load, temperature, contamination exposure, and material compatibility. For example, a multifunctional synthetic gear oil might serve both high-load reducers and air compressors if formulated with the correct additive balance and viscosity profile.

However, the risks of overgeneralization are high. Functional substitution must be data-driven and validated through rigorous testing including load capacity tests, oxidation stability, filterability, and elastomer compatibility. Furthermore, software tools and decision trees should be used to identify intersection points in lubricant performance requirements.

The economic and logistical benefits of substitution must be balanced with long-term reliability concerns. Case studies have shown that aggressive substitution without technical validation often leads to premature wear, additive incompatibility, and lost warranties.

THE CASE FOR LUBRICANT SIMPLIFICATION

Functional substitution is a strategic approach to reduce lubrication complexity in operations running multiple machine types or platforms. When a lubricant can safely serve more than one application, inventory can be streamlined, risk of cross-contamination minimized, and procurement simplified. This is increasingly relevant in facilities where storage constraints, budget pressure, or limited technician expertise makes SKU reduction a priority. Success hinges on identifying lubricants with sufficient performance overlap across targeted applications.

The selection process starts with mapping the critical parameters of each application. Load severity, operating speed, temperature fluctuations, environmental exposure, and system metallurgy must be compared. A lubricant cannot be chosen solely on matching viscosity grades. For instance, a synthetic ISO 150 gear oil with high thermal stability and a balanced EP/AW additive system might double as a compressor oil, but only if volatility, foam control, and copper corrosion metrics meet both sets of requirements.

Failure modes tied to poor substitution often trace back to unvalidated assumptions. A lubricant optimized for gearbox load may fail in a moisture-prone compressor sump if water separation is inadequate. Similarly, substitution across elastomer types can lead to swelling or embrittlement. Test protocols must include elastomer compatibility, filterability under fine micron ratings, and oxidation stability at application-specific temperatures.

Functional substitution requires a performance envelope matrix. Tools such as decision trees or algorithmic software help identify intersecting nodes in performance characteristics. Lab validation is mandatory. ASTM methods such as D2893 (oxidation), D892 (foam), and D943 (oxidation life) should be standard. Field trials should run over multiple maintenance cycles.

Economic gains depend on execution. Substitution that prevents one unscheduled outage or failed seal can offset months of fluid cost. However, shortcutting validation leads to warranty losses, reduction in MTBF (mean time between failures), or untraceable reliability drift. Engineering teams must be involved in qualification testing. Product data sheets are insufficient unless backed by real-world trials. Use of fluid condition monitoring and drain interval trending adds safety margins.

In facilities with mixed assets, gear reducers, hydraulics, or air compressors, a tiered strategy may be optimal. Grouping applications by shared functional demands, then finding a common lubricant for each tier, offers better control than blanket substitution. Custom-formulated fluids may be viable when substitution ROI justifies R&D expense.

Functional substitution works when treated as an engineered process. It fails when treated as an administrative shortcut. Decisions must be based on technical data, validated against application parameters, and continuously monitored in the field.

Defining Functional Substitution

Functional substitution refers to the engineered or validated use of a single lubricant across multiple machine applications where operating conditions, materials, and performance requirements intersect. The goal is to reduce lubricant variety without sacrificing equipment integrity or operational performance. This approach is most effective when applications share comparable demands in film strength, thermal range, contamination exposure, and elastomer compatibility.

Common application pairs where substitution is frequently viable include gear reducers and air compressors, particularly when both operate under similar speed and load regimes. In such cases, a synthetic gear oil with balanced anti-wear, EP, and oxidation resistance properties may serve both roles. Hydraulic and bearing systems are another candidate pair, especially when both systems use moderate pressure, experience low-to-moderate thermal cycling, and share exposure to the same contaminants or operating fluids. In mobile equipment, multiuse fluids are sometimes deployed to service gearboxes, hydraulic circuits, and even wet brake systems, provided all functional loads fall within the lubricant's tested operating window.

Substitution boundaries are defined by divergence in operational stressors or materials of construction. For instance, gear oils formulated with high levels of sulfur-phosphorus EP additives are incompatible with compressors using yellow metals or sensitive seals. Likewise, compressor oils with very low ash and detergent content may fail under shock load or low-speed boundary conditions typical of gear reducers. Substitution fails when thermal extremes, fluid shear degradation, or additive incompatibility introduces wear, noise, or leakage. Validation through lab testing and limited field trials is mandatory before any substitution becomes standard practice.

Identifying Crossover Requirements

Effective functional substitution begins with identifying shared mechanical and environmental conditions across the target applications. Load-bearing overlap is a critical baseline. Lubricants must accommodate the peak and sustained load ranges of both systems, with consideration for hydrodynamic, mixed, and boundary lubrication regimes. Thermal compatibility is also essential. Operating temperature bands must be examined not only for average values but for transient spikes, cooldown rates, and ambient extremes. For example, a lubricant suited for an enclosed gear reducer running at 85°C must also tolerate intermittent compressor discharge pipe temperatures exceeding 100°C without oxidation breakdown or viscosity drift.

Speed regime evaluation determines whether shear stability and film formation behavior align across systems. High-speed applications may demand low-viscosity, shear-stable fluids, while slow-turning gearboxes may prioritize load capacity and anti-wear chemistry. The substitution candidate must fall within the overlapping performance envelope. Environmental exposure further narrows options. The lubricant must resist emulsification, particulate contamination, and degradation from process chemicals if present in either system.

Material compatibility must be validated across all wetted components. Seals and elastomers in compressors and gearboxes may differ in polymer type and tolerance to base oil polarity and additive reactivity. A PAO-based oil that performs well in nitrile seals may cause shrinkage or hardening in fluorosilicone or EPDM (ethylene propylene diene monomer) components. Similarly, the lubricant's interaction with filter media, adhesives, and anti-drainback materials must be tested. Foam control and air-release behavior may also differ based on the system's circulation rate and sump geometry. Compressors often present higher churning conditions with smaller reservoirs, while gearboxes may tolerate longer air-release cycles.

System architecture, including pressure ranges, circulation speeds, and reservoir design, dictates whether a single fluid can serve both systems without compromising film integrity or creating operational lag. Filtration requirements vary as well; ISO cleanliness levels may be more stringent in hydraulic or compressor circuits than in enclosed gearboxes. Substitution must ensure that fluid properties don't compromise filter performance or shorten maintenance intervals.

Formulation Criteria for Substitutable Lubricants

Base oil selection drives compatibility across dissimilar systems. PAOs offer the most consistent performance across high-load, high-speed applications due to their oxidative stability, low volatility, and thermal resistance. Their low polarity requires tuning with cosolvents or esters to dissolve performance additives and maintain seal compatibility. Synthetic esters provide better solvency and biodegradability, often beneficial in systems with potential exposure to moisture or food processing environments. However, ester reactivity with elastomers and risk of hydrolysis must be accounted for in both compressor and gearbox architectures. PAGs provide natural resistance to varnish and excellent thermal conductivity but are limited by their immiscibility with mineral oils and potential incompatibility with common seals and filters. Suitability depends on the base oil's performance across the shared environment's volatility threshold, oxidative load, and contaminant profile.

Additive chemistry must accommodate overlapping wear, corrosion, and contamination risks. For high-load reducers and compressors, the balance between extreme pressure (EP) and anti-wear (AW) additives must be fine-tuned. Overuse of EP chemistries such as sulfur-phosphorus compounds may lead to corrosion in yellow metals common in compressors and control valves. AW systems based on ZDDP may also need adjustment for systems with filtration or emission constraints. Foam inhibitors must function under both churning air entrainment in compressors and the slower sump dynamics of gearboxes. Demulsifiers are required in both applications but at different response times; fast water separation is critical for gearboxes, while emulsified tolerance may be acceptable in sealed compressors with low exposure.

Corrosion inhibitors must address multiple metal types and condensation risks, especially in vertically mounted or intermittently cycled equipment. Formulations must avoid additive combinations that impair filterability, especially across systems with fine-micron filters or those operating under high differential pressure. Compatibility with cellulose, glass fiber, and synthetic filter media must be verified under ISO 13357 and D6316 filtration tests. Successful substitution requires achieving chemical equilibrium across diverging equipment without compromising individual system requirements.

Technical Validation Requirements

Substitutable lubricants must be validated through rigorous laboratory testing before field deployment. Load-carrying capacity is benchmarked using FZG scuffing tests for gear applications and Timken OK Load for boundary film performance. Oxidation stability is evaluated through RPVOT (ASTM D2272) for turbine and gear oils or through PDSC for rapid assessment of oxidative resistance. Wear protection must be confirmed with four-ball wear and weld point data, ensuring performance across anticipated contact regimes in both applications. Compatibility testing is nonnegotiable; seals, filter media, and coatings used in both systems must be exposed to the candidate lubricant under heat and soak conditions, typically 70°C to 150°C for elastomer swelling and degradation screening. A drop in filter pressure across standard micron ratings must be logged using ISO 4548-12 or D7621 protocols to confirm flow behavior and contaminant retention without excessive delta-P or bypass.

Field validation requires active monitoring. Temperature trends must be tracked at bearing housings, gearboxes, and compressor heads using in-situ thermocouples or thermal imaging. Wear-metal trends from inductively coupled plasma (ICP) spectroscopy are essential to detect gear or bearing distress. Particle count via ISO 4406 should be trended before and after substitution to flag any changes in system cleanliness or lubricant filtration performance. Equipment operating under substitutable lubricants must be benchmarked side by side against the previously used fluids for noise, thermal stability, and component surface condition. Maintenance teams must adjust drain intervals and relubrication schedules based on lubricant aging profiles captured through TAN/TBN (total base number) drift, oxidation/nitration peaks (FTIR), and viscosity change from baseline. Acceptance criteria must be data-driven, documented through oil analysis and component inspection to support continued use or reversion to separate fluids.

Tools for Substitution Decision-Making

Effective lubricant substitution depends on structured analysis. Decision trees built around core lubricant properties, viscosity range, temperature stability, base oil compatibility, and additive function help determine feasibility. Application-matching matrices list system requirements by asset class and compare them against candidate lubricant capabilities. These tools should flag nonnegotiables like incompatible elastomers, excessive temperature gradients, or filtration incompatibility. Red-flag criteria include load regimes beyond verified test thresholds, seal incompatibility by material class (e.g., EPDM versus PAO), and exposure to cross-system contaminants like process chemicals or water ingress that one application may tolerate but the other cannot.

Software platforms extend decision accuracy. Expert systems embed tribology databases with ASTM test results and OEM specifications, allowing users to match machine demands with validated lubricant performance profiles. Modeling tools simulate film thickness, temperature rise, and degradation under specific system stressors. AI-based suggestion engines use historical performance data, oil analysis trends, and environmental parameters to recommend viable substitution options. These systems can reduce substitution risk by flagging outliers in compatibility or predicting the degradation curve under mixed-duty conditions. The best results come from pairing algorithmic recommendations with real-world validation and domain expertise.

Benefits and Pitfalls of Functional Substitution

Properly implemented substitution reduces cost and complexity. Fewer SKUs lower procurement errors, reduce training scope, and simplify labeling, storage, and safety compliance. Inventory carrying cost decreases, and cross-contamination risk drops due to fewer incompatible lubricants in proximity. Technicians spend less time matching lubricants to equipment, especially in multi-machine facilities where overlapping lubricant duties are common. In regulated environments, substitution can streamline audits by consolidating environmental documentation and SDS requirements.

The risks, however, are significant. Additive profiles designed for one function may fall short in another. For instance, a lubricant with strong EP characteristics may not offer adequate oxidative stability for high-speed compressors, leading to varnish or deposit formation. Seal incompatibility may only appear after thermal cycles or long-term exposure. Filter plugging can occur if soot or oxidation by-products increase under different duty cycles. Failures often trace back to unverified assumptions, extrapolated from similar, not identical, applications. Substitute lubricants must be proved under load, temperature, material, and contamination conditions of both systems, or they represent a latent risk embedded in every maintenance cycle.

Table 4.1 summarizes function substitutions.

TABLE 4.1 Summary of Function Substitutions

ASPECT	SUMMARY
Definition	Use of one lubricant to serve multiple systems with overlapping requirements to reduce SKUs and simplify operations
Primary benefits	Reduces inventory, cross-contamination risk, and procurement complexity and simplifies compliance
Core risks	Additive incompatibility, seal degradation, reduced performance, warranty loss due to unverified assumptions
Critical success factors	Overlap in load, speed, temperature, contamination, and material compatibility; validated with lab and field tests
Candidate applications	Gear reducers and compressors, hydraulics and bearings, mobile machinery systems with unified thermal/mechanical loads
Selection criteria	Must match viscosity, shear stability, oxidation resistance, filterability, and elastomer compatibility across systems
Formulation considerations	Base oils (PAO, esters, PAGs) and additives must support EP/AW balance, corrosion inhibition, foam/demulsifier tuning
Validation requirements	Requires ASTM tests (FZG, D2272, four ball), seal and filter testing, in-field thermals, ICP and particle count trending
Decision tools	Use decision trees, property matrices, expert systems, and software modeling to assess crossover feasibility
Common pitfalls	Unvalidated substitutions result in wear, filter plugging, NVH issues, or reliability loss—especially under thermal or material mismatch

CASE STUDIES

Success Case: Multi-Duty Synthetic Gear Oil Across Reducers and Blowers
A food processing facility operating both high-torque gear reducers and positive displacement blowers transitioned to a single synthetic gear oil based on PAO with a balanced EP and AW additive package. The original configuration required two lubricants: a high-viscosity gear oil for reducers and a separate blower oil with superior oxidative stability. Cross-analysis showed that both applications shared similar ambient temperature ranges, low water exposure, and moderate operating loads. Lab validation confirmed adequate film thickness, copper corrosion stability, and oxidative resistance in both conditions. Filter pressure drop testing showed no increase over baseline, and seal materials across both systems were compatible with the PAO-ester blend. Field deployment reduced lubricant inventory by 50%, eliminated misapplication incidents, and extended average drain intervals by 18%.

Failure Case: Hydraulic-Gear Cross-Use Leading to Varnish and Pump Wear
A mining operation attempted to consolidate lubricants by using a detergent-laden hydraulic fluid across mobile hydraulic units and enclosed reduction gearboxes. Initial compatibility screens focused only on viscosity and base oil. Within 4 months, gearbox inspections showed varnish buildup, gear pitting, and bronze bushing discoloration. Oil analysis detected elevated oxidation products and foam tendencies under high churning conditions. The detergent-disbursing additive package, while appropriate for hydraulic contamination, destabilized under high-shear gear loading, promoting sludge. Filtration units also showed increased restriction due to insoluble residue. The failure resulted in unplanned downtime, warranty voidance, and complete drain-and-flush recovery. Post-failure review identified a lack of shear and oxidation testing under gear-specific operating temperatures.

Tiered Approach: Primary-Substitute Fallback in Remote Asset Sites
A regional water utility managing remote pumping and generator sites adopted a two-tier lubrication strategy. Primary lubricants were optimized for each application, but a vetted fallback fluid was identified through lab testing to meet minimum acceptable performance in both pump and generator lubricated systems. This substitute fluid was stored at remote sites to enable emergency service when the primary inventory was unavailable. The fallback lubricant was selected based on proven compatibility with elastomers, acceptable shear stability, and anti-

wear capacity under moderate load. Field performance logs showed that during supply chain disruptions, substitute use extended equipment uptime without accelerated wear or failure. The program reduced field service delays and avoided high-cost unscheduled maintenance, validating the use of performance-tiered substitution as a strategic resilience measure.

Best Practice Guidelines

- Start with detailed application mapping across all equipment types considered for substitution. Define operating conditions, including load spectra, ambient and internal temperature ranges, contamination risks, and material compatibility. Match lubrication regimes by viscosity, elastomer type, filtration requirements, additive sensitivity, and service interval expectations. Identify performance overlaps, and isolate outliers that may disqualify a universal solution.
- Prioritize lubricants that carry dual certification or have documented multi-application testing under relevant standards such as DIN 51517 for gears and ISO 11158 for hydraulics. Review third-party validation data showing consistent performance in both regimes. Ensure that base oil and additive systems are compatible across all targeted environments, with attention to EP/AW balance, filterability, oxidative stability, and foam control.
- Implement controlled substitution trials under conservative service intervals. Collect baseline oil analysis data and track key indicators, viscosity shift, TAN/TBN, wear-metal trend, and particle count. Use pressure drop and filter condition monitoring to detect additive fallout or incompatibility. Cross-reference runtime data with lab results for validation before full-scale deployment.
- Engage OEMs early to confirm compatibility and avoid warranty conflict. Document all correspondence, including risk mitigation strategies and lubricant certifications. Ensure maintenance teams are briefed on substitution logic, failure modes, and fallback protocols. Maintain clear records of substitution boundaries, and update CMMS entries to reflect validated lubricant mappings. Incorporate ongoing oil condition monitoring as a permanent safeguard.

DYNAMIC LUBRICANT SELECTION USING PERFORMANCE-BASED ALGORITHMS

Advancements in artificial intelligence, sensor integration, and industrial informatics have opened the door to dynamic, condition-based lubricant selection. Rather than static OEM-prescribed products, algorithms now support adaptive recommendations based on real-time and historical operating data.

Performance-based selection algorithms integrate inputs such as ambient temperature, load cycles, vibration patterns, fluid condition, and maintenance intervals to recommend or adjust lubricant choice. These tools can be integrated into CMMSs or digital twin models of machinery.

Machine learning (ML) models trained on operational and laboratory datasets can predict lubricant degradation patterns, enabling more precise drain intervals and product substitutions. This evolution from descriptive to prescriptive lubrication management helps operators adjust formulations in response to system stress or environmental shifts.

The future of lubricant selection lies in its integration with Industry 4.0 platforms. Smart reservoirs, real-time fluid analytics, and cloud-based formulation updates are expected to become increasingly commonplace, allowing the lubrication system itself to request the most appropriate fluid on demand.

The Shift from Prescriptive to Adaptive Lubrication

Static lubrication schedules and OEM-specified lubricant selections are insufficient in dynamic, variable-load environments. OEM recommendations assume stable, nominal conditions and fail to account for operational deviations such as start/stop cycles, temperature excursions, contamination exposure, or system modifications. As a result, these prescriptive models contribute to accelerated lubricant degradation, suboptimal film formation, and increased failure rates.

Field data confirms that operating conditions rarely match design assumptions. Load patterns fluctuate, thermal envelopes shift, and contamination ingress varies across duty cycles. Prescriptive lubrication does not respond to these deviations. Adaptive lubrication, driven by condition monitoring and real-time data, corrects this mismatch by selecting, dosing, and modifying lubricant properties based on actual, not theoretical, conditions.

Performance-based algorithms ingest machine data: vibration, temperature, RPM, bearing load, moisture content, and particle counts. These inputs inform lubricant selection frameworks that dynamically recommend base stock type, viscosity grade, additive class, and replenishment frequency. Machine learning algorithms trained on failure modes and tribological behavior adjust selections as field conditions evolve.

Integration with Industry 4.0 platforms enables deployment of these algorithms at scale. Edge computing nodes collect and preprocess sensor data, while centralized analytics platforms evaluate lubricant effectiveness across asset fleets. Feedback loops refine selection algorithms in real time. Systems adjust lubrication intervals, suggest viscosity modifiers, or switch between synthetic and mineral formulations as needed.

In rotational equipment, for example, torque variation and operating temperature influence minimum viscosity thresholds. The algorithm flags marginal hydrodynamic film conditions, prompting a shift to higher VI oils or synthetics with better shear stability. In hydraulic systems, pressure pulsations and fluid aeration profiles are tracked, triggering changes in anti-foam chemistry and base oil solubility characteristics.

Implementation reduces lubrication-related failures, extends mean time between lubricant changes (MTBLC), and improves component life-cycle cost (LCC) profiles. Field deployments have demonstrated up to a 30% reduction in lubricant consumption and over a 40% reduction in lubrication-induced downtime.

Static models are obsolete in variable-condition environments. Lubrication must evolve from time-based and prescriptive to condition-driven and adaptive. Performance-based selection algorithms, powered by real-time telemetry and predictive analytics, are now essential reliability tools.

Foundations of Performance-Based Algorithm Design

Effective algorithm design for lubricant selection depends on accurate, high-frequency input data streams. Core variables include real-time temperature, pressure, mechanical load, vibration, and machine duty-cycle patterns. These metrics define the lubrication environment and directly influence film thickness, oxidative stress, additive depletion, and base oil volatility.

Temperature data must be collected at critical points, bearing housings, sump reservoirs, and load zones. Transient spikes and sustained high temperatures accelerate oxidation and reduce oil viscosity. Pressure sensors identify boundary regime risks in hydraulic and circulation systems. Load fluctuations, especially in variable-torque machinery, change the minimum viscosity requirement for maintaining hydrodynamic film integrity. Algorithms adjust lubricant selection to match load severity and contact stress profiles.

Vibration signatures provide early indicators of surface distress, lubricant starvation, and contamination ingress. High-frequency envelope analysis correlates with bearing inner race spalling and can be tied to lubricant film breakdown. Duty-cycle mapping quantifies operational mode variability, cold starts, ramp-up durations, idle times, and overload events, critical for calculating shear stability and volatility resistance needs.

Historical datasets define degradation pathways and baseline wear rates. Tracking ferrous and nonferrous metals via spectroscopic analysis, along with fluid property shifts such as TAN/TBN drift and oxidation peaks, informs expected fluid life under load-specific conditions. This historical context trains the algorithm to recognize failure precursors and optimize future selections.

Maintenance intervals and operational context determine permissible lubricant change windows, downtime thresholds, and replenishment schedules. In batch production environments, algorithm constraints may prioritize maximum runtime per fill. In continuous-duty assets, protection under sustained thermal load takes precedence.

Effective algorithms integrate these inputs into dynamic decision matrices. Selection outputs include base stock type, additive package class, viscosity index range, and contamination tolerance level. The algorithmic framework continuously recalibrates as new data feeds in, aligning lubricant characteristics to real-world operating demands without human intervention.

Building a dynamic lubricant selection system using performance-based algorithms involves combining tribological science, system engineering, and AI-driven decision tools to select lubricants based on *measured performance under actual or simulated conditions*, rather than static specifications alone. Here's the **structured process** to implement such a system:

1. **Define Operational and Application Parameters**
   Start by mapping the **lubrication requirements** of each asset or machine group:

   - ➤ Load (steady, shock, cyclic)
   - ➤ Speed (RPM ranges, duty cycle)
   - ➤ Temperature (ambient and operating)
   - ➤ Contaminants (water, dust, process fluids)
   - ➤ Metallurgy and materials (seals, coatings, alloys)
   - ➤ Lubrication regime (boundary, mixed, hydrodynamic)
   - ➤ Compliance needs (food grade, biodegradable, low SAPS)

   These become the input variables for the algorithm.

   The following is an example of how to define operational and application parameters for an asset—specifically an integrated electric motor and gearbox (e-axle) used in a battery-electric delivery van.

Asset Description

Component: An e-axle unit (motor + gear reduction + power electronics)
Application: An electric delivery vehicle (urban stop/start cycles)

Table 4.2 shows the mapped operations and application parameters, and Table 4.3 shows the parameters converted to algorithmic input variables.

TABLE 4.2 Mapped Operational and Application Parameters

PARAMETER	DETAILS
Load type	Mixed—steady cruising load with frequent shock loading (start/stop torque surges)
Speed regime	2,000–18,000 RPM; variable-duty cycle; frequent acceleration/deceleration
Temperature range	Ambient: –10°C to 35°C; operating: 80°C to 115°C; hot spots up to 140°C
Contamination risk	Low particulate; moderate moisture from condensation in sealed housing
Materials/metallurgy	Aluminum housing; copper stator windings; NBR seals; sintered bronze bushings
Lubrication regime	Predominantly hydrodynamic, with brief boundary conditions during cold start or acceleration
Compliance needs	Must meet dielectric insulation > 30 kV, nonconductive, low volatility, oxidation-resistant; not food grade

TABLE 4.3 Parameters Converted to Algorithmic Input Variables

INPUT VARIABLE	VALUE OR RANGE
Load type	"Mixed shock"
RPM range	2,000–18,000
Operating temp	80°C–115°C
Peak temp	140°C
Ambient temp	–10°C–35°C
Moisture exposure	Moderate condensation
Material compatibility	["aluminum," "NBR," "copper," "bronze"]
Lubrication mode	"hydrodynamic with boundary"
Compliance requirements	["dielectric 30 kV," "low volatility," "oxidation stable"]

This structured input is what feeds into a dynamic selection algorithm. The system would then filter and rank lubricants based on their proven performance in these conditions—for example, a PAO-based ISO VG 68 fluid with dielectric additives, validated oxidation life (RPVOT > 1,000 min), and low moisture uptake.

2. **Create a Centralized Lubricant Performance Database**
   Compile lab and field data for lubricants, including:

   ➤ **ASTM/ISO test results** (e.g., FZG, four ball, RPVOT, D892, D943, D2272, ISO 4406)
   ➤ **Additive package properties** (EP, AW, antioxidants, demulsifiers, detergents)
   ➤ **Base oil characteristics** (Groups I–V, polarity, solvency, oxidative stability)
   ➤ **Compatibility data** (elastomers, filters, yellow metals, coatings)
   ➤ **OEM approvals and certifications** (NSF H1, ISO 21469, REACH, TSCA)

Organize the data in a relational database or structured knowledge graph. Example 4.1 is a **structured example** of how to create a **centralized lubricant performance database,** showing how lubricant properties are collected and stored for analysis and comparison across multiple products.

EXAMPLE 4.1   Lubricant Performance Database Entry (Relational Table Format)

FIELD	EXAMPLE ENTRY
Product name	SynFlow Ultra PAO 68
Product ID	SFU-PAO-068
Base oil group	Group IV (PAO)
Base oil polarity	Nonpolar
Solvency rating	Low
Viscosity @ 40°C (cSt)	68.4
Viscosity index	155
Pour point (°C)	–45°
Flash point (°C)	240°
RPVOT (ASTM D2272)	1,450 min
FZG load stage (ASTM D5182)	12
Four-ball wear scar (mm, ASTM D4172)	0.42
Foam stability (ASTM D892)	10/0, 10/0, 10/0

FIELD	EXAMPLE ENTRY
Oxidation stability (ASTM D943)	> 2,000 hours
ISO cleanliness (ISO 4406)	16/14/11
EP additive type	Sulfur-phosphorus
AW additive type	ZDDP
Antioxidants	Aminic + phenolic blend
Demulsifier present	Yes
Detergent/dispersant	Calcium sulfonate
Elastomer compatibility	NBR—pass; FKM—caution; EPDM—fail
Filter media compatibility	Compatible with cellulose, glass fiber
Yellow metal compatibility	Caution—limited exposure recommended
OEM approvals	Siemens AG Drive Systems, Flender, Bosch Rexroth
Certifications	NSF H2, ISO 21469-compliant, TSCA listed, REACH registered
Application fit score (e-axle)	88/100 (post-screening)

Back-End Implementation Tips

➤ **Relational database schema (SQL)** could include:
  - Products table (product ID, name, base oil group)
  - Test results table (product ID → FZG, four ball, RPVOT, etc.)
  - Additives table (product ID → AW, EP, etc.)
  - Compatibility table (product ID → materials, filters, seals)
  - Certifications table (product ID → NSF, ISO, REACH, etc.)
➤ For **knowledge graph** (e.g., RDF/OWL):
  - Nodes = lubricant components (base oil, additives, OEMs)
  - Edges = relationships like "has compatibility with," "meets requirement of," "tested by"

EXAMPLE 4.2　Lubricant Performance Database Comparison

PROPERTY	SYNFLOW ULTRA PAO 68	BIOGEAR ESTER 150	HYDROSAFE H1 46
Product name	SynFlow Ultra PAO 68	BioGear Ester 150	HydroSafe H1 46
Product ID	SFU-PAO-068	BGE-EST-150	HSH1-046
Base oil group	Group IV (PAO)	Group V (ester)	Group III (hydrotreated)
Base oil polarity	Nonpolar	Polar	Low polarity
Viscosity @ 40°C (cSt)	68.4	150	46
Viscosity index	155	175	130
Pour point (°C)	−45°	−36°	−30°
Flash point (°C)	240°	270°	220°
RPVOT (min)	1,450	1,300	800
FZG load stage	12	10	11
Four-ball wear scar (mm)	0.42	0.38	0.5
Foam stability (D892)	10/0, 10/0, 10/0	20/0, 10/0, 10/0	10/0, 10/0, 10/0
Oxidation stability (hours)	2,000	1,800	1,200
ISO cleanliness code	16/14/11	17/15/12	18/16/13
EP additive type	Sulfur-phosphorus	Ashless EP	None
AW additive type	ZDDP	Phosphate ester	Food-grade AW
Antioxidants	Aminic + phenolic	Phenolic only	Phenolic only
Demulsifier present	Yes	Yes	No
Detergent/ dispersant	Calcium sulfonate	Ashless dispersant	None
Elastomer compatibility	NBR—pass; FKM—caution; EPDM—fail	FKM—pass; NBR—pass; EPDM—caution	All pass (NSF H1)
Filter compatibility	Cellulose, glass fiber	Glass fiber only	Cellulose, glass fiber
Yellow metal compatibility	Caution	Pass	Pass
OEM approvals	Siemens, Flender, Rexroth	Buhler, Voith	NSF H1, USDA-approved
Certifications	NSF H2, ISO 21469, REACH	EU Ecolabel, Blue Angel	NSF H1, ISO 21469, TSCA
Application fit score (e-axle)	88	70	55

3. **Develop a Lubricant Performance Envelope Model**
   Construct performance envelopes using:

   ➤ Load versus speed versus temperature tolerance mapping
   ➤ Shear stability curves
   ➤ Oxidation degradation models
   ➤ Compatibility thresholds (materials, filters, seals)
   ➤ Safety/environmental limits (flash point, toxicity, biodegradability)

   This is used to *mathematically describe what a lubricant can handle.*

   Examples 4.3A–F show how to develop a lubricant performance envelope model for "SynFlow Ultra PAO 68" using real-world parameters. This model describes the *functional operating window* of a lubricant, enabling algorithmic selection, substitution decisions, or failure prevention.

EXAMPLE 4.3A   Operating Envelope Mapping: Load Versus Speed Versus Temperature*

PARAMETER	MIN	OPTIMAL RANGE	MAX
Load (MPa)	0.5	1.5–3.5	5.0
Speed (RPM)	100	1,000–15,000	20,000
Temp (°C)	–40°	10°–110°	135°

*Operates optimally in high-speed, medium-load systems with moderate thermal cycling.

EXAMPLE 4.3B   Shear Stability Curve (ASTM D6278)*

TEST CONDITION	INITIAL VISCOSITY (CST)	FINAL VISCOSITY (CST)	SHEAR LOSS (%)
100 cycles @ 30,000 psi (CEC L-45-A-99)	68.4	64.1	6.3%

*Shear-stable for high-speed electric motors and compressors.

EXAMPLE 4.3C   Oxidation Degradation Model*

TEST	VALUE	THRESHOLD	TIME TO FAILURE
RPVOT (ASTM D2272)	1,450 min	≥1,000 min	~3–5 years typical use
PDSC (°C onset)	235°	>200°	Good resistance
FTIR oxidation index	<1.0 (start)	1.5 (limit)	Monitored over 18 months

*Exhibits slow oxidation rate under heat and pressure, making it suitable for extended drain intervals.

EXAMPLE 4.3D   Compatibility Thresholds*

COMPONENT	COMPATIBILITY RATING	NOTES
NBR	Pass	No swelling or hardening after 1,000 hours @ 100°C
FKM (Viton)	Caution	Minor shrinkage observed
EPDM	Fail	Severe embrittlement
Cellulose filters	Pass	Stable differential pressure (ΔP) at 10 μ
Glass fiber filters	Pass	No media degradation
Yellow metals (brass)	Caution	Mild tarnish after extended exposure

*Requires elastomer review before use in EPDM-sealed components.

EXAMPLE 4.3E   Environmental and Safety Limits*

PROPERTY	VALUE	THRESHOLD	COMPLIANCE
Flash point (°C)	240°	> 200°	✔ Meets safety margin
Biodegradability	40% (28 days)	≥ 60% for EAL status	✔ Not EAL-compliant
LC50 (OECD 203, fish)	> 1,000 mg/L	> 100 mg/L	✔ Nontoxic aquatic
REACH/TSCA listed	Yes	Required	✔ Fully registered

*Environmentally safe, but not classed as biodegradable under OECD 301B.

EXAMPLE 4.3F   Example of Mathematical Use

CATEGORY	MIN	OPTIMAL RANGE	MAX	NOTES
Load (MPa)	0.5	1.5–3.5	5	Operates under high load
Speed (RPM)	100	1,000–15,000	20,000	High-speed tolerance
Temperature (°C)	–40°	10°–110°	135°	Moderate thermal range
Initial viscosity (cSt)		68.4		CEC L-45-A-99 test
Final viscosity (cSt)		64.1		Shear stability after 100 cycles
Shear loss (%)		6.3%		Low shear loss
RPVOT (min)		1,450		Oxidation resistance
PDSC onset (°C)		235°		Thermal stability
FTIR oxidation index	< 1.0	Monitored	Limit: 1.5	Stable over 18 months
NBR compatibility	Pass			No swelling at 100°C
FKM compatibility	Caution			Minor shrinkage
EPDM compatibility	Fail			Embrittlement observed

CATEGORY	MIN	OPTIMAL RANGE	MAX	NOTES
Flash point (°C)	240°	> 200°		Safe margin
Biodegradability (28 days)	40%	≥ 60% for EAL		Not EAL-compliant
LC50 (mg/L)	>1,000	>100		Nontoxic
REACH/TSCA	Yes	Registered		Fully listed

4. **Implement a Rule-Based and Machine Learning Algorithmic Engine**
    Combine two layers:

    ➤ **Rule-based filters.** Eliminate incompatible options based on nonnegotiables (viscosity mismatch, chemical incompatibility, regulation).
    ➤ **Performance scoring algorithm.** Use weighted criteria to rank lubricants by suitability, using a scoring matrix (e.g., on a 0–100 scale).

    Then optionally layer in:

    ➤ **Supervised ML models** trained on past lubricant selection outcomes, oil analysis data, and MTBF history
    ➤ **Bayesian inference** to update recommendations based on new field data

    Table 4.4 is an updated comparison table showing five lubricant candidates evaluated using a weighted performance scoring system. Each lubricant's suitability score considers dielectric strength, oxidation stability, and biodegradability.

TABLE 4.4 Comparison of Five Lubricant Candidates

LUBRICANT	VISCOSITY (CST)	DIELECTRIC STRENGTH (KV)	OXIDATION STABILITY (HRS)	BIODEGRADABLE	SUITABILITY SCORE
Lube X	68	42	1,400	TRUE	88.6
Lube Y	46	35	800	FALSE	48
Lube Z	100	50	1,600	TRUE	100
Lube W	68	38	1,200	TRUE	80.4
Lube Q	68	45	1,000	FALSE	61

5. **Build a User Interface and Decision Tree Input Tool**
    Allow users to enter key operational inputs through:

    ➤ Dropdown menus for asset type
    ➤ Numeric fields for load, speed, and temperature

> ➤ Checkboxes for environmental or compliance constraints
> ➤ The interface queries the algorithm and returns:
> ➤ Top three to five lubricants by score
> ➤ Application notes and caveats
> ➤ Expected change in performance indicators (e.g., extended drain interval)

Table 4.5 summarizes the steps to develop a user interface and decision tree input tool for dynamic lubricant selection.

TABLE 4.5   Steps to Develop a User Interface and Decision Tree Input Tool

STEP	DESCRIPTION	TOOLS/ TECHNOLOGIES	DELIVERABLES
1. Define requirements	Identify functional goals (input, filter, recommend lubricants) and target users	Stakeholder input Requirements document	Functional specification document
2. Map input/output variables	List and define all input parameters (load, speed, temperature, compliance) and expected outputs	Domain expertise ISO/API application standards	Variable definitions and user input schema
3. Develop decision logic	Build logic that filters incompatible options and scores remaining lubricants	Rule-based filters Scoring matrices Optional ML models	Decision tree and scoring algorithm design
4. Design user interface	Create a user-friendly front end for data entry and result display	React/Vue.js/HTML/ CSS Figma (for design)	UI wireframes and functional mock-ups
5. Do back-end integration	Build a back end to receive inputs, run logic, and return results	Python (FastAPI/Flask) Node.js PostgreSQL or SQLite	API end points and logic integration
6. Interpret output	Build messaging logic for performance caveats, compliance flags, and application notes	Rule tagging Warning generation logic	Annotated results with performance insights
7. Test and validate	Perform input validation and unit tests, and compare against known field cases	PyTest Postman Manual test cases	Test cases, validation reports, pass/fail logs
8. Deploy	Host the app and ensure secure access and uptime	Vercel, Netlify (front end) AWS/Heroku (back end) S3 or Cloud DB	Live web app, database connection, SSL security
9. Monitor and update	Add analytics and user feedback, and schedule database updates	Google Analytics Admin dashboard	Feedback loops, updated lubricant database

6. **Validate Recommendations Through Lab and Field Trials**

For each algorithmic recommendation:

➤ Run **ASTM validation tests** for oxidation, wear, and filterability.
➤ Perform **soak tests** for seals and materials.
➤ Conduct **short field trials** with temperature tracking, oil analysis (ICP, FTIR), and particle counts.

Use this real-world data to train or refine the model.

Table 4.6 illustrates the process of validating lubricant recommendations through lab and field trials, complete with example methods and goals for each step.

TABLE 4.6   Steps to Validate Lubricant Recommendations Through Lab and Field Trials

STEP	ACTIVITY	METHOD/ STANDARD	PURPOSE	EXAMPLE OUTCOME
1. Lab oxidation testing	Assess resistance to oxidative breakdown	ASTM D943 (TOST), ASTM D2272 (RPVOT)	Verify long-term stability under thermal stress	Lubricant A exceeds 1,000 hours in D943 test, indicating extended drain capability
2. Wear protection testing	Evaluate anti-wear performance under boundary conditions	ASTM D4172 (four-ball wear), ASTM D2783 (weld point)	Confirm protection against metal-to-metal contact	Lubricant B shows wear scar diameter < 0.4 mm; passes OEM spec
3. Filterability testing	Ensure lubricant maintains flow through fine filters	ISO 13357-1 (filterability), ASTM D7621	Confirm compatibility with system filtration	Lubricant C maintains flow with < 10% pressure drop across 10-μm filter
4. Elastomer soak testing	Check material compatibility (swelling, cracking, hardening)	ASTM D471 (rubber property change)	Prevent seal degradation or leakage	Lubricant D causes < 5% volume change in Viton and NBR seals
5. Field trial setup	Install candidate lubricant in limited assets for real-world evaluation	OEM procedure + instrumentation	Collect data in operational environment	Three compressors charged with Lubricant E, monitored over 500 hours
6. Field monitoring: temperature	Measure running temps at bearings, gearboxes, sumps	Infrared thermography or thermocouples	Detect thermal load differences and heat transfer efficiency	Average temp drop of 3°C using Lubricant E vs. control
7. Field monitoring: oil analysis	Analyze used oil samples over trial period	ICP (wear metals), FTIR (oxidation/ nitration), ISO 4406 (particles)	Track lubricant health and component wear	Lubricant A shows no significant iron/copper rise; oxidation index stable
8. Model feedback loop	Use test and field data to refine scoring model or machine learning weights	Manual input + ML training (if used)	Improve algorithmic prediction accuracy	Oxidation stability adjusted to a higher value in a sealed system

This structured validation ensures that algorithm recommendations are not only theoretically sound but also practically verified in relevant operational contexts.

7. **Continuously Monitor and Retrain**
Integrate the system with:

- ➤ **CMMS** or **EAM** (enterprise asset management) platforms for field data
- ➤ **Oil analysis labs** for result feeds
- ➤ Maintenance logs and warranty claims

Retrain algorithms monthly or quarterly to improve accuracy and adapt to new chemistries or machine behavior.

Optional Advanced Enhancements

- ➤ Digital twin simulations of lubricant behavior under variable-load and thermal conditions
- ➤ Predictive modeling of oxidation, wear-metal generation, or TAN increase
- ➤ Integration with procurement systems to auto-suggest compliant, in-stock lubricants

Table 4.7 is a structured example of how to implement continuous monitoring and algorithm retraining in a lubricant selection system, including optional advanced features.

TABLE 4.7 Steps to Implement Continuous Monitoring and Algorithm Retraining

STEP	FUNCTION	DATA SOURCE	PROCESS	OUTCOME
1. Data integration	Real-time field data ingestion	CMMS/EAM platforms (e.g., SAP, Maximo), SCADA	Sync lubrication-related events (run hours, fault codes, component changes)	Update system with operational context for each asset
2. Oil analysis feed	Analytical results for lubricant health	Partner labs (ICP, FTIR, viscosity, TBN/TAN) via API or batch upload	Automatically populate wear trends, oxidation status, contamination alerts	Trigger retraining flags if trends deviate from expected lubricant life
3. Maintenance log parsing	Connect lubricant use with asset performance	Technician entries, service records, warranty claims	Apply natural language processing (NLP) or structured input to associate lubricant choice with failures or extended performance	Weight scoring engine with reliability feedback
4. Algorithm retraining	Improve model accuracy	Aggregated performance data, oil lab results, runtime metrics	Monthly or quarterly batch retraining of supervised ML models; adjust weights in rule-based systems	Increased recommendation accuracy and relevance to changing conditions
5. Alert system	Detect anomalies and trigger retraining	Outlier detection in wear rates, sudden oil property changes	Alert engineers to potential model drift or new operating conditions	Preemptive model tuning or validation test scheduling
6. Report and dashboard update	User insight into system updates	Internal UI with change logs and model versioning	Show users what factors influenced recent updates or new recommendations	Transparency builds trust and encourages field adoption

Table 4.8 offers optional enhancements.

TABLE 4.8 Optional Advanced Enhancements

FEATURE	PURPOSE	METHOD	EXAMPLE USE CASE
Digital twin simulation	Model lubricant behavior in synthetic system	Multiphysics model incorporating fluid properties and component geometry	Simulate varnish risk in variable-speed gear reducers under tropical climate
Predictive degradation modeling	Forecast oxidation, TAN rise, wear-metal generation	Time-series modeling with historical input (ML regression or physics-informed models)	Predict filter clogging or lubricant breakdown before next preventive maintenance (PM)
Procurement system integration	Suggest approved, in-stock lubricants in real time	ERP tie-in (e.g., Oracle, Coupa) linked to lubricant SKU database	Recommend top two products that meet both spec and warehouse availability

The continuous loop enables a **self-improving lubricant recommendation engine**, leveraging both machine intelligence and field reality.

Integration with Asset Management Systems

Computerized maintenance management systems serve as operational hubs for lubrication control. When configured with performance-based rulesets, CMMS platforms transition from static recordkeepers to active lubricant decision engines. Work order triggers, failure logs, runtime counters, and asset criticality rankings feed into selection algorithms. The system dynamically generates lubrication tasks based on condition severity, not calendar intervals.

Digital twins replicate machine behavior in virtual space, using real-world sensor input to simulate thermal, mechanical, and tribological conditions. These models predict how lubricants perform under fluctuating loads, speeds, and temperatures. The twin forecasts lubricant breakdown curves and alerts when viscosity loss, oxidation, or film failure thresholds are approaching. Algorithms can simulate multiple formulations and recommend the best-fit fluid based on virtual test results before application in the field.

PLCs and SCADA systems provide the real-time data flow required for adaptive lubrication strategies. PLCs relay sensor signals, bearing temperatures, shaft loads, and filter differential pressures directly into the lubrication control matrix. SCADA platforms visualize asset status across the facility and send exception flags when lubrication conditions deviate from control limits. This enables immediate intervention or autonomous corrective action.

Cloud integration allows centralized storage, analytics, and decision feedback loops across multiple facilities. Data from edge devices is aggregated and analyzed in real time, with recommendations fed back into local systems. If a hydraulic press in one plant shows a favorable response to a viscosity shift under certain thermal conditions, that insight can be pushed to similar assets enterprise-wide within minutes.

Integrated systems eliminate lag between condition change and lubrication response. Instead of waiting for manual inspection or scheduled PMs, lubrication control becomes continuous, automated, and self-correcting. The results are reduced lubricant consumption, increased machine uptime, and data-driven lubrication precision at scale.

Algorithmic Structure and Decision Logic

Algorithmic frameworks for lubrication selection evolve through three analytic stages: descriptive, predictive, and prescriptive.

Descriptive analytics ingests historical and real-time data. Wear-metal concentrations from ferrography or ICP, viscosity shifts from kinematic testing, oxidation index values, and acid number trends form the foundation. These values quantify what has occurred in the fluid system, offering empirical baselines for comparison and deviation tracking.

Predictive analytics projects lubricant and component behavior into the future. Models estimate the remaining useful life (RUL) of the lubricant based on current degradation rates, load exposure, and thermal stress. For example, rate-of-change calculations in oxidation or additive depletion can be regressed against past trendlines to forecast the point of functional failure. Machine learning models refine these predictions with each new data input.

Prescriptive analytics generates action. Change intervals, product substitutions, filtration deployment, or operating procedure adjustments are recommended in real time. Algorithms select the most viable action from a predefined decision tree, factoring constraints such as cost ceilings, criticality, maintenance windows, and product compatibility.

Multivariable optimization models drive the decision logic. Inputs include thermal load profile, fluid shear stability, contamination exposure (particulate, moisture, coolant), and economic metrics such as fluid cost, disposal cost, and downtime risk. Algorithms solve for the lubricant that best balances these competing variables.

Weighting matrices guide the optimization. High-criticality assets may weight reliability at 0.8 and cost at 0.2. For noncritical utilities, cost and efficiency may have equal weight. Adjusting matrix coefficients allows tailoring of lubricant decisions to specific operational objectives.

Fuzzy logic systems address uncertainty and partial truths, such as sensor drift, intermittent contamination, or conflicting test results. Rather than binary thresholds, fuzzy sets define ranges—for example, "low oxidation," "moderate wear," "high load"—and assess condition severity with gradient rules. Probabilistic approaches further enhance resilience in noisy or incomplete data environments by assigning likelihoods to failure modes and recommending risk-mitigated lubrication responses.

This logic stack—descriptive, predictive, prescriptive—drives lubricant strategy from observation to autonomous control. The system continuously recalibrates based on new data, refining its decision logic and optimizing fluid selection in line with changing field conditions.

Machine Learning Applications in Lubricant Management

Model training in machine learning–based lubricant management relies on diverse and high-integrity datasets. Laboratory testing provides controlled baseline data for oxidation stability, shear resistance, EP performance, and thermal degradation. These structured results establish the chemical and physical response of lubricant formulations under standard test conditions.

In-field data sources supply the dynamic context missing from lab tests. Real-time wear particle sensors, in-line oil condition monitors (measuring dielectric, moisture, TAN, and viscosity), and vibration analysis platforms capture operating conditions that influence lubricant behavior. These datasets offer continuous feedback on lubricant effectiveness under load, temperature, and contamination variability.

Failure mode databases and historical maintenance logs allow supervised learning models to correlate specific lubricant conditions and machine profiles with actual failure events. Metadata, such as equipment type, failure severity, root cause, and lubricant formulation, serve as labeled input for classification and regression tasks.

Predictive models classify operating environments by profile clusters—low-speed/high-load, high-speed/clean, thermally variable/contaminated. Within each cluster, wear rate modeling predicts component degradation based on lubricant properties and machine state. Models calculate the probability of film loss, additive depletion, or surface scoring, triggering early alerts.

RUL forecasting uses time-series regression, recurrent neural networks, or ensemble tree models to estimate fluid performance longevity under evolving conditions. Drain interval recommendations shift from fixed schedules to data-driven intervals adjusted in real time.

Reinforcement learning introduces a feedback loop into lubricant scheduling. Algorithms observe the outcome of lubrication actions, such as early drain, additive boost, and product switch, and refine future decision policies based on long-term equipment health and downtime penalties. Rewards are assigned for extending fluid life without inducing wear, while penalties train the model to avoid conditions that increase failure probability.

As models evolve, cross-validation with fresh field data improves accuracy. Predictive accuracy increases as sensors proliferate and machine histories deepen. The result is a continuously improving system that selects, monitors, and maintains lubrication strategies tailored to asset-specific behavior and mission-critical reliability goals.

Dynamic Lubricant Recommendation Systems

Formulation adjustment is triggered by operational thresholds that exceed defined lubricant performance margins. Heat spikes beyond viscosity stability limits initiate recommendations for higher-viscosity base oils or synthetic alternatives with elevated oxidative resistance. Load anomalies, such as short-term torque surges or chronic overloading, trigger adjustments toward fluids with higher film strength or increased anti-wear additive concentrations. Fluid aging markers, such as elevated acid number, viscosity shift, or loss of dispersancy, initiate a lubricant switch or rejuvenation protocols based on historical degradation curves.

Contamination detection plays a critical role in formulation change. Water ingress detected via dielectric loss or Karl Fischer titration initiates a shift toward demulsifying formulations or water-tolerant base stocks. Fuel dilution flags volatility incompatibility and demands a reformulation with lower NOACK volatility and adjusted solvency. Elevated ISO particle counts or the presence of hard contaminants activates filtration events or recommends a switch to higher-detergency lubricants with strong soot dispersal capacity.

System-driven lubricant requests are handled by smart reservoirs equipped with condition sensors and integrated flow control. When onboard analysis identifies a deviation beyond formulation tolerance, the system initiates a lubricant change, top-off, or purge without human input. Intelligent reservoirs communicate with CMMS and SCADA to trigger maintenance events, send formulation codes, and execute changeover sequences.

Cloud-based lubricant libraries store detailed chemical and physical property data, indexed by OEM, formulation ID, ISO VG grade, viscosity index, and additive compatibility. These libraries cross-reference the asset's operational profile, environmental conditions, and historical lubricant performance to identify suitable replacements.

Compatibility cross-checks prevent formulation conflicts, additive antagonism, or seal material incompatibility during product substitution.

Connected supply chain integration automates the restock and reformulation process. When a system requests a specific formulation or detects depletion, it triggers a reorder through integration of ERP (enterprise resource planning). If the required lubricant isn't in stock, cloud logic initiates a reformulation request with supplier parameters and sends the adjusted spec to blending or fulfillment centers. This closes the loop between machine condition, lubricant performance, supply inventory, and production scheduling, creating a just-in-time, condition-responsive lubrication ecosystem.

FIELD IMPLEMENTATION AND CASE STUDIES

Industrial Gearboxes in Mining Applications

Mining operations impose extreme load variability on gearbox systems, haul truck differentials, conveyor drives, and crushing plant gear reducers cycle through high shock loads, idling, and sustained torque. Static lubrication strategies fail to accommodate these fluctuations, leading to premature fluid breakdown, foaming, and micropitting.

Field deployment of dynamic lubricant recommendation systems in a copper mine's overland conveyor gearboxes demonstrated load-based switching between conventional high-viscosity mineral gear oils and synthetic shear-stable formulations. Vibration sensors and torque monitors were networked with SCADA and cloud-based lubricant analytics platforms. When gearboxes experienced prolonged high-load conditions above baseline torque thresholds, the system issued a recommendation to switch to PAO-based gear oils with higher VI and shear resistance.

Viscosity retention and thermal stability were verified through in-line oil condition sensors. When the load dropped below the dynamic threshold and no shearing risk was present, the system reverted to mineral-based options optimized for cost and drain interval efficiency. Smart reservoirs handled the automated blend transition without shutting down operations.

Results over a 12-month period showed a 27% reduction in unscheduled gearbox maintenance events. Oil sampling verified a 45% improvement in viscosity retention under thermal stress and a 38% reduction in insoluble contamination, indicating reduced oxidation and varnish formation. Lubricant usage dropped by 22% due to longer intervals and reduced volume of corrective flushes.

The adaptive strategy also allowed predictive scheduling of fluid changeovers based on torque history and temperature cycling instead of calendar intervals. Mean time between failures improved by 31%, and lubricant disposal costs were cut in half.

This case confirms that dynamic lubricant algorithms, when aligned with real-world load conditions, outperform static OEM prescriptions. Shear-stable synthetics are applied precisely when required, minimizing waste while extending equipment life.

FIELD IMPLEMENTATION AND CASE STUDIES: ELECTRIC BUS FLEETS WITH REAL-TIME OIL ANALYTICS

In electric bus powertrains, dielectric fluids serve dual roles: cooling and electrical insulation. Fluid breakdown leads to loss of thermal transfer efficiency and potential voltage tracking or arcing. Static drain intervals do not account for variable urban duty cycles, regenerative braking profiles, or ambient temperature swings that directly influence fluid stress.

A metropolitan electric bus fleet deployed real-time oil analytics across traction motor inverters and e-axle cooling circuits. Sensors tracked fluid dielectric strength, moisture content, and conductivity. These inputs fed predictive modeling algorithms that correlated fluid property degradation with load cycles, inverter switching frequency, and ambient thermal excursions.

Predictive models established RUL curves based on actual field data, not fixed mileage or time. Once dielectric strength dropped below defined thresholds or moisture trended toward saturation, the system issued alerts and scheduled fluid replacement during low-demand maintenance windows.

The analytics system also tracked transient heat spikes during regenerative braking and flagged buses with atypical thermal patterns for targeted diagnostics. This allowed isolation of underperforming thermal paths and early identification of inverter cooling inefficiencies.

After 1 year of implementation across 220 of these electric vehicles (EVs), the fleet documented a 35% reduction in maintenance hours tied to cooling system service. Fluid replacement intervals increased by 60% without compromising dielectric performance. Unscheduled inverter maintenance related to thermal failure dropped by 41%.

Component life extension was verified through infrared thermography and post-service inspection. Inverter modules operated at consistently lower temperatures, and fluid oxidation was significantly reduced. No electrical tracking or insulation failures were recorded in units running on algorithm-optimized service intervals.

Real-time fluid analytics enabled a shift from reactive to predictive maintenance in electric bus fleets. By modeling dielectric fluid performance dynamically, downtime was minimized, and fluid utilization was optimized, extending the operational life span of critical EV components.

Offshore Wind Turbines with Remote Lubrication Systems

Offshore wind turbines operate under volatile environmental conditions, rapid temperature shifts, fluctuating wind loads, and high-humidity exposure. These variables directly affect lubricant viscosity, film formation, and oxidation stability in main bearings, gearboxes, and pitch/yaw systems. Static lubrication schedules and manual refills are infeasible due to turbine inaccessibility and exposure risk.

A remote lubrication system was implemented across a North Sea wind farm, integrating real-time sensors with cloud-based lubrication algorithms. Inputs included nacelle temperature, wind torque variability, gearbox load profiles, and humidity levels within the nacelle and hub enclosures. These variables triggered formulation-specific lubricant selection and delivery adjustments.

As temperatures dropped below −5°C, the system switched to synthetic low-temperature greases with pour points below −50°C and stable base oil viscosity under low shear. During high-load wind events, the algorithm selected high-film-strength greases with increased EP additive content to mitigate sliding wear in pitch bearings. When humidity exceeded the 80% threshold for extended periods, the system increased fill frequency with water-tolerant, corrosion-inhibited lubricants.

Automated refill cycles were governed by torque cycles and thermally induced viscosity changes rather than operating hours. Refill volumes were precisely metered to avoid overlubrication, and fill schedules were shifted to periods of low energy generation to minimize the impact on turbine efficiency.

Over a 24-month deployment across 48 turbines, lubricant-related maintenance visits dropped by 58%. Gearbox failures associated with lubricant starvation or improper fill volume were eliminated. Average lubricant usage decreased by 21% due to algorithmic control of refill cycles and energy-efficient lubricant selection.

Lubricant performance was optimized continuously based on the actual operating environment, reducing wear and extending component service intervals. Remote delivery reduced technician exposure and vessel deployment costs. Adaptive lubricant management in offshore environments ensures mechanical reliability without compromising access safety or energy output.

Infrastructure and Cyber-Physical Systems

Sensor calibration is critical. Measurement drift in temperature, pressure, vibration, or dielectric sensors compromises algorithm output and leads to lubrication misapplication. Field calibration schedules must be enforced quarterly or per OEM recommendation, using traceable standards. On-site calibration should include in-situ validation against known conditions. Calibration data must be stored locally and pushed to the cloud for audit traceability.

Data fidelity depends on time-stamped, high-resolution acquisition. Sampling intervals must align with the dynamic range of the equipment. For rotating assets with variable-load profiles, subsecond resolution is required. Filtering algorithms must reject noise without suppressing transient load spikes or thermal surges, which drive critical lubrication decisions. Lossless compression and checksum validation ensure data integrity across transmission nodes.

Network security governs the viability of system-level integration. TLS encryption, end-point authentication, and access partitioning are required. Onboard processors must operate on hardened firmware with write-protected memory regions. Modbus, OPC UA (Unified Architecture), and MQTT (Message Queuing Telemetry Transport) traffic must be monitored for spoofing and injection attempts. Isolated VLANs and edge firewalls mitigate unauthorized access to lubricant delivery controls.

Firmware updates must be remotely deployable but locked to authenticated push protocols. Over-the-air updates require checksum verification, rollback capability, and cryptographic signing. Field units should support redundant boot images to prevent bricking from interrupted updates. Firmware revisions must be regression-tested against all sensor types and lubricant control logic trees before deployment.

Redundancy in autonomous lubrication systems includes dual-sensor configurations, secondary reservoir lines, and fail-closed solenoid valves. Sensor redundancy should include diversity—for example, both thermocouple and RTD for temperature—to avoid single-point signal failure. Logic controllers must execute fallback routines if input data fails checksum or exceeds confidence thresholds. In such cases, the system defaults to conservative fill parameters based on last-known good state.

Fail-safes include mechanical overrides, manual fill ports, and alarm-state lockouts. If algorithmic decision latency exceeds predefined limits, the lubrication cycle proceeds on static safety protocol. Event logs, decision trees, and control status must be archived and synchronized to the cloud for forensic review.

Cyber-physical integration requires hardened infrastructure, authenticated control pathways, and validation at every layer, sensor, algorithm, transmission, and actuation. Reliability depends not only on fluid properties, but also on system resilience, security posture, and verified data lineage.

Future Outlook and Design Trends

Fluid-on-demand systems are being integrated with cloud-hosted lubricant databases to enable precise formulation delivery based on real-time operating conditions. These systems use asset-specific data, such as load, temperature, or duty cycle, to request a matched lubricant from a centralized repository. The request triggers autonomous mixing, dispensing, or additive adjustment at the point of use. Inventory is minimized, and lubricant logistics shift from bulk storage to just-in-time microdosing.

Smart components are embedding AI modules directly into field devices. Valves equipped with onboard diagnostics modulate flow rates based on local wear trends and predictive fluid behavior. Pumps adjust stroke frequency and delivery pressure in response to feedback from viscosity and temperature sensors. Reservoirs equipped with embedded processors manage volume control, detect stratification, and trigger contamination alarms without external computation layers.

Custom-formulated lubricants are now being synthesized at the equipment site. AI models evaluate operational data against historical failure profiles, environmental conditions, and OEM constraints. Using microblending units, base oils and additive packages are dosed in real time to match evolving needs. This eliminates overengineering of lubricant stock and ensures compatibility with immediate operating demands. Formulations can vary by machine, shift, or weather pattern, with no human intervention.

Regulatory oversight is increasing. AI-managed lubrication systems must comply with ISO 21434 for cybersecurity and IEC 62443 for industrial automation security. Regulatory bodies are drafting frameworks for AI validation in predictive maintenance, including explainability of decision logic and audit trails for autonomous lubricant dosing events. Any system influencing lubricant selection or delivery must log algorithmic actions, inputs and override events in machine-readable format.

Cybersecurity hardening is mandatory. AI decision engines must operate on segmented networks, and all firmware modules require encryption, authentication, and rollback controls. Systems managing lubricant dosing are classified as critical infrastructure components in defense, aviation, and energy sectors. Attack surface must be minimized— no open ports, no default credentials, no unsecured cloud bridges.

Future design trends are converging around autonomous, adaptive, and accountable lubrication infrastructure, driven by embedded intelligence, governed by cybersecurity, and optimized through in-field data fusion.

CHAPTER 5

SELECTION OF LUBRICATION EQUIPMENT

S electing the appropriate lubrication system is foundational to achieving operational efficiency, equipment reliability, and maintenance consistency. Each application, whether bearings, gears, chains, hydraulics, or slideways, has unique requirements in terms of delivery rate, distribution pattern, cleanliness, and frequency.

Application precision requires matching the lubrication method to the machine's mechanical demands. Drip-feed systems, for example, may suffice for low-speed chains but are unsuitable for high-speed precision bearings, which require spray, mist, or oil-air systems. Grease-lubricated components may benefit from single-point lubricators or centralized grease systems to ensure accurate, timed delivery without overapplication.

Critical selection parameters include lubricant type, flow rate, pressure, temperature conditions, and system control options. Precision in lubrication directly impacts energy efficiency, temperature regulation, and wear minimization. The correct system must also align with maintenance capabilities, service frequency, and accessibility constraints.

CHOOSING THE RIGHT LUBRICATION SYSTEM FOR APPLICATION PRECISION

Lubrication system precision is defined by the accuracy, repeatability, and response time with which the system delivers lubricant relative to component demand. Application type determines the required precision class. High-speed spindle bearings, robotics joints, and medical-grade actuators require microdosing systems with volumetric control down to microliters per cycle. Heavy-duty gearboxes, open gears, and off-road equipment require high-volume delivery systems capable of compensating for environmental ingress, evaporation, and leakage.

Functional mismatch results in measurable degradation. Underlubrication due to imprecise delivery or delayed actuation leads to metal-to-metal contact, wear acceleration, and eventual seizure. Overlubrication results in increased fluid churn, seal degradation, and thermal loading. Excess grease in bearing housings causes churning losses, elevates temperatures, and can induce early fatigue spalling. Mistimed delivery cycles in high-speed equipment introduce variability in oil film thickness, increasing surface stress and fatigue risk.

System selection must align delivery rate, actuation speed, and lubricant type with equipment demand profiles. For example, centralized single-line resistance systems are suitable for low-cyclic, steady-state machines. Dual-line or progressive divider systems are better suited for high-output, variable-cycle industrial equipment where precise sequencing is required. Air-oil and oil-mist systems match environments requiring minimal lubricant quantity with high placement accuracy, such as textile spindles or CNC tool changers.

Demand profiling involves mapping thermal load, duty cycle, vibration patterns, and seal type to determine the lubricant consumption rate and replenishment frequency. Equipment with intermittent duty cycles may need prelube functionality, which some systems cannot support. Systems must be rated for viscosity range, pressure delivery consistency, and compatibility with monitored feedback systems, temperature, pressure decay, or in-line flow verification.

Balancing equipment capabilities with lubrication precision involves ensuring that the control logic, metering devices, reservoir sizing, and distribution network can accommodate the range of operating conditions. Overspecifying leads to cost and complexity. Underspecifying leads to increased wear and premature failure. System design must be based on failure mode analysis, environmental factors, and real-time control integration potential.

CLASSIFICATION OF LUBRICATION SYSTEMS

Manual Systems

Manual lubrication systems include hand-operated devices such as grease guns, drip oilers, and squeeze bottles. These systems rely on operator input for both timing and quantity of lubricant delivery. Delivery is nonautomated and varies by user technique, ambient conditions, and lubricant properties.

Grease guns, whether lever, pistol-grip, or pneumatic, are used primarily for bearings, joints, and couplings requiring semisolid lubricants. They deliver high-viscosity grease under pressure, typically between 3,000 and 10,000 psi. Grease volume per stroke must be known to avoid overlubrication or underlubrication. The guns are effective for low-duty, low-frequency lubrication points where automation is not cost-justified or physically feasible.

Drip oilers provide gravity-fed lubrication through adjustable needle valves or sight glass reservoirs. Common in older textile machinery and chain applications, they offer simple flow control but limited responsiveness to dynamic operating conditions. Flow rate depends on viscosity, temperature, and wick condition. Drip oilers are not suitable for variable-speed or high-shock-load equipment.

Squeeze bottles allow point application of fluid lubricants for small-scale or infrequent maintenance. They offer no metering control or pressure and are limited to applications such as hinges, small bushings, and sliding surfaces without significant load or speed. Contamination risk is high if bottles are reused or exposed.

Limitations of manual systems include inconsistent delivery, reliance on human judgment, high contamination risk, and inability to synchronize with machine load cycles. Missed intervals, excessive application, and lubricant incompatibility are common failure modes. Manual systems lack feedback or verification, making them unsuitable for critical equipment, high-frequency lubrication points, or environmentally sensitive zones.

Use cases are limited to noncritical assets, legacy machinery, or locations with limited access to power and infrastructure. Where precision, repeatability, or contamination control is required, manual systems should be replaced with automated or monitored alternatives.

Single-Point Automatic Systems

Single-point automatic lubrication systems deliver lubricant directly to 1 lubrication point using a self-contained mechanism. These systems are designed for low-volume, consistent delivery over extended periods without manual intervention. They are commonly used in hard-to-reach locations, elevated assets, or environments where regular manual access is restricted or hazardous.

Spring-driven and gas-driven lubricators rely on stored energy to push lubricant through a discharge piston. In spring-loaded types, mechanical compression provides constant force over time. Gas-driven systems generate pressure from an electrochemical reaction (commonly hydrogen or nitrogen) that expands a chamber and displaces the lubricant. These systems typically operate at low pressure (15–75 psi) and deliver lubricant continuously or intermittently over a set duration ranging from 1 month to 1 year.

Battery-operated single-point units utilize programmable logic to control lubricant discharge through an internal motor or solenoid. These devices offer higher delivery precision and customizable settings for discharge volume and interval. They often include status LEDs, failure alarms, or Bluetooth connectivity for remote condition checks. Common discharge rates range from 0.5 cc/day to 5 cc/day, depending on configuration.

Typical applications include electric motor bearings, pillow block housings, fan shafts, and conveyor rollers, especially those located behind guarding, at elevation, or in hazardous zones (e.g., high heat or chemical exposure). These systems minimize the need for manual intervention and ensure consistent lubrication where neglect or inaccessibility is common.

Limitations include single-outlet design, limited lubricant capacity (typically 60–125 cc), and restricted adaptability to sudden load or environmental changes. High-viscosity greases may not discharge uniformly, particularly in cold environments. Most single-point systems are not pressure-rated for long or complex distribution lines.

These units are best suited for isolated or low-criticality components requiring steady, low-rate lubrication where manual access is impractical. They are not recommended for high-speed, high-load, or dynamically variable equipment unless integrated into broader condition-monitoring infrastructure.

Centralized Lubrication Systems

Centralized lubrication systems supply multiple lubrication points from a single pump or reservoir. These systems are designed for medium- to high-point-count machines operating under continuous or severe-duty conditions. Lubricant delivery is automated, metered, and distributed through a fixed network of lines and valves, reducing manual intervention and ensuring consistent application.

Single-line resistive systems operate by pressurizing a main supply line that branches into multiple feed lines with flow restrictors or metering orifices. Each point receives lubricant simultaneously, but volume is dependent on line length, resistance, and pressure. These systems are low-cost and best suited for small to midsize equipment operating under consistent conditions. They cannot compensate for pressure variation or point-specific demand and are typically used with oil, not grease.

Dual-line systems utilize two alternating pressure lines feeding a series of metering valves. During each cycle, one line delivers lubricant while the other vents, allowing pressure buildup and delivery in a staggered pattern. These systems support long line runs, high back pressure, and the use of grease up to NLGI #2. Suitable for large-scale machinery, steel mills, paper machines, and mining equipment, dual-line systems can operate reliably with hundreds of lubrication points and long cycle intervals.

Progressive systems use a central pump to pressurize lubricant into a master metering valve, which sequentially feeds a network of secondary valves. Each valve discharges a fixed volume before triggering the next. Blockage at any point halts the entire circuit, providing built-in fault detection. These systems offer precise volume control and are commonly applied in CNC machines, injection molding systems, packaging equipment, and mobile hydraulics. Grease and oil variants are available, and feedback sensors can be added for confirmation of delivery.

Each system uses a centralized pump to deliver lubricant under pressure to the distribution network. Metering devices, not the pump, control the volume per point. Timers, PLCs, or condition-based sensors initiate cycles. Flow monitors, pressure switches, and progressive cycle indicators provide verification and system diagnostics.

Centralized systems reduce labor, improve reliability, and ensure accurate lubricant delivery across all points, especially in high-duty or high-risk operations. Proper system sizing, line routing, and metering configuration are critical to maintain consistent delivery and avoid starvation or overload at downstream points.

Circulating Oil Systems

Circulating oil systems provide continuous lubrication by recirculating oil through a closed-loop system that includes a reservoir, a pump, filters, and heat exchangers. These systems are designed for high-speed, high-load, and thermally sensitive equipment where lubrication, cooling, and contaminant removal must occur simultaneously and without interruption.

Oil is drawn from the reservoir by a pump, typically gear or screw type, and delivered under pressure to lubrication points, bearings, gears, or seals. After passing through the equipment, oil returns via gravity or drain lines to the reservoir. During this cycle, the oil passes through filtration units to remove wear debris and external contaminants, and through a cooler to dissipate heat accumulated from mechanical friction and shear.

Flow rates and pressures are engineered based on load zones and component geometry. Common operating pressures range from 10 to 100 psi, with flow rates sufficient to ensure turbulent flow in critical areas, maximizing heat transfer and flushing capability. Reservoir sizing is typically 3–5 times the system flow rate to allow for proper de-aeration and thermal stability.

Circulating oil systems are mandatory in equipment where static oil pools would result in excessive thermal buildup or inadequate lubrication film; this would include equipment such as large gearboxes, steam turbines, rolling mills, paper machine dryers, and compressor trains. The oil in circulating systems also acts as a heat transfer medium, carrying thermal energy away from high-friction interfaces and preventing lubricant oxidation.

System instrumentation includes differential pressure sensors across filters, temperature probes at supply and return lines, flow meters, and oil condition sensors (moisture, particle count, TAN). Redundant pumps and duplex filter assemblies are common for mission-critical applications to maintain uptime during maintenance or component failure.

Circulating systems support condition-based maintenance, allowing in-line sampling and real-time monitoring of lubricant condition. They also enable centralized oil management, including additive replenishment, contamination control, and viscosity correction without requiring shutdowns.

These systems are capital-intensive but necessary where lubrication must be continuous, clean, cool, and chemically stable under severe load and environmental conditions. Proper system design, component redundancy, and routine monitoring are essential for performance and reliability.

Mist, Air/Oil, and Spray Systems

Mist, air/oil, and spray lubrication systems deliver lubricant in atomized or finely metered form, providing controlled application to high-speed, low-tolerance components. These systems are designed for environments where minimal lubricant quantity, low viscous drag, and high delivery precision are required.

Mist systems generate an oil aerosol by mixing compressed air with oil at a central reservoir. The aerosol is transported through piping and directed at the lubrication point, where it condenses upon contact with rotating or moving surfaces. These systems are commonly used in enclosed gearboxes or bearing housings of high-speed spindles and motors. Delivery is continuous, with typical air pressures between 15 and 30 psi. Proper line routing and system balance are essential to avoid oil pooling or insufficient mist density at distal points.

Air/oil systems deliver discrete, metered oil pulses carried by a continuous stream of compressed air. Each pulse lubricates a precise point, such as a ball screw, spindle bearing, or linear guide. These systems prevent excess lubricant accumulation and are suited for high-speed applications where even a thin oil film can cause energy loss or thermal instability. Pulse frequency and volume are programmable through a controller, and in-line sensors verify delivery. Air/oil systems are ideal for CNC equipment, textile machinery, and robotic actuators.

Spray systems atomize lubricant externally at the point of application, using air-assisted nozzles. These are commonly deployed for open gears, chains, and exposed drive elements. Spray patterns can be adjusted for coverage area, droplet size, and direction.

Delivery can be timed, condition-based, or controlled via PLC. Typical configurations include solenoid-actuated nozzles, flow monitoring, and airflow interlocks to ensure safe and accurate delivery.

Advantages of these systems include low lubricant consumption, minimal frictional drag, and reduced contamination risk from overapplication. They also allow lubrication of moving or rotating components without requiring direct contact or enclosure penetration.

System limitations include sensitivity to air quality, line pressure variation, and nozzle clogging. Mist and air/oil systems require dry, filtered air to prevent fluid contamination and maintain consistent delivery. For spray systems, overspray control and drift minimization are necessary to protect surrounding components and reduce airborne residue.

These systems provide critical lubrication control in speed-sensitive, precision-driven applications where overlubrication or drag can degrade performance or cause failure. Proper configuration, regular maintenance, and air system integrity are key to long-term reliability.

Table 5.1 is a decision table comparing different lubrication system types based on application suitability, lubricant type, delivery precision, automation level, and known limitations.

APPLICATION-SPECIFIC EQUIPMENT MATCHING

Bearings

Bearing lubrication must be matched to the rotational speed, load, operating environment, and bearing type. An improper lubrication selection or delivery method leads to film breakdown, heat generation, and surface fatigue.

High-Speed Bearings

For high-speed applications, spindles, electric motors, turbochargers, and oil-air and mist systems provide controlled delivery with minimal drag and thermal buildup. Oil-air systems meter small pulses of lubricant into a continuous stream of compressed air. Delivery is precise and synchronized with operating cycles. This method maintains a consistent, thin film without generating excess friction. Mist systems deliver oil as an aerosol, allowing vaporized lubricant to condense on bearing surfaces, making mist systems suitable for enclosed high-speed units with integrated ventilation. Both mist and air-oil systems prevent overlubrication, reduce churning, and manage heat dissipation in speed-sensitive components.

TABLE 5.1 Decision Table Comparing Different Lubrication System Types

SYSTEM TYPE	BEST FOR APPLICATION	LUBRICANT TYPE	DELIVERY PRECISION	AUTOMATION LEVEL	LIMITATIONS
Manual	Low-duty, infrequent lubrication points	Grease/oil	Low	Manual	Inconsistent, contamination risk
Single-point automatic	Hard-to-reach, low-criticality bearings or rollers	Grease	Medium	Self-contained	Limited volume, not load-adaptive
Centralized (single line)	Small to midsize machines with stable loads	Oil	Medium	Basic auto	No pressure compensation
Centralized (dual line)	Large equipment, high point counts, long runs	Grease	High	High auto	Complex, costlier setup
Centralized (progressive)	Precise volume control with fault detection needs	Grease/oil	High	High auto	Stops entire circuit if one valve clogs
Circulating oil	High-load, high-speed, thermal-sensitive machinery	Oil	Very high	Full auto	High capital cost; requires monitoring
Mist	High-speed spindles in enclosed gearboxes	Oil	High	Full auto	Mist density control, air quality–sensitive
Air/oil	CNC (computer numerical control) machines, robotics, ball screws	Oil	Very high	Full auto	Air pressure dependence, filter clogging
Spray	Open gears, chains, exposed drives	Oil	High	High auto	Overspray and drift issues

Oil viscosity must align with the speed factor (DN value). Incorrect viscosity or excessive film thickness causes temperature rise and fluid shear. Mist and oil-air systems require clean, dry compressed air, monitored for pressure stability and flow rate. Delivery lines must be balanced to prevent uneven distribution.

Low-Speed or Sealed Bearings

Low-speed bearings, such as those on conveyors, gear reducers, and pivot arms, require higher-viscosity grease with sufficient mechanical stability to maintain film integrity under low motion. These are typically lubricated via centralized systems or single-point lubricators. Progressive systems ensure consistent grease volume to each point. Single-point automatic units suit isolated or inaccessible bearings with low relubrication frequency.

Sealed bearings are typically prelubricated and not intended for relubrication. For field-sealed designs with lubrication ports, grease intervals must be based on calculated relubrication schedules using bearing diameter, speed, and operating temperature. Overgreasing is a known cause of seal rupture and bearing overheating. Monitor for back pressure or purge at vent ports during greasing.

In contaminated environments, use grease with enhanced water resistance, corrosion inhibitors, and sealing compatibility. Lubricant selection and delivery system must consider ingress protection level (IP rating) and exposure to washdown, dust, or particulates.

Matching bearing type and duty cycle to the lubrication method ensures proper film thickness, thermal stability, and protection against wear, minimizing failure risk and extending service life.

Table 5.2 summarizes the bearing types and applications.

TABLE 5.2 Summary of Bearing Types and Applications

BEARING TYPE	LUBRICATION METHOD	KEY CRITERIA	RISKS AND REQUIREMENTS
High-speed bearings	Oil-air systems	Precise dosing; low drag; minimal heat	Requires clean, dry air; line balancing critical
High-speed bearings	Mist systems	Aerosol delivery; enclosed systems; vapor condensation	Pressure/flow monitoring required; air quality sensitive
Low-speed bearings	Centralized grease systems	High-viscosity grease; consistent volume; multiple points	Grease selection critical; potential for cross-contamination
Low-speed bearings	Single-point lubricators	Remote locations low relube frequency	Limited capacity; unsuitable for high load/ high frequency
Sealed bearings	None (prelubricated)	Factory-sealed; no maintenance required	No lubrication access; failure = replacement
Sealed bearings	Grease port lubrication	Scheduled based on speed, size, temperature	Risk of overgreasing; must monitor vent purge or back pressure

GEARS

Gear lubrication requirements depend on enclosure type, gear geometry, load profile, and operating speed. Incorrect lubricant selection or delivery method leads to wear, pitting, scuffing, and thermal failure.

Enclosed Gears

Enclosed gearboxes, such as those found in speed reducers, planetary drives, and helical or bevel gear units, typically use splash or circulating oil systems. Splash lubrication relies on gear rotation to distribute oil via partial immersion of the gear teeth in the sump, which is effective in moderate-speed, moderate-load applications with consistent operating angles. The sump level must be maintained within OEM-specified limits to ensure adequate tooth engagement without causing fluid aeration or thermal rise.

Circulating oil systems are required when splash lubrication is insufficient due to thermal load, contamination, or system complexity. These systems continuously deliver filtered, cooled oil to contact surfaces and are used in high-speed, high-torque applications, turbines, rolling mills, and large reducers. Oil is pumped through filters and heat exchangers before being returned to the gearbox sump or reservoir. Flow rates are engineered based on gear mesh losses, specific sliding velocity, and thermal load. These systems support condition monitoring and are compatible with in-line oil analysis sensors for moisture, wear metals, and viscosity.

For enclosed units operating in low-temperature or intermittent service, synthetic oils with high VI and low pour point prevent viscosity breakdown and cold-start damage. EP additives are required for high-load gear meshes, particularly in hypoid or worm gear configurations.

Exposed Gears

Exposed gears, such as those in open gear sets on mills, rotary kilns, and slewing drives, require specialized application methods due to environmental exposure and lack of containment. Spray systems or brush-on methods apply high-viscosity adhesive greases or asphaltic compounds. Lubricant must maintain adhesion under centrifugal force, resist wash-off, and provide cushioning under heavy loads.

Spray systems use air or hydraulic pressure to atomize lubricant and direct it onto gear flanks and root zones. Application must ensure uniform coverage without overspray or drift. Nozzle alignment, spray timing, and delivery pressure require periodic calibration. Film thickness must be verified with visual inspection or ultrasonic measurement.

Grease is appropriate for low-speed, high-load gear applications where fluid retention is not possible and relubrication access is limited. Manual application should follow a consistent interval and be monitored for buildup or insufficient coverage. Use of open gear lubricants with solid additives (e.g., graphite or molybdenum disulfide) improves load-carrying capacity and wear protection under boundary conditions.

Gear lubrication strategy must align with enclosure type, load profile, and environmental exposure. Selecting the wrong system results in excessive wear, overheating, and mechanical failure. Lubricant delivery, condition monitoring, and maintenance intervals must be adapted to the gear drive's functional profile.

Table 5.3 summarizes gears and their applications.

TABLE 5.3   Summary of Gears and Applications

GEAR TYPE	LUBRICATION METHOD	KEY CRITERIA	RISKS AND REQUIREMENTS
Enclosed gears	Splash lubrication	Moderate load/speed; partial immersion; consistent alignment	Oil level critical; aeration risk if overfilled
Enclosed gears	Circulating oil systems	High load/speed-filtered/circulated; thermal management	Complex system; requires monitoring and maintenance
Exposed gears	Spray systems	Adhesive grease; uniform spray; environmental resistance	Overspray/drift risk; requires nozzle calibration
Exposed gears	Manual or brush-on grease	Solid additives manual access; consistent intervals	Risk of buildup or missed coverage; labor-intensive

CHAINS AND CONVEYORS

Chain and conveyor lubrication must ensure penetration to internal friction surfaces, specifically pins and bushings, where metal-to-metal contact occurs. Surface coating alone provides insufficient protection. Lubrication method selection depends on chain type, operating speed, exposure to contaminants, and thermal environment.

Drip Systems

Drip lubrication delivers oil via gravity from adjustable flow nozzles positioned directly above the chain. It's suitable for slow- to moderate-speed chains in clean environments.

The drip rate must be calibrated to ensure oil reaches the pin-bushing interface without excessive runoff. Positioning is critical—the drip point must align with the slack side of the chain to allow capillary action into internal components. Drip systems are low cost but sensitive to viscosity, ambient temperature, and misalignment.

Brush Systems

Brush lubrication systems apply oil or light grease via contact with rotating chain elements. They are typically used on conveyors and accumulation systems in dry or dusty environments. Brushes must be positioned on the return side of the chain to minimize debris accumulation and provide time for lubricant migration into wear zones. Brush material must be compatible with lubricant and chain material. Periodic inspection is required to prevent bristle clogging and to verify wear surface contact.

Brush systems deliver consistent film but have limited penetration under high-speed or high-load conditions. They are inappropriate where contamination risk, chemical washdown, or elevated temperatures degrade brush integrity or lubricant film.

Spray Systems

Spray lubrication uses compressed air or pressurized oil to atomize lubricant and direct it precisely onto moving chain surfaces. Spray systems are effective at medium to high chain speeds, offering fine control of volume and coverage. Systems may be timed or sensor-triggered and can be integrated with PLCs for cycle-based application.

Nozzle orientation and delivery pressure must be configured to target the load side of the chain and achieve pin and bushing penetration. Chain guards and shields may be required to contain overspray and reduce lubricant loss. Air/oil spray systems are often used in food-grade conveyors or high-speed packaging lines, where minimal lubricant quantity and clean operation are critical.

Functional Requirements

Lubricant viscosity must support migration into tight clearances without excessive sling-off. For elevated temperature chains (e.g., ovens or dryers), synthetic or ester-based oils with low evaporation loss and high thermal stability are required. In corrosive or wet environments, rust-inhibited lubricants with water-repelling additives prevent seizing and premature wear.

Consistent pin and bushing lubrication prevents elongation, chain pitch variance, and roller freeze-up. Chain lubrication schedules should be based on wear rate, operating hours, and exposure severity, not calendar time. Visual inspection, elongation measurement, and thermal imaging support verification.

Chain and conveyor reliability depends on internal lubrication, not visible coating. The delivery method must ensure targeted application, sufficient penetration, and compatibility with the operating environment and chain construction.

Table 5.4 summarizes chain applications.

SLIDEWAYS AND GUIDES

Slideways and linear guides require consistent, low-friction lubrication to prevent stick-slip motion and ensure positional accuracy. Lubrication must form a stable film under boundary and mixed-lubrication regimes, especially during low-speed or start/stop motion. The primary failure modes include stiction, wear at reversal points, loss of geometric accuracy, and migration of lubricant into adjacent components.

Intermittent Oil Metering

Intermittent oil metering delivers controlled volumes of lubricant at timed intervals or motion-based triggers. This approach minimizes excess buildup while maintaining film strength at the contact surface. Flow is controlled via positive displacement injectors, metering valves, or electronically actuated pumps. Delivery frequency must be matched to stroke length, travel frequency, and load.

Common systems include air/oil units, micro-oilers, and progressive divider blocks connected to a central pump. Lubricants must exhibit high film strength, good wetting properties, and resistance to emulsification if exposed to coolant or washdown. Oils with tackifiers are often used to improve adhesion to vertical or inverted slideways.

Slideway oils should meet or exceed ISO 6743-13 (CKC/CKD) or DIN 51502 CGLP classifications for anti-stick-slip properties. Viscosity selection depends on slide size, operating temperature, and expected load, typically ISO VG 68 to 220.

TABLE 5.4 Summary of Chain Applications

APPLICATION TYPE	TYPICAL USE CASE	LUBRICATION DELIVERY METHOD	KEY SELECTION CRITERIA	LIMITATIONS	MAINTENANCE NEEDS
Drip systems	Slow- to moderate-speed chains in clean environments	Gravity-fed oil drip onto slack side of chain	Viscosity, ambient temperature, nozzle alignment	Misalignment, oil runoff, temperature sensitivity	Check drip rate and alignment regularly
Brush systems	Conveyors in dry or dusty environments with low to moderate load	Oil or grease applied via rotating brush contact	Brush material compatibility, location on return side, dust resistance	Limited penetration in high-speed/load, brush clogging	Inspect brush wear, clogging, and surface contact
Spray systems	Medium- to high-speed chains, including food-grade or packaging lines	Compressed air or pressurized oil spray with nozzles	Nozzle orientation, delivery pressure, spray pattern control	Overspray risk, complex setup, air/oil quality dependence	Monitor nozzle condition, flow rates, and containment

Distribution Uniformity

Uniform distribution across the slide or guide surface is critical. Uneven lubrication leads to skewing, chatter, and alignment loss in multi-axis systems. Manifolds, flow restrictors, or metering screws are used to balance flow between lubrication points. Oil grooves in the slideway surface must be properly machined and free of debris or scoring.

In precision machine tools, such as grinders, machining centers, and CNC mills, motion accuracy is directly tied to lubrication uniformity. Film failure at one end of a guideway can produce torsional strain, impacting repeatability and surface finish.

Cross-contamination between way lubrication and hydraulic systems must be avoided. Shared reservoirs require separation by barrier flow valves or isolation tanks to prevent dilution of critical slideway oil properties.

Maintenance routines must include line flushing, filter checks, and confirmation of metering actuator function. Oil film presence can be verified through swipe tests or contact sensors in high-precision systems.

Effective slideway lubrication prevents mechanical hysteresis, supports geometric control, and protects high-value guide surfaces from scoring and microwelding under dynamic load. System performance depends on proper timing, volume control, and uniform distribution.

Table 5.5 summarizes slideway and guide applications.

TABLE 5.5   Summary of Slideway and Guide Applications

PARAMETER	INTERMITTENT OIL METERING SYSTEMS	MANUAL LUBRICATION	CIRCULATING OIL SYSTEMS
Delivery method	Timed or motion-triggered metered injection (pumps or injectors)	Operator-applied (brush, syringe, or oil can)	Continuous oil circulation via pump, filters, and reservoir
Application suitability	CNC (computer numerical control) machines, grinders, machining centers, robotics	Light-duty guides, occasional-use equipment	Large industrial slideways with high load/temperature
Lubricant type	ISO VG 68–220 with tackifiers, CGLP-grade oils	Light spindle or general-purpose oils	ISO VG 68–150 with oxidation inhibitors and foam control
Film formation needs	Must support boundary/mixed regime with minimal stiction	Often inconsistent, risk of underlube or overlube	Stable hydrodynamic film under continuous flow

PARAMETER	INTERMITTENT OIL METERING SYSTEMS	MANUAL LUBRICATION	CIRCULATING OIL SYSTEMS
Flow control	High precision (e.g., micro-oilers, divider blocks, restrictors)	No metering—operator-dependent	Controlled by pump pressure and restrictors
Uniformity of coverage	Achieved through manifold balancing and flow calibration	Highly variable	Excellent, if system is properly balanced
Contamination risk	Low; enclosed metering and sealed lines	High; exposed tools, open oil containers	Moderate; depends on filter maintenance and seals
Cross-system compatibility	Must avoid mixing with hydraulic oils; isolate systems if needed	Uncontrolled; high risk of cross-contamination	Typically isolated or shared with filtration safeguards
Maintenance needs	Inspect injectors; verify pulses; check for line leaks	Frequent reapplication, difficult to monitor	Filter change, flow check, viscosity/top-off monitoring
Failure modes prevented	Stick-slip, microwelding, chatter, wear at stroke reversals	None reliably—manual errors common	Overheating, wear from lack of oil film, misalignment
Best use cases	High-precision, multi-axis systems; CNCs; robotic actuators	Low-use tools, legacy machines, simple slides	Continuous-duty high-value machines with critical geometry

HYDRAULIC SYSTEMS

Hydraulic systems rely on the working fluid to perform both power transmission and lubrication. The fluid must maintain film strength across pumps, valves, actuators, and bearings while delivering consistent response under pressure. Proper lubrication within a hydraulic system depends on the fluid's base stock, additive package, cleanliness level, and temperature-dependent viscosity behavior.

Lubrication via Working Fluid Versus Dedicated Additives

In most systems, the hydraulic fluid itself provides lubrication through its base oil properties and additive content. Anti-wear hydraulic fluids (e.g., ISO 6743-4 HM or DIN HLP types) contain ZDDP or ashless phosphorus-based additives that form protective boundary films in vane and piston pumps. These additives activate under high contact stress, minimizing scuffing, galling, and adhesive wear.

In systems with high-pressure or high-cycle components, such as servo valves or axial piston pumps, fluids with enhanced boundary layer additives or friction modifiers are required. Where precision or extreme loads are present, fluid selection must ensure compatibility with elastomers, seals, and internal coatings.

In systems where fluid lubricity alone is insufficient, such as exposed actuators or where fluid dilution from process chemicals occurs, external lubrication or grease-packed interfaces may be required to supplement internal protection.

Filtration and Viscosity Control Hardware

Lubrication effectiveness in hydraulic systems is tightly coupled to contamination control. Solid particles, water ingress, and oxidation by-products compromise film integrity and accelerate wear. Systems must include high-efficiency filtration ($\beta x \geq 200$) at both pressure and return lines. Kidney-loop filtration improves off-line fluid conditioning and extends service life. Water removal elements (coalescers or desiccant breathers) are required in high-humidity or thermally cycling systems.

Viscosity control is critical for maintaining minimum film thickness in hydrodynamic and elastohydrodynamic regions. Viscosity must remain within equipment-defined ISO VG limits across the entire operating temperature range. Excessively low viscosity leads to leakage and boundary wear; viscosity that is too high causes sluggish actuation and cavitation risk. VI improvers are common in multigrade fluids, but their stability must be verified under shear stress.

Heat exchangers and reservoir sizing are key to managing fluid temperature. Overheated fluid accelerates oxidation, reduces viscosity, and depletes additive performance. In-line thermocouples and temperature-compensated pressure regulators maintain system integrity and lubrication performance.

Hydraulic system lubrication hinges on fluid cleanliness, chemical integrity, and temperature stability. Component longevity depends not just on power transmission but on maintaining a continuous and protective fluid film under all operating conditions.

CRITICAL SELECTION PARAMETERS

Proper lubrication system design begins with aligning component specifications and operating conditions to critical selection parameters. Failure to match lubricant type, flow characteristics, system pressure, and environmental constraints results in film breakdown, increased wear rates, and premature system failure.

Lubricant Type

Selection depends on application demands, component geometry, and operating environment.

➤ **Grease** is used for enclosed or slow-moving components where lubricant retention is critical. Viscosity, base oil type, thickener compatibility, and NLGI grade must align with speed and load.

➤ **Mineral oils** are suitable for general-purpose applications, providing adequate film strength and additive compatibility at moderate temperatures and loads.

➤ **Synthetic oils** (e.g., PAO, esters) offer superior oxidation resistance, thermal stability, and viscosity index, making them suitable for high-speed, high-temperature, or extended-drain applications.

➤ **Food-grade lubricants** (NSF H1/H2) are required where incidental contact with product is possible. Formulations must be nontoxic, inert, and thermally stable.

Flow Rate

Lubricant volume per point must match the consumption rate, film formation needs, and operating conditions. Flow rate calculations depend on component speed, load, and contact area.

➤ **Bearings and guides.** Volume must support full film replenishment before starvation.

➤ **Gears.** Flow must ensure flushing of wear debris and heat transfer.

➤ **Mist/air-oil.** Flow must be low-volume and high-frequency and be controlled in microliters per cycle.

The system must support adjustable metering and periodic verification through flow sensors or stroke counters.

Operating Pressure and Pump Compatibility

Lubrication systems must deliver lubricant at pressures compatible with line length, metering device resistance, and fluid viscosity.

➤ **Grease systems (progressive, dual-line).** Require pump pressures from 2,000–5,000 psi.
➤ **Oil systems (single line, circulating).** Operate between 15 and 100 psi.
➤ **Mist/air-oil.** Require precise pressure regulation for air and oil phases.

The pump must handle the required pressure and viscosity range without cavitation or bypass. High-viscosity lubricants may require heated reservoirs or positive displacement pumps.

Temperature Extremes and Thermal Stability

Lubricant performance must remain stable across the operating temperature envelope.

➤ Cold starts require fluids with low pour point and sufficient flow at low temperatures.
➤ High-temperature applications demand oxidation-resistant base stocks and thermally stable thickeners or additives.
➤ Greases must resist bleeding or hardening under sustained thermal load.

System design may require in-line heaters, coolers, and insulated lines in extreme environments.

Duty Cycle and Lubrication Frequency

Lubrication must match operating cycle patterns—continuous, intermittent, or variable load.

➤ High-frequency or continuous operation requires timed or sensor-triggered delivery.
➤ Intermittent or batch cycles benefit from prelube functionality and delayed delivery post-start.
➤ Grease and oil film degradation rates must be factored into scheduling logic to avoid overlubrication or underlubrication.

The delivery method (continuous versus intermittent) must match equipment speed and downtime allowance.

Each parameter must be evaluated based on actual application conditions and equipment design constraints. Lubricant type, flow control, pressure capability, and environmental tolerance must align to ensure continuous, effective film formation under real-world operating conditions.

INTEGRATION WITH CONTROL SYSTEMS

Lubrication systems must align with plant-level control architecture to ensure delivery accuracy, real-time monitoring, and maintenance efficiency. Control integration defines how lubrication actions are initiated, monitored, and verified, directly affecting system reliability and asset uptime.

Manual Versus Automated Control

Manual control systems operate on fixed schedules or operator-initiated cycles. These are stand-alone units with basic timers or push-button actuation. They are suitable for small-scale or low-criticality assets, but limited by lack of real-time feedback, high risk of skipped cycles, and no condition-based adjustment.

Automated systems use internal controllers or external signals to initiate lubrication events. Cycle frequency, delivery volume, and fault responses are programmable. Stand-alone logic units handle simple time-based sequencing. More advanced configurations use signal input from equipment load, runtime, or thermal sensors to trigger cycles dynamically.

PLC/SCADA Integration

PLC integration enables full synchronization with machine operation. Lubrication logic is embedded into equipment start/stop routines, interlocks, and alarm states. SCADA integration provides centralized visibility across systems, allowing remote monitoring, diagnostics, and cycle override. This is essential for large facilities managing multiple lubrication zones and for systems operating in hazardous or inaccessible locations.

Condition-based lubrication is achieved by feeding sensor data, such as temperature, vibration, cycle count, and fluid pressure, into the PLC. The controller adjusts lubrication timing, duration, or fluid type based on real-time asset condition. This minimizes overlubrication and responds to early wear indicators without manual intervention.

Alarm, Feedback, and Delivery Confirmation Features

Integrated systems must include feedback mechanisms to confirm lubricant delivery. Critical components include:

- ➤ **Pressure switches** to verify line pressurization
- ➤ **Cycle indicators** to confirm progressive divider operations
- ➤ **Flow meters** to detect lubricant volume per point
- ➤ **Limit switches or proximity sensors** on actuator-driven systems

Alarm logic must be configured to flag low reservoir levels, blocked lines, failed delivery, air ingress, or abnormal pressure drops. Alarms should trigger automatic shutdowns or hold states in high-criticality equipment, with notifications sent to SCADA or CMMS for maintenance response.

Data logging of cycle counts, delivery confirmation, and alarm events supports traceability and system auditing. Integration with CMMS platforms enables automated work order generation based on system status, cycle thresholds, or fault conditions.

Robust control integration elevates lubrication systems from passive subsystems to active reliability mechanisms. System architecture must support real-time feedback, fail-safe response, and condition-based delivery to meet the demands of high-uptime operations.

ENVIRONMENTAL AND MAINTENANCE CONSIDERATIONS

Effective lubrication system performance depends not only on system design but also on how well it aligns with the physical environment and maintenance infrastructure. Field conditions, technician access, and contamination exposure all influence long-term reliability.

Accessibility of Lube Points

Lube point location determines serviceability and system layout. Hard-to-reach points, such as elevated bearings, guarded driveshafts, and enclosed linear rails, necessitate remote or centralized delivery systems. Manual access to these locations increases the risk of missed lubrication, improper dosing, and technician exposure to mechanical or fall hazards. The routing of supply lines must account for thermal expansion, movement, and mechanical interference.

Accessibility drives system selection: single-point lubricators for isolated or infrequent points; centralized systems for distributed points on mobile or continuous-duty machinery.

Operator Skill Level and Maintenance Bandwidth

System complexity must match the available skill set. Facilities with limited technical bandwidth or high turnover require systems with minimal setup, automated cycle initiation, and built-in delivery verification. Progressive or single-line systems with visual indicators, low-maintenance metering valves, and minimal configuration burden are preferred in low-skill environments.

Facilities with trained technicians can deploy more complex systems, such as dual line, PLC-integrated, or condition-responsive, with programmable dosing, feedback sensors, and fault diagnostics. Skill level also affects the ability to troubleshoot pump priming, pressure loss, or line blockage.

Contamination Control: Dust, Washdown, Chemicals

Environmental exposure dictates lubricant formulation, sealing strategy, and delivery method. In dusty conditions, such as mining, cement, and grain, lines and fittings must be sealed with compression-type connectors and nonbreathing reservoirs. Desiccant breathers or membrane vents prevent airborne particulate ingress.

Washdown environments, such as food processing, bottling, and pharma, require IP-rated enclosures, corrosion-resistant components, and food-grade lubricants. Air/oil and mist systems must avoid overspray, pooling, or lubricant migration to adjacent equipment. Spray nozzles require shields or retractable mounts to prevent chemical wash-off.

Chemical exposure, including exposure to acids, solvents, or cleaning agents, demands chemically resistant hoses, fittings, and seal materials. Compatibility with elastomers and gaskets must be confirmed during specification.

Refill Intervals and Tank/Reservoir Sizing

Reservoir capacity must align with relube volume, cycle frequency, and service interval expectations. Undersized tanks result in frequent refilling, and as a consequence, increase labor hours and the risk of lubricant starvation. Oversized tanks may lead to fluid degradation due to long residence times, especially for water-sensitive or oxidation-prone lubricants.

Remote fill ports with level sensors allow top-up without removing guards or covers. Systems should include low-level alarms tied to control logic or CMMS alerts. Oil return systems (in circulating oil setups) must ensure proper de-aeration, sediment separation, and residence time to maintain lubricant quality.

System reliability improves when reservoir size, lubricant selection, and refill procedures are matched to operating hours, environmental conditions, and maintenance shift availability. Field constraints must drive practical system specification, not theoretical design assumptions.

CHAPTER 6

SELECTION TOOLS AND SPECIFICATION AIDS

System selection requires structured analysis using technical tools to match lubrication demands with equipment design, operational constraints, and criticality. Selection aids reduce specification errors, standardize component choices, and support documentation during procurement or audits.

OEM LUBRICATION CHARTS AND SYSTEM RECOMMENDATIONS

Most equipment manufacturers provide lubrication charts detailing recommended lubricant types, quantities, and intervals per point. These charts also specify acceptable delivery methods (e.g., grease via manual gun versus centralized grease system). OEM data includes critical information such as:

- Lubrication point type (bearing, gear, slide, chain)
- Lubricant volume per cycle or per hour
- Maximum pressure and flow limits for the component
- Environmental exposure requirements (IP ratings, chemical compatibility)

These charts form the baseline for system configuration and must be reviewed against actual operating conditions, as OEM assumptions often reflect ideal conditions not present in the field.

Lubrication Demand Calculators (Volume/Point/Hour)

Quantitative demand calculators convert machine data into flow rate and replenishment schedules. Input variables include:

- ➤ Shaft diameter and speed (for bearings)
- ➤ Gear pitch line velocity and load
- ➤ Chain pitch, width, and operating cycle
- ➤ Slideway stroke length and frequency

Outputs include:

- ➤ Lubricant volume per point per hour
- ➤ Relube interval based on film loss or mechanical motion
- ➤ Total system flow rate for pump sizing

Calculators also support viscosity selection by estimating required film thickness under specified operating conditions. This aids in confirming whether an oil, grease, or air/oil system is appropriate for the duty cycle.

Decision Matrix Based on Application Criticality and Point Count

A structured matrix helps select appropriate lubrication systems based on two primary axes: equipment criticality (high, moderate, low) and number of lubrication points.
As an example:

- ➤ **Low criticality/low point count** → Manual lubrication or single-point auto-lubricators
- ➤ **Moderate criticality/medium point count** → Single-line or progressive centralized systems
- ➤ **High criticality/high point count** → Dual-line or circulating systems with PLC/SCADA integration

Other matrix variables include access difficulty, duty-cycle severity, environmental exposure, and available maintenance bandwidth.

Selection matrices standardize system decisions across plant assets and help prioritize upgrades or retrofits based on reliability impact and ROI. These tools guide appropriate system architecture, metering design, and integration planning for efficient and risk-aligned lubrication strategy deployment.

FIELD CASE EXAMPLES

Real-world applications of lubrication system upgrades highlight the performance gains, risk reduction, and maintenance efficiency achieved by aligning system architecture with equipment demands and environmental constraints.

Upgrade from Manual Greasing to a Progressive System in a Steel Plant

In a hot strip mill, mill stands and roller bearings were originally lubricated with manual grease guns during shift change. Missed points, inconsistent grease volume, and technician exposure to high-temperature environments led to premature bearing failures, excessive downtime, and overlubrication in accessible areas.

A progressive centralized lubrication system was installed, servicing 138 points across the finishing stand. A single electrically driven pump with a 30-liter reservoir fed metering valves through rigid and flexible lines. Grease volume per point was calculated based on bearing size, duty cycle, and ambient temperature.

Cycle initiation was tied to roller motor runtime via PLC. Pressure switches and cycle indicators confirmed full system function. After installation, bearing-related downtime dropped by 47% within 6 months. Grease consumption was reduced by 34%, and thermal imaging confirmed more consistent operating temperatures across all bearing housings.

Switch from Oil Bath to Circulating System in High-Temp Gearboxes

In a cement kiln application, gearboxes on induced draft (ID) fans were lubricated with static oil baths. Operating temperatures reached 95°C, causing oil oxidation, foam formation, and reduced viscosity. This led to repeated gear scuffing, bearing wear, and emergency shutdowns.

A circulating oil system was installed with a dedicated pump, plate heat exchanger, dual-stage filtration, and 100-liter reservoir. Flow was targeted to gear mesh zones and bearing pockets. Return oil was routed through a high-efficiency filter and temperature sensor before re-entering the reservoir.

Operating temperatures were stabilized at 65°C, and oxidation-related oil changes dropped from every 3 months to once annually. Gear condition monitoring showed no additional scuffing, and oil cleanliness levels consistently met ISO 4406 17/15/12. Vibration levels on the ID fan gearbox dropped by 22%, and MTBF increased significantly.

Adoption of an Air-Oil Mist System for Spindle Lubrication in CNC Machining

In a precision CNC machining facility, high-speed spindles were originally lubricated with periodic manual oil injection. Stick-slip behavior and elevated bearing temperatures at high RPMs (24,000+) limited production consistency and reduced spindle life.

An air-oil system was retrofitted on each spindle unit, delivering 3 μL of oil per pulse via a controlled air stream. Pulse frequency was tied to spindle RPM and adjusted in real time via machine PLC. The system included in-line flow sensors and feedback alarms to detect nozzle clogging or oil starvation.

Post-upgrade analysis showed bearing temperatures stabilized 10°C lower across full speed range. Stick-slip was eliminated, enabling smoother surface finishes and reduced scrap rate. Spindle life extension was confirmed via bearing inspections, with replacement intervals increasing from 18 to 36 months. Oil consumption was reduced by 87%, supporting both performance and environmental targets.

These field examples demonstrate measurable improvements in uptime, component life, and lubricant use when lubrication systems are engineered to match operational requirements. System design, implementation, and validation must be aligned with load, speed, thermal profile, and accessibility to achieve sustainable reliability gains.

FINAL DESIGN AND IMPLEMENTATION GUIDELINES

Lubrication system design must support long-term equipment reliability, efficient maintenance access, and future system scaling. Field implementation requires alignment among mechanical, electrical, and operations teams to ensure functionality and maintainability from start-up through continuous operation.

PRIORITIZE RELIABILITY, ACCESSIBILITY, AND SCALABILITY

System architecture must deliver consistent lubricant volume, timing, and coverage under all expected load and environmental conditions. Component selection, including components such as pumps, metering valves, reservoirs, and fittings, must meet pressure, temperature, and compatibility requirements. Use vibration-resistant lines and industrial-grade fittings rated for the lubricant type and viscosity.

Accessibility is critical. Position reservoirs, filters, and refill ports where technicians can service them without disassembly or safety compromise. Avoid routing lines through moving assemblies or high-heat zones without shielding or protection.

Scalability must be designed in from the outset. Include spare ports on manifolds, expandable PLC input/output blocks, and oversized reservoirs where point counts may increase. Progressive systems should include bypass options or expansion zones for future loads.

Validate Performance Through Commissioning and Runtime Monitoring

Commissioning must verify full system function under operating conditions. Validate:

- ➤ Pressure rise and decay during cycle
- ➤ Flow rate at all lubrication points
- ➤ Timing synchronization with machine operation
- ➤ Function of cycle indicators, pressure switches, and flow sensors

Baseline data, such as bearing temperature, vibration, and wear particle count, should be collected before and after lubrication system activation. Runtime monitoring via PLC or SCADA ensures that lubricant delivery occurs at correct intervals, volumes, and system pressures. Alarm thresholds should be configured for reservoir level, delivery failure, and metering faults.

Maintain As-Built Documentation, Component Specs, and Training Plans

As-built drawings must document reservoir location, pump specs, line routing, metering block configurations, and wiring diagrams. Maintain a full bill of materials (BOM), including OEM part numbers for pumps, valves, filters, hoses, and sensors. Label all lubrication lines and control panels clearly.

Develop technician training plans covering operation, fault diagnosis, component replacement, and cycle verification. Include inspection intervals, sensor calibration schedules, and flushing procedures for contamination events.

System effectiveness depends not only on design but also on how the system is maintained and operated over time. Documented procedures, routine validation, and scalable architecture are required to ensure long-term lubrication system integrity.

AUTOMATED VERSUS MANUAL DELIVERY: A RISK AND ROI EVALUATION

Manual lubrication is still prevalent due to its perceived simplicity and low capital cost. However, automated systems often yield superior ROI when evaluated over the total life cycle. Key risks associated with manual delivery include inconsistent application, overlubrication/underlubrication, safety concerns during access, and human error.

Automated systems, including single-line, dual-line, progressive, and injector-based designs, offer consistent volume and timing, improved safety, and reduced labor. Risk-based assessments help justify automation investment, particularly for high-value, mission-critical assets or inaccessible components.

ROI is calculated through metrics such as equipment uptime, labor savings, lubricant consumption, reduced failures, and extended service intervals. However, implementation costs, training requirements, and system reliability must also be considered.

Table 6.1 is a decision matrix comparing automated versus manual lubrication delivery systems across key evaluation factors, with weighted risk and ROI considerations. This can be used to guide capital investment, reliability planning, or retrofit decisions.

TABLE 6.1   Automated Versus Manual Lubrication: Risk and ROI Decision Matrix

EVALUATION FACTOR	WEIGHT (%)	MANUAL DELIVERY	AUTOMATED DELIVERY	JUSTIFICATION/ NOTES
Consistency of lubricant volume	15%	Low—user-dependent	High—programmable and repeatable	Automation ensures metered dosing regardless of personnel
Lubrication frequency accuracy	10%	Low—risk of missed or delayed intervals	High—schedule-based or sensor-triggered	Timing precision reduces wear and extends asset life
Labor cost/ resource allocation	15%	High—manual rounds required	Low—reduced labor for routine tasks	Automation reallocates personnel to higher-value tasks

EVALUATION FACTOR	WEIGHT (%)	MANUAL DELIVERY	AUTOMATED DELIVERY	JUSTIFICATION/ NOTES
Contamination risk	10%	High—open tools, reused containers	Low—sealed reservoirs and enclosed systems	Automated systems reduce exposure to particulates and moisture
Overlubrication/ underlubrication risk	10%	High—inconsistent application	Low—system-calibrated dosing	Prevents bearing overheating (over) or failure (under)
Upfront investment cost	10%	Low—basic tools required	High—pumps, injectors, PLC integration	CapEx hurdle must be evaluated against long-term savings
Maintenance and repair	5%	Low—simple devices	Moderate—pumps, timers, nozzles need service	Maintenance of automation requires technical oversight
Suitability for remote assets	10%	Low—hard to access frequently	High—enables unattended or elevated service	Automated systems shine where manual access is dangerous or infrequent
Impact on equipment uptime	10%	Medium—increased failure risk	High—fewer breakdowns, more predictable cycles	Predictive lubrication avoids unplanned downtime
ROI over 2–5 years	5%	Low—ongoing labor cost, increased risk	High—failure reduction, fewer repairs	Especially favorable in high-cycle, high-value equipment environments

Table 6.2 is a companion table that summarizes scoring.

TABLE 6.2 Scoring Summary (Weighted %)

SYSTEM	TOTAL WEIGHTED SCORE
Manual	~48%
Automated	~82%

Automated lubrication systems offer superior **long-term ROI** and **risk mitigation**, especially in environments with high equipment criticality, frequent service needs, or safety/accessibility concerns. Manual delivery may remain appropriate for **noncritical**, low-frequency, or budget-constrained applications.

Automated Versus Manual Delivery—A Risk and ROI Evaluation of Lubrication Delivery Modes

Manual lubrication involves human-operated tools such as grease guns, oil cans, squeeze bottles, or drip feeders. Delivery frequency, quantity, and method depend entirely on technician availability, access, and technique. These systems offer no automated confirmation of delivery, no cycle validation, and no synchronization with machine duty cycle. Accuracy varies significantly based on operator skill, shift workload, and access to lube points.

Automated lubrication uses programmable pumps, metering valves, flow regulators, and control logic to deliver lubricant in consistent, predefined volumes at timed or condition-based intervals. Lubricant is delivered to multiple points through fixed distribution lines. Delivery confirmation can be tied to pressure switches, cycle counters, or PLC input. Systems are scalable, configurable, and capable of operating in hazardous or remote locations without human intervention.

Historically, industrial plants depended on manual methods due to simplicity, low upfront cost, and the absence of automation infrastructure. As equipment complexity increased, manual lubrication introduced variability in lubrication quality and consistency. Missed points, overlubrication, and lubricant cross-contamination became leading contributors to premature bearing and gear failures. Downtime for lubrication-related issues increased, especially on high-speed or critical assets.

Modernization drivers include rising asset uptime expectations, reduced maintenance staffing, higher environmental control requirements, and integration with digital maintenance systems. Lubrication delivery is now recognized as a controllable failure point. Automation reduces variability, eliminates missed cycles, and supports real-time monitoring of lubricant delivery and reservoir levels. Integration with PLC and SCADA systems allows for alarm-based escalation, preventive maintenance scheduling, and precise lubricant usage tracking for cost control.

The shift from manual to automated systems is driven by the need to eliminate human error, reduce lubrication-related failure risk, and provide a data-backed ROI through reduced lubricant waste, extended component life, and lower labor costs. Evaluation of system type must include downtime history, labor availability, asset criticality, and existing control system infrastructure.

Risks and Limitations of Manual Lubrication

Manual lubrication presents inherent reliability risks due to inconsistent volume delivery and timing. Grease or oil is applied based on technician interpretation, often without measurement tools. Variability in stroke pressure, nozzle alignment, and point access results in overlubrication, underlubrication, or complete omission. Film starvation at friction interfaces causes surface fatigue, while overapplication leads to seal damage, fluid churning, and elevated operating temperatures.

Human error is frequent. Missed lubrication points occur during shift transitions, during high workload periods, or when components are obscured or difficult to access. Cross-contamination results when lubricants are applied with unclean fittings or mixed between incompatible greases. Application of the wrong lubricant, such as an incorrect NLGI grade, base oil type, or additive package, alters film characteristics, increases wear rate, or causes chemical attack on seals and bearings.

Manual lubrication introduces safety hazards. Performing lubrication on running equipment exposes technicians to rotating machinery, pinch points, and elevated temperatures. Confined spaces, overhead lubrication points, and equipment requiring climbing increase the risk of falls, burns, or crush injuries. Lockout/tagout procedures are often bypassed to save time, elevating risk further.

Labor-intensive lubrication routes consume valuable maintenance hours. Repetitive, low-value manual tasks divert resources from diagnostic and corrective maintenance. Understaffed facilities with large asset footprints struggle to maintain lubrication schedules, increasing failure probability due to skipped or delayed lubrication cycles.

Manual methods offer no delivery confirmation, pressure validation, or volume tracking. There is no embedded traceability. Lubrication records are often paper-based, unnverifiable, and disconnected from actual machine conditions. Failure investigations are limited by lack of documentation on when lubrication was last performed or whether it met required standards.

Without monitoring, automation, or performance feedback, manual lubrication becomes a significant reliability liability, especially in high-duty, inaccessible, or high-speed applications where failure consequences are severe.

Automated Lubrication Technologies Overview

Automated lubrication systems are engineered to deliver precise lubricant volumes to multiple points using centralized pumps, metering components, and programmed control logic. System architecture is selected based on point count, lubricant type, required volume, and operating pressure.

Single-Line Resistive Systems

These systems use a central pump to pressurize a main supply line. Flow restrictors or calibrated orifices meter lubricant to each point. Delivery is simultaneous but dependent on line resistance, making it suitable only for low-viscosity oils in low-pressure, short-distance systems. This results in a lack of point-specific control, common in light-duty machinery and compact industrial assets.

Dual-Line Systems

Dual-line systems are designed for large installations with long lines and high point counts. Two supply lines alternate between pressure and vent cycles. Each metering device receives lubricant when its line is pressurized. The devices operate at high pressure (up to 5,000 psi) and can handle grease up to NLGI #2. They are suited for steel mills, mining equipment, and heavy industrial plants, offering high reliability, scalability, and fault tolerance.

Progressive Divider Block Systems

A central pump feeds lubricant into a master metering block, which divides and sequences flow through secondary metering valves. Each outlet delivers a fixed volume. A blockage at any point stops the sequence, providing built-in fault detection. The architecture is ideal for medium-sized systems with moderate point counts and the need for precise control. Progressive divider block systems are common in packaging, plastics, and automated machinery.

Injector-Based (Positive Displacement) Systems

Each lubrication point in an injector-based system has a dedicated injector delivering an exact volume on each cycle. Injectors operate independently and can accommodate varied flow requirements across points. The systems are suited for applications requiring high accuracy and flow variability. They operate under moderate to high pressure and are common in mobile equipment, precision machinery, and applications with dissimilar lubrication demands.

Control Types

Control types include:

> **Time-based.** Operates on preset intervals, regardless of machine condition. Low complexity, moderate effectiveness.

> **Cycle-based.** Triggers lubrication after a fixed number of machine cycles or operational strokes. Matches application better than time-based systems.
> **Load-based.** Activates lubrication based on measured mechanical load or runtime. Aligns delivery with demand.
> **Condition-based.** Uses sensor data (temperature, vibration, torque) to initiate lubrication based on the actual system condition. Highest precision, used in critical systems.

Smart Systems

Modern systems include feedback sensors such as pressure switches, cycle indicators, flow meters, and grease stroke counters. Alarms notify operators of failed delivery, low reservoir levels, line blockages, or abnormal pressures. PLCs adjust delivery based on feedback, runtime, or environmental variables.

Integration with CMMS and SCADA

Automated lubrication systems can interface with CMMS and SCADA platforms. This enables real-time status monitoring, alarm escalation, work order generation, and historical logging of lubrication events. Integration supports preventive maintenance strategies and simplifies root cause analysis for lubrication-related failures.

Automated systems increase precision, reduce variability, and provide continuous assurance of lubricant delivery under changing operating conditions. System selection and control strategy must align with equipment criticality, operating profile, and maintenance structure.

Comparative Risk Analysis

Failure modes in lubrication delivery directly impact mechanical reliability. Inconsistent delivery, whether from skipped manual cycles, underdosing, or overapplication, results in boundary lubrication, film collapse, overheating, and accelerated wear. These failures manifest as bearing seizure, gear scuffing, chain elongation, and slideway stick-slip. In manual systems, inconsistency is operator-dependent. In automated systems, inconsistency arises from unverified delivery, clogged lines, or pump failure without feedback.

Critical Asset Vulnerability and System-Level Consequences

High-speed rotating equipment, continuous-process machinery, and safety-critical actuators are vulnerable to lubrication faults. A single underlubricated point in a centralized drive or process line can cascade into full-line stoppage. For example, seizure in a single bearing on a production conveyor can trigger emergency shutdowns, backed-up material flow, and unplanned production loss.

Systems with multiple lubrication points, such as gear trains, CNC beds, and packaging equipment, are particularly sensitive to uniformity. Failure at a single node within a progressive system halts downstream delivery. In manual systems, points requiring disassembly or elevated access are frequently skipped, creating repeat failure zones that go undocumented.

Downtime Costs Due to Lubrication-Related Faults

Lubrication-induced failures incur high downtime costs, particularly on automated production lines, batch processing systems, or capital-intensive rotating equipment. Recovery from a bearing or gearbox failure includes teardown, part replacement, labor hours, and requalification or alignment procedures. Unscheduled downtime often results in missed production targets, delayed shipments, and overtime labor.

In industries like food and beverage, pharmaceuticals, and energy, unplanned downtime can also trigger regulatory consequences, spoilage, or systemwide energy loss. Lubrication-related downtime is often underreported due to poor traceability in manual systems, masking the true cost impact.

Safety Risk Mitigation via Automation

Automated systems eliminate the need for manual lubrication during operation, reducing technician exposure to rotating equipment, hot surfaces, confined spaces, or elevated work areas. This directly lowers fall risk, burn potential, and pinch point incidents. Lockout/tagout violations associated with manual lubrication are also mitigated.

Automated systems with feedback controls detect delivery failures in real time, enabling preventive shutdown or alarm escalation before mechanical damage occurs. This preempts unsafe conditions such as overheated bearings, hydraulic stall, or mechanical seizure that could lead to uncontrolled motion, ejection, or fire.

By reducing dependency on operator consistency, providing closed-loop control, and removing exposure to hazardous service points, automated lubrication systems significantly lower the operational risk profile across critical assets.

ROI JUSTIFICATION PARAMETERS

Return on investment for automated lubrication systems is established through measurable reductions in direct costs and quantifiable improvements in asset reliability and availability. Evaluation includes lubricant consumption, labor utilization, component lifespan, and unplanned downtime metrics.

Direct Cost Savings

Reduced Lubricant Waste

Automated systems deliver lubricant in calibrated volumes, eliminating overapplication common in manual methods. Grease guns and squeeze bottles often exceed required quantities, leading to seal damage, excess purging, and environmental cleanup costs. Automated metering valves match volume to bearing or gear specifications, reducing total lubricant use by 25–40% depending on application type and frequency.

Lower Labor Hours for Lubrication Tasks

Manual lubrication consumes technician time, particularly in large plants with high point counts. Route-based greasing can account for hundreds of labor hours annually. Automated systems eliminate routine manual cycles, reducing maintenance time allocation for lubrication by 50–80%. Labor is reallocated to diagnostic, corrective, or higher-value tasks. Travel, climb, and lockout times associated with hard-to-reach points are eliminated.

Indirect Gains

Extended Component Life

Consistent lubrication reduces surface fatigue, corrosion, and adhesive wear. Automated systems maintain continuous film integrity under all duty cycles, preventing underlubrication during load spikes or overlubrication during idle periods. Bearings, gears, chains, and guides see a documented 30–60% increase in service life, reducing replacement frequency and inventory carrying cost.

Increased Equipment Uptime

Automated lubrication prevents unplanned stoppages from heat buildup, bearing seizure, or lubrication starvation. Predictable cycle timing aligns with runtime, allowing for precise control of film formation and replenishment. Equipment remains in service longer between maintenance intervals. Uptime improvements range from 2–5% annually in batch process and high-duty environments.

Fewer Emergency Maintenance Events

Emergency work orders due to lubrication-related failures decline as systems provide early fault detection through feedback sensors and alarms. Lubrication events become preventive and scheduled. MTBF increases, and corrective work transitions into planned maintenance. Plants using automated systems report 40–70% reductions in lubrication-related emergency interventions.

Table 6.3 is an example of a decision matrix comparing automated and manual lubrication.

Quantitative ROI Modeling over 1-, 3-, and 5-Year Intervals

ROI calculations include:

- System cost (hardware, installation, commissioning)
- Annual lubricant cost reduction
- Labor hour savings (converted to wage cost)
- Component replacement cost deferment
- Downtime cost avoidance (based on production rate and average failure downtime)

TABLE 6.3 Automated Versus Manual Lubrication Decision Matrix Example

EVALUATION FACTOR	WEIGHT (%)	MANUAL LUBRICATION (SCORE 1 TO 5)	AUTOMATED LUBRICATION (SCORE 1 TO 5)	MANUAL WEIGHTED SCORE	AUTOMATED WEIGHTED SCORE
Consistency of lubricant volume	10%	2	5	20	50
Lubrication timing accuracy	10%	2	5	20	50
Labor requirements	10%	2	5	20	50
Contamination risk	7%	2	5	14	35
Overlubrication/underlubrication risk	8%	2	5	16	40
Delivery verification and feedback	8%	1	5	8	40
Safety risk	7%	2	5	14	35
Downtime due to lube failures	10%	2	5	20	50
Component life extension potential	7%	2	4	14	28
Lubricant usage efficiency	6%	2	5	12	30
System cost (CapEx)	5%	5	2	25	10
ROI timeline (2–5 years)	5%	2	5	10	25
Integration with digital systems	4%	1	5	4	20
Suitability for remote/high-access points	2%	1	5	2	10
Maintenance complexity	1%	3	4	3	4
Total weighted score				202	477

Example

Installation cost: $42,000
Annual lubricant savings: $7,200
Labor savings: $18,400
Downtime reduction: $28,000
Component life extension: $9,000

Year 1 savings: $62,600
ROI year 1 = 149%
Cumulative ROI at year 3 = 447%
Cumulative ROI at year 5 = 745%

Cost recovery typically occurs within 6–18 months depending on system scale and criticality. Systems with feedback integration and SCADA connectivity provide the highest return due to added failure prevention and labor displacement benefits.

Implementation of Cost Considerations

Implementation of an automated lubrication system requires detailed cost planning across capital, labor, and operational domains. The upfront investment varies based on system type, number of lubrication points, environmental conditions, and integration requirements.

Capital Cost of Components

Core hardware includes centralized pumps, reservoirs, distribution lines, metering valves, fittings, and control units. Examples of costs are:

- ➤ **Pump/reservoir assemblies.** $1,500–$10,000, depending on pressure class, capacity, and redundancy
- ➤ **Metering valves** (progressive, injector, or dual line). $50–$400 per point
- ➤ **Tubing and fittings.** Cost scales with point count and run length, typically $5–$15 per foot installed
- ➤ **Controllers/PLCs.** $1,000–$8,000, depending on features, communication protocols, and input/output requirements

Sensor packages, pressure switches, and cycle indicators add cost but are necessary for systems requiring feedback or SCADA integration.

Design and Engineering for Retrofit or OEM Integration

Retrofit applications require site surveys, mechanical clearance validation, and line routing design. The system must be matched to existing lubrication specifications and access constraints. Custom brackets, routing protections, and structural penetrations may be needed.

OEM integration is more cost-efficient per point, as design is embedded in the build phase. However, it may still require controller compatibility checks and lubrication system isolation from process control networks.

Engineering design and commissioning costs typically range from 10–20% of hardware costs for complex systems.

Installation Labor and Operational Disruptions

Labor includes mounting of pump units, installation of tubing, termination at lube points, and system wiring. Access equipment (lifts, scaffolds), confined space permits, or machine downtime must be factored in.

Labor costs are typically $75–$125 per point for standard accessibility. Shutdown coordination and post-install verification (pressure tests, functional checks) must be included in planning. For critical production assets, installation may need to occur during planned outages to avoid production loss.

Ongoing Maintenance and Spare Parts

Automated systems reduce lubrication labor but still require maintenance. Reservoirs must be refilled, filters replaced, pumps serviced, and lines inspected for leakage or damage.

Typical spares inventory includes:

➤ Metering blocks
➤ Fittings and line repair kits
➤ Pressure switches
➤ Pump motor components or rebuild kits

Annual maintenance budget would be 5–10% of system capital cost, depending on the system complexity and environment (dust, vibration, chemical exposure).

Training and Upskilling Maintenance Personnel

Technicians must be trained on:

- ➤ System operation and cycle verification
- ➤ Alarm response procedures
- ➤ Basic troubleshooting and pressure diagnostics
- ➤ Safe refill and inspection practices

Training time is 4–8 hours per technician. Training cost is the cost of internal (on-the-job), vendor-led, or third-party certification. High-turnover environments may require recurring onboarding sessions.

Cost of implementation must be balanced against asset criticality, failure history, and available maintenance bandwidth. Shortcuts in design, installation, or training result in poor system performance and negated ROI. Proper execution requires upfront capital and technical alignment across operations, maintenance, and engineering.

APPLICATION-BASED DECISION MATRIX

Selection of a lubrication delivery system requires structured evaluation of application-specific parameters. The decision matrix is based on asset criticality, accessibility, environmental conditions, lubricant characteristics, high-frequency and low-volume requirements, and system scalability. Each factor directly influences the system type, control method, and component specification.

Asset Criticality and Accessibility

- ➤ **High-criticality assets** (e.g., mainline conveyors, turbine gearboxes, CNC spindles). Require automated, monitored systems with delivery confirmation, fault detection, and SCADA or PLC integration
- ➤ **Low-criticality assets** (e.g., auxiliary chains, infrequent-use motors). May tolerate single-point lubricators or manual lubrication if access is safe and schedules are enforced

➤ **Hard-to-reach or guarded locations.** Favor centralized or remote systems to eliminate manual access risks

➤ **Mobile assets** (e.g., cranes, loaders). Require vibration-resistant, self-contained automated systems with refill ports accessible from ground level

Environmental Conditions: Dust, Water, Vibration

➤ **Dust-heavy environments** (e.g., mining, grain, cement). Require sealed fittings, shielded lines, pressurized reservoirs, and grease-compatible components to prevent ingress.

➤ **Wet or washdown areas** (e.g., food processing, chemical plants). Require IP67 or higher-rated enclosures, corrosion-resistant tubing and fittings, and food-grade lubricants where required.

➤ **High vibration zones** (e.g., stamping presses, mobile equipment). Require flexible hoses, crimped ends, and vibration-isolated pump mounts. Use progressive or injector systems with rigid feedback to detect blocked lines.

Lubricant Type: Grease Versus Oil, Viscosity Range

➤ **Grease (NLGI 1–2).** Best suited to progressive, dual-line, or injector-based systems. High-pressure capability (2,000–5,000 psi) is essential for delivery over distance or against back pressure.

➤ **Oil (ISO VG 32–220+).** Compatible with single-line resistive, circulating, and mist/air-oil systems. Viscosity influences pump type (gear versus piston), metering method, and filter specification.

➤ **Specialty lubricants** (synthetics, food grade, biodegradable). Require chemical compatibility of seals, metering valves, and reservoir materials.

Frequency and Volume Requirements

➤ **High-frequency, low-volume applications** (e.g., spindles, linear guides). Use air/oil or mist systems with pulse controllers and micrometering.

- ➤ **Low-frequency, high-volume points** (e.g., large bearings, slow-turning shafts). Use centralized grease systems with adjustable metering for longer intervals.
- ➤ **Variable load or runtime cycles.** Require load-based or condition-based control with runtime counters or sensor input. Match system logic to operating profile.

System Scalability and Future Expansion

- ➤ **Progressive systems.** Use add-on blocks and extensions if possible within volume and pressure constraints.
- ➤ **Dual-line systems.** These should be easily scalable and support long distances and high point counts.
- ➤ **Single-line resistive systems.** They are not easily scalable; flow balancing becomes unstable with added points.
- ➤ **Expansion.** Systems should be specified with future load in mind. Include spare I/O, open metering ports, and sufficient reservoir volume to accommodate expansion.

The decision matrix supports objective selection aligned with field constraints. Failure to account for any one of these variables results in compromised delivery, increased maintenance cost, and reduced reliability. Each lubrication point must be evaluated based on duty, environment, and operational consequence.

Table 6.4 is an example of a decision matrix assessing application-based lubrication.

TABLE 6.4 Application-Based Lubrication Decision Matrix Example

APPLICATION FACTOR	MANUAL LUBRICATION (1 TO 5)	SINGLE-POINT AUTO (1 TO 5)	PROGRESSIVE SYSTEM (1 TO 5)	DUAL-LINE SYSTEM (1 TO 5)	AIR/OIL OR MIST SYSTEM (1 TO 5)
Asset criticality	2	3	4	5	5
Accessibility	2	3	4	5	3
Environmental condition: dust	1	2	4	5	2
Environmental condition: water/ washdown	1	2	4	5	3
Environmental condition: vibration	1	2	4	5	3
Lubricant type: grease	2	3	5	5	1
Lubricant type: oil	3	3	4	4	5
Lubricant type: specialty	1	2	4	4	4
Lubrication frequency: high	1	2	5	5	5
Lubrication Frequency: low	3	4	4	4	2
System scalability	2	2	4	5	3

BEST PRACTICES

Pilot automated lubrication systems on high-risk, failure-prone assets where downtime or repair cost is highest. Select systems with known lubrication-related failure history, for example, gearboxes, critical bearings, or high-speed spindles, for initial deployment. Measure baseline conditions (temperature, vibration, lubricant consumption) to establish preinstall benchmarks.

Conduct lubrication-specific failure mode and effects analysis (FMEA). Identify failure modes linked to underlubrication, contamination, overlubrication, and incompatible lubricant application. Rank severity, occurrence, and detection capability. Use FMEA output to guide system selection, metering point priority, and monitoring strategy.

Include delivery monitoring and feedback systems from day 1. Integrate pressure switches, flow indicators, cycle counters, and reservoir level alarms. Tie feedback into PLC or SCADA where possible for real-time alerts and historical data logging. Configure alarm logic to escalate lubrication faults before mechanical damage occurs.

Audit delivery effectiveness and lubricant usage at regular intervals. Perform line checks, pressure verification, and visual inspection of lube points. Review system log files for missed cycles, sensor faults, and refill frequency. Monitor lubricant consumption per machine and compare against calculated demand. Investigate variances to detect leaks, bypassed metering, or degradation in system performance. Use audit findings to refine delivery timing, adjust metering volumes, and schedule preventive maintenance.

CHAPTER 7

SPECIFICATION TOOLS FOR CENTRALIZED LUBRICATION SYSTEMS

Centralized lubrication systems distribute controlled amounts of lubricant to multiple points from a single source, typically with programmable control. Specification tools for these systems include system sizing calculators, component selectors, piping layout software, and delivery diagnostics.

A typical specification process involves:

➤ Identifying all lube points and required delivery rates
➤ Determining system type (progressive, injector, or dual line)
➤ Sizing pumps and reservoirs based on delivery rate and cycle time
➤ Designing the layout for optimal line lengths and minimal pressure drops

Advanced configuration tools allow simulation of cycle behavior, detection of flow anomalies, and optimization for redundancy and maintainability. Properly specified systems prevent cross-contamination, optimize lubricant consumption, and enable real-time monitoring of lubricant flow to critical assets.

CENTRALIZED LUBRICATION SYSTEMS

Centralized lubrication systems deliver controlled amounts of lubricant from a single pump to multiple lubrication points through a fixed distribution network. These systems eliminate manual application variability, improve film consistency, and enable precise timing and volume control across distributed assets.

System architecture includes a reservoir and pump unit, supply lines, metering components, and delivery lines terminating at each lubrication point. Control methods vary by system type and may include timers, PLC signals, runtime counters, or sensor-driven logic. Feedback devices, such as pressure switches, flow meters, and cycle indicators, confirm delivery.

Progressive Systems

In progressive systems, lubricant is delivered sequentially through metering valves that divide flow into fixed-volume discharges. Each valve section feeds a downstream point and triggers the next. A blockage halts the sequence, providing immediate fault isolation. These systems operate at moderate pressures (300–1,500 psi) and are suited for grease or oil delivery. Progressive systems are ideal for medium point-count applications where consistent volume and self-monitoring are critical. They are used extensively in packaging lines, injection molding machines, mobile equipment, and light industrial conveyors.

Injector-Based Systems

Injector systems use individual metering devices, each fed from a pressurized manifold. Each injector delivers a precise volume of lubricant independently, allowing flexible volume assignment per point. The injectors operate at higher pressures (up to 5,000 psi), making them suitable for grease delivery in high-load or variable-volume environments. Injector systems are commonly used in off-road equipment, cranes, and installations where lubrication points differ significantly in flow demand.

Dual-Line Systems

Dual-line systems alternate pressure between two supply lines. When one line is pressurized, a metering valves discharges lubricant through it, while the other line vents.

These systems support long line lengths, hundreds of lubrication points, and high-viscosity lubricants. The pressure range is typically 1,500–5,000 psi. Dual-line systems are suited for steel mills, mining operations, cement plants, and any large-scale facility with dispersed lubrication needs. They tolerate harsh environments and provide delivery redundancy.

Single-Line Resistive Systems

These systems use a central pump to pressurize a single line feeding all lubrication points through restrictors or calibrated orifices. Delivery occurs simultaneously, but metering is nonadjustable, and volume depends on line resistance and flow path geometry. The systems operate at low pressure (15–100 psi), limiting them to low-viscosity oils and short distribution runs. Single-line resistive systems are commonly used in small machine tools, textile lines, and compact industrial units where simplicity outweighs precision.

Common Industrial Applications

Common industrial applications include:

- ➤ **Mining.** Dual-line and injector systems for haul trucks, loaders, crushers, and conveyors. High point counts, long runs, and heavy grease volumes.
- ➤ **Pulp and paper.** Progressive systems for calendar rolls, dryer sections, and press bearings. Lubricant distribution must operate in wet, hot, and high-vibration environments.
- ➤ **Food processing.** Progressive or single-line systems using NSF H1-approved lubricants. Stainless steel tubing, washdown resistance, and precise dosing at frequent intervals.
- ➤ **Heavy equipment.** Injector-based systems dominate in mobile platforms requiring robust, vibration-tolerant, high-pressure grease delivery.

Each system type must be matched to the equipment layout, environmental exposure, lubricant characteristics, and access constraints. Specification accuracy directly influences system reliability, lubricant efficiency, and maintenance workload.

Table 7.1 summarizes the centralized lubrication systems.

TABLE 7.1　Centralized Lubrication Systems Summary

SYSTEM TYPE	PRESSURE RANGE (PSI)	SUITABLE LUBRICANTS	APPLICATION TYPES	KEY FEATURES
Progressive	300–1,500	Grease or oil	Medium point counts, packaging, injection molding, mobile equipment	Sequential delivery, blockage detection, self-monitoring
Injector-based	Up to 5,000	Grease	High load, variable volume, off-road equipment, cranes	Independent metering, flexible volume per point
Dual line	1,500–5,000	High-viscosity grease/oil	Large-scale industrial (mining, steel, cement); long runs	Supports long lines, high point counts, redundancy
Single-line resistive	15–100	Low-viscosity oil	Compact systems, small tools, textile machines	Simple design, simultaneous delivery, limited precision

CORE STEPS IN THE SPECIFICATION PROCESS

Lube Point Identification and Requirement Mapping

Begin with a complete inventory of all lubrication points on the asset or system. Record component type (bearing, gear, chain, slideway), lubrication method (oil or grease), and mounting orientation. Document operating conditions at each point, including speed, load, ambient temperature, exposure to contaminants, and duty cycle.

Determine the required flow rate for each point based on bearing size, load factor, and relubrication interval. Reference OEM guidelines, or use engineering calculations (e.g., ISO 281 for bearings). Define lubricant type per point—NLGI grade for grease and ISO VG rating for oil—and confirm compatibility with seals and materials. Identify required lubrication frequency: continuous, intermittent, or per operating cycle.

System-Type Selection

Select system architecture using a decision matrix with the following criteria:

➤ **Progressive systems.** Suited for 10–150 points per pump, moderate distances (< 15 meters per outlet), fixed-volume delivery, and low to moderate load sensitivity. Not ideal for inaccessible or high-criticality zones without feedback monitoring.

➤ **Dual-line systems.** Best for 50–400+ points, long line lengths (> 25 meters), heavy grease, and applications requiring redundancy. Works under high-pressure, large-scale configurations. Allows point isolation and staged delivery.

➤ **Injector systems.** Optimal for varying point demands or where lubricant volume per point differs. Independent metering per outlet. Handles grease at high pressure, suitable for mobile and high-shock environments.

➤ **Single-line resistive.** Only applicable for small installations with consistent point loads, using low-viscosity oil. Limited control and unsuitable for critical or large-scale use.

Match system type to asset layout, lubricant characteristics, accessibility, failure impact, and expansion potential.

Pump and Reservoir Sizing

Select pump based on required total system flow rate, pressure, and lubricant type. Calculate maximum back pressure based on the farthest point plus metering device resistance and line friction losses. Ensure the pump output exceeds peak delivery demand during the shortest expected lubrication interval.

The reservoir volume must support the full lubrication cycle plus a buffer margin (typically 25–50%) to avoid air entrainment or suction loss. For grease systems, use vertical or agitated reservoirs to reduce channeling. Sizing must also accommodate refill intervals based on available maintenance bandwidth and refill logistics.

Account for lubricant viscosity at minimum and maximum ambient temperatures. High-viscosity grease may require heating elements or booster pumps. Oil systems may require temperature-compensated flow regulation or reservoir cooling if ambient temperature exceeds 40°C.

Line Layout and Piping Optimization

Design piping routes to minimize pressure drop, avoid unnecessary bends, and ensure equalized delivery timing. Use equal-length branches from metering valves where possible. For progressive systems, limit each block to a consistent point count and group points by proximity and load class.

Calculate hydraulic resistance using pipe diameter, lubricant viscosity, length, and expected flow rate. Use pressure drop formulas or simulation tools to verify that delivery pressure remains above the minimum metering threshold across the system. Consider a dual-line system for long runs or high-viscosity grease delivery beyond 20 meters.

Ensure that line routing is accessible for inspection, leak detection, and service. Use color-coded or labeled lines. Protect tubing from mechanical damage and vibration. Include junction boxes, isolation valves, and drain fittings in exposed or elevated areas.

A properly executed specification process ensures reliable lubrication delivery, extends component life, and simplifies long-term maintenance. Every parameter must align with asset duty, environment, and operational constraints to ensure system integrity and performance.

Specification Tools and Software Platforms

Precision in centralized lubrication system design depends on the use of validated engineering tools that model flow behavior, pressure dynamics, component compatibility, and installation layout. These tools reduce specification errors, prevent underdelivery or overdelivery, and ensure mechanical and control system integration from concept to commissioning.

Line Sizing and Pressure Loss Calculators

Hydraulic resistance and pressure drop must be calculated for each branch and distribution line. Use pressure loss calculators specific to lubrication applications, factoring line diameter, lubricant viscosity, flow rate, elevation change, and fitting losses. Grease systems require correction for non-Newtonian flow behavior, particularly with NLGI 1–2 products.

Tools should output:

➤ Required pump pressure at max system resistance
➤ Flow uniformity across longest and shortest branches
➤ Line velocity constraints to prevent cavitation or churning
➤ Minimum pressure at each metering device to ensure function

Standard tools include SKF LubeSelect, Lincoln QuickCalc, and in-house spreadsheets based on the Darcy-Weisbach and Hagen-Poiseuille flow equations modified for grease.

Component Selection Configurators

OEM-specific configurators (e.g., Bijur Delimon, DropsA, Groeneveld-BEKA, SKF, Lincoln) allow selection of pumps, metering blocks, injectors, reservoirs, filters, and sensors based on input variables such as:

- ➤ Number of lubrication points
- ➤ Flow per point
- ➤ Lubricant type and viscosity
- ➤ Required operating pressure and environmental conditions

Third-party configurators offer broader compatibility across brands but may lack precision for proprietary components. Output includes part numbers, data sheets, and bills of materials for procurement. Many also export to STEP, DWG, or native CAD formats for integration with layout drawings.

CAD-Integrated Piping Layout and Modeling Tools

3D piping tools within platforms like AutoCAD Plant 3D, SolidWorks Routing, or Inventor Tube & Pipe allow integration of lubrication lines into machine design. Use for:

- ➤ Precise routing within equipment constraints
- ➤ Clash detection with mechanical components
- ➤ Support bracket placement
- ➤ Integration with electrical and pneumatic routing plans

Custom libraries with lubrication components (fittings, junction boxes, metering blocks) can be imported for BOM accuracy and install visualization.

Digital Twins and Simulation of Cycle Timing and Pressure Dynamics

Simulation tools model lubrication system behavior under dynamic load and pressure conditions. Digital twin environments replicate:

- ➤ Cycle timing and sequence logic for progressive or dual-line systems
- ➤ Pressure buildup and decay across long runs
- ➤ Sensor response lag, flow imbalance, and blockage effects
- ➤ Impact of environmental temperature on flow and metering

Simulation platforms such as Siemens NX, EPLAN Fluid, and advanced SKF and Lincoln system modeling tools allow real-time cycle validation before hardware installation. These platforms are useful for validating pump sizing, sensor placement, and metering balance across variable-demand lubrication zones.

Use of validated specification tools eliminates assumptions, enforces engineering discipline, and improves coordination among mechanical, electrical, and maintenance functions. Simulation and digital modeling reduce commissioning delays, prevent misroutes, and ensure system scalability.

Diagnostic and Optimization Utilities

Centralized lubrication systems require continuous validation to ensure delivery accuracy, detect faults, and maintain uptime. Diagnostic and optimization utilities enable real-time monitoring, event-based maintenance, and system fine-tuning. These tools provide assurance that lubricant is reaching each point within volume and pressure tolerances—this is critical in high-speed, high-load, or safety-critical applications.

Table 7.2 shows the steps in the centralized lubrication system.

TABLE 7.2   Centralized Lubrication System Specification Steps

STEP	DETAILS
Lube point identification and requirement mapping	Inventory all lubrication points by component type, location, and method. Document speed, load, temperature, contamination, and duty cycle. Calculate flow rate and relubrication interval using OEM specs or ISO 281. Define lubricant type, grade, and compatibility with materials.
System-type selection	Use a decision matrix: progressive (10–150 points), dual-line (50–400+ points), injector (independent flow per point), single-line resistive (small, low-precision systems). Match to layout, criticality, and lubricant requirements.
Pump and reservoir sizing	Select pump for required pressure and volume. Account for max back pressure and delivery interval. Reservoir size should cover full cycle + 25–50% margin. Use heaters or boosters for high-viscosity greases; ensure compatibility with ambient temp range.
Line layout and piping optimization	Design equal-length branches, minimize pressure drops, and use calculation/simulation tools. Group lubrication points by location and duty class. Protect lines, provide access for inspection, and include fittings for serviceability.
Specification tools and software use	Use pressure drop calculators, OEM configurators, CAD layout tools, and digital twins. Validate system behavior under simulated loads, align electrical/mechanical integration, export BOMs, and validate system scalability.
Diagnostic and optimization utilities	Install pressure and flow sensors, pulse/stroke counters, and real-time feedback loops. Enable predictive fault detection for blocks, leaks, and air pockets. Use redundancy tools for critical systems (dual pumps, filters, logic controllers).

Key Diagnostic and Optimization Tools

Flow and Pressure Sensors with Real-Time Feedback

In-line flow sensors and pressure transducers monitor system dynamics during each lubrication cycle.

- ➤ **Flow sensors** verify that lubricant volume matches expected discharge per point or block.
- ➤ **Pressure sensors** installed before and after metering valves detect pressure rise, hold, and decay trends.
- ➤ **Sensors can be analog** (4–20 mA) or **digital** (CAN, Modbus, IO-Link), feeding into PLC or SCADA.
- ➤ **Alarm thresholds** can be configured for low delivery pressure, high back pressure, or zero flow conditions.

These sensors enable live verification of lubricant movement and are critical for systems operating under variable load, with long line lengths, or with NLGI 1–2 greases.

Pulse Monitoring and Stroke Verification Tools

Progressive systems require cycle verification to confirm that all outlets have discharged.

- ➤ **Cycle indicators** (mechanical or inductive) track the movement of metering pistons.
- ➤ **Pulse counters** log completed cycles and enable alarms when expected cycles fail to complete.
- ➤ **Digital stroke sensors** installed on injectors or pump plungers confirm lubricant delivery per actuation.
- ➤ **Monitoring tools also include LED indicators** or **signal outputs** for integration into control logic.

These tools detect partial delivery, blocked outputs, or pump priming issues. They are essential for systems without direct flow confirmation at each point.

Predictive Fault Detection (Blocked Lines, Air Pockets, Leaks)

Predictive analytics and sensor feedback enable early fault identification:

- ➤ **Blocked lines.** Detected by an abnormal pressure rise with no flow signal or uncompleted metering cycle.
- ➤ **Air pockets.** Identified by erratic flow readings or incomplete stroke movement. Often caused by low reservoir level, improper priming, or suction line leaks.
- ➤ **Leaks.** Diagnosed by pressure decay or drop in reservoir level inconsistent with logged flow volume.

Advanced control systems use trend analysis to differentiate between mechanical wear and temporary anomalies. Feedback enables corrective action before component damage occurs.

Redundancy Configuration Tools for Critical Systems

For high-availability systems, such as turbines, kilns, and mining gearboxes, redundancy is required at the component level:

- ➤ **Dual-pump arrangements** with switchover logic to maintain pressure on failure
- ➤ **Parallel filtration units** to allow element replacement without shutdown
- ➤ **Dual-reservoir or multi-outlet manifolds** to segregate circuits and isolate faults
- ➤ **Fault-tolerant PLC logic** to override or reroute failed lubrication zones

Tools include pump logic controllers with built-in diagnostics, modular hardware blocks for fast swap-out, and relay bypass systems to maintain lubrication during sensor failure.

In critical applications, diagnostics and optimization tools are not optional; they define system effectiveness. Monitoring must be continuous, actionable, and integrated into asset management protocols to prevent lubrication-induced failures and maximize service continuity.

Table 7.3 summarizes the key diagnostic and optimization tools for centralized lubrication systems.

TABLE 7.3   Summary of Key Diagnostic and Optimization Tools for Centralized Lubrication Systems

TOOL CATEGORY	KEY FUNCTIONS	COMMON TECHNOLOGIES	CRITICAL USE CASES
Flow and pressure sensors	Monitor real-time lubricant flow and pressure across system; confirm volume and cycle delivery	Analog (4–20 mA), CAN, Modbus, IO-Link; PLC/SCADA integration	Systems with long runs, NLGI 1–2 grease, or variable-load demands
Pulse monitoring and stroke verification	Verify metering sequence completion; detect partial delivery and pump stroke issues	Cycle indicators, digital stroke sensors, pulse counters, LED indicators	Progressive systems requiring complete discharge confirmation
Predictive fault detection	Identify abnormal pressure/flow behavior; detect leaks, air pockets, or blockages	Sensor trend analysis, pressure/flow signal interpretation, reservoir-level tracking	Applications with tight tolerance for downtime and reliability
Redundancy configuration tools	Ensure lubrication delivery continuity via fault-tolerant system architectures	Dual pumps, switchover logic, parallel filters, relay bypass, fault-tolerant PLCs	High-availability environments (e.g., turbines, kilns, mining gearboxes)

COMPLIANCE AND PERFORMANCE ASSURANCE

Performance assurance in centralized lubrication systems depends on verified volume accuracy, contamination control, and seamless integration with plant control infrastructure. Failure to meet delivery tolerances or maintain system hygiene results in component wear, seal failure, and unplanned downtime. Continuous validation and system-level logging are required to meet reliability, regulatory, and safety expectations.

Ensuring Correct Volume per Point per Cycle

Each lubrication point must receive a calibrated volume matched to its load, speed, and relubrication interval.

- ➤ **Progressive systems** deliver fixed-volume discharges per cycle, verified by stroke sensors or cycle counters.
- ➤ **Injector systems** require confirmation of actuator stroke and refill timing; differential pressure sensors flag failed injector function.
- ➤ **Flow meters** or in-line volume counters are used for high-criticality points where delivery must be confirmed per cycle.

Any deviation from specified volume—underdelivery or overfeed—must trigger local alarms or centralized flags. System configuration must include point-specific metering sizing and periodic functional verification using test ports, purge cycles, or pulse count validation.

Avoiding Cross-Contamination and Overlubrication

Cross-contamination between incompatible lubricants (e.g., lithium and polyurea greases, or food-grade and non-food-grade oils) causes additive breakdown, grease hardening, or film collapse.

To avoid cross-contamination:

➤ Use color-coded fittings, dedicated fill ports, and clearly labeled reservoirs.
➤ Implement lockable covers and lubricant-specific transfer equipment.
➤ Flush all lines, valves, and reservoirs when changing lubricant types.

Overlubrication leads to seal purge, bearing heating, and unnecessary lubricant waste. Automated systems must be configured with accurate cycle intervals and volume limits per point. Progressive and injector systems prevent accidental overfeed by design, but improper programming or metering selection can bypass safeguards. Delivery must stop once metering capacity is reached; verify with back-pressure limits and system feedback.

Integration with CMMS or Control Systems for Logging and Alarms

Lubrication system performance must be logged, traceable, and auditable. Integration with CMMS or plant control systems enables:

➤ Automatic logging of cycle completions, delivery confirmation, and fault conditions
➤ Generation of maintenance work orders based on stroke count, reservoir levels, or alarm triggers
➤ Dashboard visibility of active lubrication system status across assets and locations
➤ Alarm escalation protocols for failure to deliver, empty reservoirs, blocked metering, or abnormal pressure trends

Standard integrations use OPC UA, Modbus TCP, or digital I/O to communicate with SCADA or DCS platforms. CMMS interfaces provide time-stamped records of lubrication events and condition trends. Systems with cloud-based analytics support fleetwide performance benchmarking and exception reporting.

To assure compliance, lubrication systems must be treated as instrumentation, subject to calibration, validation, and data logging. Precision in volume, isolation of lubricant classes, and end-to-end visibility of system operation are foundational to sustaining reliability in centralized lubrication programs.

SPECIFICATION PITFALLS AND FIELD CORRECTION

Incorrect specification of centralized lubrication systems leads to delivery failures, system imbalance, and accelerated component wear. These issues originate from design-phase assumptions that do not reflect actual field conditions. Early identification and correction are critical to restoring system integrity and avoiding chronic lubrication-related faults.

Underestimating Back Pressure and Line Expansion

Failure to account for pressure losses over long runs or through undersized tubing leads to underdelivery at distal points. Grease systems operating over 10 meters with NLGI 1–2 lubricants require high-pressure pumps (3,000–5,000 psi) and properly sized feed lines.

- Progressive systems stall when pressure falls below the metering threshold (~200 psi), causing incomplete cycles.
- Injector systems fail when differential pressure is insufficient for full stroke completion.
- Dual-line systems compensate with alternating pressure but require balanced resistance across branches to avoid premature valve actuation.

Field correction involves verifying actual pressure at remote points using in-line gauges or pressure transducers. Replace undersized lines with larger-diameter tubing, minimize bends and fittings, or segment the system into shorter, pressure-optimized circuits. Add pressure boosters if required for remote zones.

Mismatch of Metering Elements and Actual Load Requirements

Incorrect sizing of metering blocks or injectors results in overlubrication or underlubrication. Progressive systems with fixed-volume outlets cannot adapt to varying load demands per point. A bearing operating under higher radial load requires more lubricant per cycle than a lightly loaded adjacent component.

➤ Using identical metering elements across all points leads to unequal lubrication effectiveness.
➤ In injector systems, stroke adjustment screws or variable-volume models may be incorrectly set or left uncalibrated.

Correction includes remapping lubrication demand per point based on actual load, speed, and operating conditions. Replace or recalibrate metering devices to match volume demand. Use flow indicators or collection tests to validate output per cycle.

Failure to Account for Lubricant Behavior at Low/High Temperatures

Changes in lubricant viscosity caused by temperature affect system performance.

➤ In cold environments, grease hardens, increasing resistance and delaying flow initiation.
➤ In high-temperature zones, oil or grease may thin, increasing flow rate unpredictably or causing separation.
➤ Mist and air/oil systems suffer from condensation or loss of atomization in temperature gradients.

Systems designed without temperature compensation will fail in seasonal or thermally variable environments. Corrective measures include:

➤ Installing reservoir heaters or trace-heated lines for cold conditions
➤ Selecting lubricants with appropriate VI, pour point, and thermal stability
➤ Adding thermal cutoffs, thermostats, or flow adjustments in control logic
➤ Verifying lubricant condition post-delivery to ensure additive retention and phase stability

Specification errors must be corrected through field validation, line pressure testing, cycle confirmation, thermal behavior assessment, and flow measurement. A properly specified and tuned system maintains consistent delivery across all operating conditions, extending component life and reducing maintenance load. Field correction is not optional; it's required to align system output with real-world operating demands.

BEST PRACTICES IN SYSTEM SPECIFICATION

Effective centralized lubrication system design begins with accurate field data and incorporates predictive modeling, verified sizing, and long-term maintainability. Each decision must be grounded in mechanical demand, environmental constraints, and operational continuity requirements.

Start with Detailed Site and Asset Mapping

Document every lubrication point with location, equipment ID, component type, load classification, duty cycle, and lubricant specification. Map physical access constraints, elevation changes, and mounting limitations. Identify heat sources, vibration zones, contamination risk areas, and washdown exposure.

These include:

➤ Distance from the central pump to each lubrication point
➤ Grouping of lubrication zones by function or accessibility
➤ Points requiring differing lubricant types or volumes
➤ Any assets requiring food-grade or specialty lubricants

Field mapping ensures accurate material takeoffs, metering assignments, and control logic development.

Simulate Systems Under Worst-Case Operating Conditions

Model system performance under the most demanding conditions, including:

➤ Lowest ambient start-up temperature (viscosity spike)
➤ Maximum line length at furthest point (pressure drop)

➤ Simultaneous cycle demand across all points (flow rate load)
➤ Thermal expansion and mechanical vibration on line support

Simulation includes:

➤ Pressure versus distance curves per lubricant type
➤ Injector or divider block function at viscosity extremes
➤ Pump start-up curves under cold grease loads
➤ Worst-case voltage and air supply drop scenarios for electrically or
   pneumatically actuated systems

Stress testing the model before procurement prevents post-installation failures and unplanned design changes.

Validate Sizing with Manufacturer or Third-Party Tools

Use OEM-specific configurators or neutral engineering platforms to confirm:

➤ Pump pressure and flow compatibility with system resistance
➤ Correct metering device selection for each point
➤ Reservoir capacity against refill interval targets
➤ Compatibility of sensors, controllers, and connectors with environmental
   and safety requirements

Submit the design for peer review or factory support validation when point count exceeds 50 or when deploying dual-line or mixed-media systems. Include tool-generated BOMs and data sheets in the spec package to support procurement and commissioning.

Build in Service Access and Modular Replacement Paths

Position reservoirs, junction boxes, metering blocks, and valves with direct technician access, no tool-required covers, elevated-only fills, or nonstandard fittings. Install quick disconnects at critical nodes for line purging, flushing, and segment isolation.
This is designed for:

➤ Modular expansion (additional point loops or metering blocks)

➤ Redundant pump mounting brackets
➤ Standardized line sizes to simplify spares stocking
➤ Independent zone isolation for diagnostics without full system shutdown

Service pathways must account for expected wear points, heat-affected zones, and component replacement under runtime pressure. Lockable fill ports, labeled lines, and accessible pressure ports reduce maintenance time and improve fault detection accuracy.

These best practices anchor system performance in field realities, ensuring lubrication precision, control integration, and long-term system reliability across the asset life cycle.

CHAPTER 8

INTEGRATION OF LUBRICATION EQUIPMENT WITH IIOT INFRASTRUCTURE

The advent of IIoT (Industrial Internet of Things) has transformed lubrication from a static support function to a dynamic, data-rich process. Modern lubrication equipment integrates sensors for flow rate, temperature, pressure, and lubricant level. These parameters feed into cloud platforms or on-site SCADA systems for predictive maintenance and performance optimization.

Wireless connectivity enables condition monitoring of lubricators and delivery points. Lubrication equipment integrated with IIoT can self-report anomalies, predict refill requirements, and adapt delivery based on machine condition. Edge computing allows localized decision-making to adjust frequency and dosage.

This connectivity also supports compliance by logging historical data, validating PM execution, and alerting operators to deviations. Smart lubrication platforms align with Industry 4.0 by reducing unplanned downtime, extending component life, and synchronizing with ERP systems for supply chain alignment.

OVERVIEW OF IIOT-ENABLED LUBRICATION

Lubrication systems are transitioning from fixed-interval delivery to dynamic, condition-based control driven by IIoT infrastructure. Legacy systems rely on time-based cycles or manual routes with no correlation to equipment load, duty profile, or real-time condition. Failures due to overlubrication or underlubrication persist, especially in variable-load or high-duty environments.

IIoT integration enables adaptive lubrication based on sensor input, performance data, and machine state. Systems operate using embedded controllers that receive and process live signals from vibration sensors, thermocouples, pressure switches, flow meters, cycle counters, and oil condition monitors. Inputs are used to trigger lubrication events, adjust volume per point, or suspend cycles under no-load conditions. Lubrication becomes synchronized with machine operation, not isolated from it.

Data acquisition is handled through edge devices connected to local PLCs or wireless transmitters. Data includes lubricant flow rates, pressure profiles, metering cycle counts, injector stroke confirmation, reservoir levels, and fault status. This data is time-stamped, processed locally, and pushed to supervisory platforms via protocols such as OPC UA, MQTT, or Modbus TCP/IP.

Real-time communication enables direct feedback to plant DCS, SCADA, or CMMS platforms. Lubrication data becomes part of the broader asset health dataset. Lubrication alarms escalate with the same priority as thermal or vibration alarms. Event histories are archived, and maintenance triggers (e.g., refill alerts, system bypass, cycle miss) generate work orders automatically.

IIoT-aligned lubrication systems meet Industry 4.0 objectives by embedding intelligence into mechanical reliability infrastructure. Lubrication no longer functions as a passive subsystem but as a digitally monitored and responsive asset protection mechanism. Smart lubrication supports predictive maintenance, reduces manual labor dependency, and improves data-driven asset optimization across distributed operations.

KEY SENSOR TECHNOLOGIES IN LUBRICATION SYSTEMS

Integration of sensor technologies into centralized lubrication systems provides direct feedback on system status, delivery accuracy, and lubricant condition. Sensor data enables real-time adjustment, alarm generation, and predictive diagnostics within IIoT frameworks.

Flow and Delivery Sensors

> **Pulse counters** are installed on progressive divider blocks or injectors to confirm metering piston movement per lubrication cycle. Each pulse corresponds to a completed discharge event. Missed pulses indicate blockages, line rupture, or actuator failure.

➤ **Gear flow meters** measure actual lubricant volume per unit of time by tracking gear rotation within the flow body. They offer high accuracy for oil or low–NLGI grease. The meters are used where volumetric validation per point is required. They can detect deviation from programmed flow set points.

➤ **Ultrasonic flow sensors** are nonintrusive sensors mounted externally on pipes. These sensors measure lubricant velocity and flow consistency using time of flight or Doppler shift. They are effective for high-viscosity or opaque fluids where mechanical meters are impractical. They are suitable for harsh or sanitary environments.

All flow sensors must be calibrated to lubricant type and operating pressure. Deviations from expected flow rates trigger alarms, enable shutdown interlocks, or escalate service requests via CMMS integration.

Pressure and Temperature Sensors

➤ **Pressure sensors** are installed at pump outlets, metering blocks, or injector manifolds. These sensors are used to verify system pressurization during cycles and to detect blockages (excess pressure), line breaks (pressure drop), or air entrainment (erratic pressure profile). Differential pressure across filters indicates clogging or bypass.

➤ **Temperature sensors** monitor ambient and lubricant temperature at reservoirs and key lines. Thermal data is used to:
   • Adjust cycle timing or flow volume based on viscosity shift
   • Detect overheating due to churning or overlubrication
   • Verify heater function in low-temperature environments

Integration of pressure and temperature sensors allows dynamic system control and provides early warning of mechanical degradation, lubricant phase change, or flow path failure.

Reservoir Level and Grease Quality Monitoring

Level Sensors

➤ **Ultrasonic sensors** are noncontact. They are reliable in grease or oil tanks. They require internal baffling for accuracy.

➤ **Capacitive sensors** are installed vertically in reservoirs and detect oil-air or grease-air interfaces.
➤ **Weight-based load cells** provide continuous monitoring in mobile or remote systems where internal sensors are impractical.

Low-level alarms are configured based on calculated refill intervals and usage rate. Integrated level monitoring prevents air entrainment, cavitation, and pump starvation.

Grease Quality and Oil Degradation Sensors

➤ **Dielectric sensors** detect changes in fluid polarity, indicating oxidation, thermal breakdown, or water contamination.
➤ **Particle counters** classify contamination levels in circulating oil systems by ISO 4406 standards.
➤ **Viscosity and TAN/TBN sensors** are available for high-value oil circuits and can be mounted in-line or in bypass loops.

These sensors validate lubricant integrity in real time and prevent continued operation with degraded or contaminated product. Integration into IIoT platforms supports trend analysis, lubricant change scheduling, and contamination source tracking.

Sensor selection must be matched to system type, lubricant properties, and environmental exposure. All feedback data should be linked to system alarms, logged in SCADA or CMMS, and used to support root cause diagnostics.

Table 8.1 summarizes the key sensor technologies in lubrication systems.

TABLE 8.1 Summary of Key Sensor Technologies in Lubrication Systems

SENSOR CATEGORY	SENSOR TYPE	FUNCTION/APPLICATION
Flow and delivery sensors	Pulse counters	Confirm metering piston movement per cycle; detect blockages or failures
Flow and delivery sensors	Gear flow meters	Measure lubricant volume per time; detect flow deviations
Flow and delivery sensors	Ultrasonic flow sensors	Nonintrusive velocity measurement for high-viscosity/opaque fluids
Pressure and temperature sensors	Pressure sensors	Detect system pressurization, blockages, line breaks, air entrainment
Pressure and temperature sensors	Temperature sensors	Monitor ambient/lubricant temperature; adjust cycle timing and detect thermal issues

SENSOR CATEGORY	SENSOR TYPE	FUNCTION/APPLICATION
Reservoir level and grease quality monitoring	Level sensors (ultrasonic, capacitive, weight-based)	Monitor lubricant levels to avoid pump starvation; prevent air entrainment
Reservoir level and grease quality monitoring	Dielectric sensors	Detect oxidation, breakdown, water contamination in grease/oil
Reservoir level and grease quality monitoring	Particle counters	Classify particulate contamination (ISO 4406); detect wear debris
Reservoir level and grease quality monitoring	Viscosity and TAN/TBN sensors	Measure viscosity and acid/base number shifts in high-value oil circuits

Connectivity Architecture

Reliable data transmission among lubrication equipment, plant control systems, and enterprise-level platforms is essential for IIoT-enabled lubrication. System architecture must account for communication range, data rate, power availability, integration compatibility, and environmental constraints.

Wireless Protocols

These protocols include:

> **LoRaWAN.** Long-range, low-power communication for remote or hard-to-reach equipment. Ideal for large industrial sites with minimal existing communication infrastructure. Suited for battery-powered lubrication sensors that transmit infrequently (e.g., reservoir level or pressure alerts). Limited bandwidth—used for status updates, not continuous high-speed data.

> **Zigbee.** Mesh networking protocol suitable for mid-range communication in clustered machinery. Enables distributed sensor nodes to relay data across plant zones. Used in lubrication networks for flow confirmation, metering stroke feedback, and temperature alarms.

> **Bluetooth Low Energy (BLE).** Short-range communication for proximity diagnostics, maintenance access, and portable sensor data retrieval. Used for handheld troubleshooting tools or technician tablets to interface with local lubrication sensors without physical connection.

> **Wi-Fi.** High-bandwidth option for areas with strong signal coverage and constant power. Enables real-time monitoring and configuration of

smart lubrication controllers. Susceptible to interference and not ideal in high-vibration or metal-dense environments without repeaters.

Battery-operated smart sensors leverage these protocols with sleep-wake cycles and event-driven reporting to preserve energy. Power optimization and protocol selection must align with expected transmission interval, payload size, and physical installation constraints.

Network Integration

SCADA and DCS Interoperability

Lubrication systems must exchange data with SCADA or DCS (distributed control system) for centralized monitoring and control. Integration typically uses Modbus TCP/IP, EtherNet/IP, or Profinet. Critical data includes pump status, delivery confirmation, reservoir level, and active alarms. Integration enables lubrication faults to be escalated alongside process faults.

MQTT and OPC UA

For IIoT and hybrid environments, MQTT supports lightweight, publish-subscribe communication with cloud brokers. OPC UA enables secure, platform-agnostic communication between lubrication controllers and industrial software. These protocols are preferred for cloud-based analytics, cross-platform visibility, and API-driven integrations with CMMS or ERP systems.

Tag naming, data normalization, and time stamp synchronization are critical for scalable integration across mixed-vendor environments.

Edge Versus Cloud Computing

Edge Analytics

Edge controllers process data locally at the lubrication system, reducing latency and minimizing bandwidth use. Logic includes:

➤ Condition-based lubrication triggering (based on sensor input)
➤ Fault detection (missed strokes, pressure anomalies)
➤ Local actuation of alarms, pump start/stop, and system isolation

Edge processing ensures fail-safe operation even if upstream communication is lost. It's recommended for high-criticality assets or isolated equipment.

Cloud Platforms

Cloud platforms enable centralized fleet monitoring, historical data aggregation, and large-scale trend analysis. Cloud dashboards display lubricant usage, system efficiency, fault frequency, and compliance to maintenance schedules. AI and machine learning models identify patterns linked to lubricant degradation, wear onset, or premature system failure.

Data from multiple plants or mobile equipment units can be accessed globally for decision-making, benchmarking, and automated alert routing to maintenance teams.

Connectivity architecture must balance real-time control needs with long-term data access and analytical depth. Network configuration, protocol selection, and system redundancy determine the responsiveness and reliability of smart lubrication infrastructure across industrial operations.

Functional Capabilities Enabled by IIoT Integration

Integration of centralized lubrication systems into IIoT infrastructure unlocks advanced control, diagnostic, and predictive capabilities that are not achievable with conventional stand-alone systems. These functions enhance reliability, reduce unplanned downtime, and support data-driven maintenance decisions.

Real-Time Anomaly Detection and Alerting

Sensor data, such as pressure, flow, cycle confirmation, reservoir level, or temperature, is continuously monitored against predefined thresholds. Anomalies such as delayed pressure rise, missed metering strokes, excessive cycle duration, or sudden flow drop trigger immediate alarms.

➤ Alerts are generated locally at the edge controller or escalated via SCADA or CMMS.

➤ Alarms can be tiered: informational (low reservoir), warning (flow variance), or critical (cycle failure).

➤ Event logs include time stamp, asset ID, and fault code for traceability.

The ability to detect and alert reduces mean time to detection (MTTD) of lubrication-related failures and supports proactive response before mechanical damage occurs.

Adaptive Delivery Rates Based on Vibration, Temperature, or Operating Cycles

IIoT-connected lubrication systems can dynamically adjust lubricant delivery based on live operating conditions.

> ➤ **Vibration inputs** from accelerometers or condition monitoring sensors can increase frequency or volume to high-load points.
> ➤ **Temperature sensors** adjust delivery to compensate for viscosity shifts in grease or oil.
> ➤ **Cycle counters or runtime data** from PLCs or encoders can trigger lubrication based on shaft revolutions, actuator strokes, or process batch counts.

This adaptive logic ensures that lubricant film is maintained under variable load and prevents overlubrication during idle or no-load periods. Adaptive delivery reduces waste, improves film integrity, and aligns lubricant application with real mechanical demand.

Self-Diagnostics for Pump or Valve Failures

IIoT-enabled systems continuously self-monitor pump pressure, motor current draw, metering stroke confirmation, and line resistance.

> ➤ **Pump diagnostics** detect cavitation, loss of prime, seal wear, and thermal overload.
> ➤ **Metering diagnostics** detect blocked injectors, unresponsive progressive valves, and downstream line leakage.
> ➤ **System logic** isolates failed branches, flags degraded components, and triggers backup pump (in redundant configurations).

Self-diagnostics prevent silent system failures and enable condition-based component replacement rather than reactive repair.

Predictive Maintenance Scheduling via Data Patterns

Collected sensor and performance data is analyzed to detect early indicators of wear or degradation.

> ➤ Trending pressure curve changes may indicate line restriction buildup.
> ➤ Increased motor amperage suggests pump resistance or hardened grease.
> ➤ Deviations in delivery time or volume predict injector wear or contamination.
> ➤ Correlation with ambient data and runtime supports prediction of seasonal or usage-based failure patterns.

Data feeds into CMMS platforms to automatically schedule inspections, lubricant changes, or component replacements. The process enables migration from fixed PM intervals to true predictive maintenance based on field conditions and machine behavior.

IIoT functionality transforms lubrication systems from passive service mechanisms into active reliability tools, enabling higher asset utilization, reduced maintenance overhead, and real-time visibility of system health.

SYSTEM INTEGRATION WITH ENTERPRISE PLATFORMS

IIoT-enabled lubrication systems must interface with enterprise-level platforms to support full life-cycle asset management, automated logistics, and regulatory compliance. Integration ensures that lubrication activities are no longer isolated events but are tracked, validated, and acted upon within the broader operational and quality frameworks.

Linking Lubrication Activity with CMMS for Closed-Loop Maintenance Logging

Lubrication controllers transmit cycle completions, fault codes, reservoir levels, and metering confirmations to the computerized maintenance management system.

> ➤ Each lubrication event is time-stamped and linked to a specific asset tag.
> ➤ Auto-generated maintenance records confirm lubricant delivery occurred as planned.

➤ Missed lubrication cycles or abnormal system behavior automatically generate corrective work orders.

➤ Usage-based triggers (e.g., total metering cycles, run hours, lubricant consumption) can initiate inspection tasks or pump service schedules.

This closed-loop structure eliminates manual logging, ensures traceability, and supports asset-level performance tracking based on lubrication reliability metrics.

ERP Integration for Automated Lubricant Restocking and Supply Chain Coordination

Consumption data from metering logs and reservoir level sensors is aggregated and analyzed to determine real-time lubricant usage rates.

➤ Threshold-based alerts are sent to ERP or inventory systems when lubricant volumes reach reorder points.

➤ Automatic purchase requisitions or restocking notifications are generated.

➤ Integration ensures that correct lubricant type, batch, and container size are aligned with usage location and application.

➤ Inventory rotation (FIFO compliance) and expiration tracking can be built into the system for food-grade and synthetic products.

ERP linkage enables lubricant procurement to shift from reactive to predictive, reduces stockouts, and minimizes obsolete inventory.

Audit Trails and Compliance Documentation (e.g., Food Safety, ISO 21469)

For industries governed by hygiene and safety standards, such as food, beverage, and pharmaceutical, lubrication systems must support full audit transparency.

➤ Each lubrication event is recorded with lubricant type, volume, time, asset ID, and operator (if manual override is allowed).

➤ Sensors confirm delivery to prevent dry contact with moving parts.

➤ System access, changes in settings, and fault events are logged in secure, non-editable formats.

➤ Documentation can be exported for compliance with standards such as ISO 21469, NSF H1 registration, and FDA GMPs.

Audit trails demonstrate that correct lubricants were applied to the correct points at the correct times, with validated delivery, supporting both internal QA and external regulatory inspections.

Enterprise integration ensures that lubrication data contributes directly to uptime, cost control, traceability, and compliance. It closes the loop among physical asset behavior, maintenance execution, inventory control, and external auditability.

Implementation Considerations and Challenges

Successful deployment of IIoT-integrated lubrication systems requires engineering alignment with environmental, operational, and cybersecurity constraints. Hardware, data infrastructure, and workforce readiness all factor into functional reliability and long-term system ROI.

Sensor Selection Based on Lubricant Type and Environmental Conditions

Sensor performance is directly influenced by lubricant viscosity, conductivity, and chemical makeup.

➤ **Flow meters** must be matched to lubricant grade; gear meters may clog with NLGI 2 grease; ultrasonic sensors require stable, nonaerated oil for accuracy.

➤ **Pressure sensors** need seals compatible with synthetic esters, PAOs, or food-grade fluids; diaphragms must resist hardening or swelling.

➤ **Dielectric sensors** may return false positives in fluids with high additive loads or water contamination.

Environmental variables like vibration, washdown, EMI, and ambient temperature extremes further dictate sensor housing, IP rating, and cable shielding requirements. Selection must be validated per application, not assumed from catalog specs.

Power Supply and Maintenance of Wireless Nodes

Battery-powered wireless sensors present deployment flexibility but require realistic planning for power longevity and access.

➤ Nodes must operate under low-duty transmission cycles with power management modes.
➤ Signal strength, update interval, and data payload size directly impact battery life.
➤ For hard-to-reach installations, consider energy harvesting (thermal or vibration), solar backup, or hardwiring to 24VDC if available.

Each sensor should be included in the asset maintenance plan with replacement interval, access method, and criticality ranking.

Data Security and Network Compatibility

Lubrication system data often routes through shared OT (operational technology) networks or bridges into IT (information technology) domains.

➤ End points must support encrypted protocols (TLS, WPA2, VPN tunneling) to prevent interception or injection attacks.
➤ Firewalls, VLANs, and access control lists must isolate lubrication data from critical process controls.
➤ OPC UA, MQTT, and Modbus TCP configurations must match existing SCADA or DCS settings to avoid handshake failures.

Unsecured or incompatible nodes risk not only data loss but also potential attack vectors into broader plant networks.

Change Management and Staff Training for Digital Adoption

Technicians accustomed to manual lubrication or timer-based systems require upskilling to interact with sensor diagnostics, network alerts, and cloud dashboards.

➤ Training must cover system architecture, basic troubleshooting, and CMMS alert workflows.

➤ Resistance may arise due to perceived job replacement, added complexity, or alert fatigue.

➤ Start with pilot systems on critical assets, with clear performance benchmarks and technician involvement in rollout.

Document all configuration, wiring, IP assignments, and sensor IDs in as-built schematics. Support adoption with visual SOPs, alert code lookups, and embedded help via HMIs (human-machine interfaces) or mobile apps.

Deployment success depends not only on technology selection but on system compatibility, workforce readiness, and maintenance infrastructure. Early planning and phased implementation reduce integration risk and accelerate digital reliability gains.

ROI AND OPERATIONAL VALUE

IIoT-integrated lubrication systems deliver measurable return on investment through asset reliability gains, waste reduction, and improved operational coordination. Value is realized across equipment uptime, resource efficiency, and organizational alignment.

Reduction in Unplanned Downtime from Lubrication-Related Failures

Improper lubrication, for example, missed points, incorrect volume, or degraded fluid, is a primary contributor to bearing, gear, and slideway failures. IIoT-enabled systems mitigate this by:

➤ Delivering lubricant precisely when and where needed based on load, temperature, or duty cycle

➤ Verifying each cycle via flow, pressure, or pulse sensors

➤ Alerting in real time to blocked lines, low reservoirs, or actuator faults

Downtime reduction is quantifiable through incident tracking: fewer emergency work orders, shorter mean time to repair, and extended mean time between failures. Asset-level data supports root cause attribution and optimization of lubrication strategy per machine class.

Decreased Lubricant Waste and Environmental Compliance Violations

Adaptive delivery replaces blanket application, reducing lubricant consumption per cycle. Real-time adjustment minimizes overlubrication, which leads to runoff, component fouling, and wasted product.

- ➤ Fluid condition monitoring extends change intervals through condition-based scheduling.
- ➤ Reservoir level tracking avoids overfills and uncontrolled releases.
- ➤ Automated systems reduce technician exposure to hazardous areas and chemicals.

Waste oil generation, cleanup costs, and EPA or ISO 14001 violations are reduced. Food, pharma, and clean manufacturing environments benefit from traceable, validated H1 and ISO 21469-compliant lubrication with full audit trail.

Enhanced Visibility Across Maintenance, Reliability, and Procurement Teams

Lubrication data becomes actionable across departments:

- ➤ **Maintenance** sees real-time asset lube status, fault alerts, and refill triggers.
- ➤ **Reliability engineering** trends performance data to optimize intervals and select higher-performing products
- ➤ **Procurement** aligns ordering with actual usage, avoiding both overstock and lubricant stockouts

Cross-platform integration (CMMS, ERP, SCADA) creates a shared dataset for decision-making, resource planning, and compliance reporting. Operational silos are removed; lubrication becomes a managed, visible, and optimized process.

When benchmarked over a 1-, 3-, and 5-year span, IIoT lubrication systems deliver positive ROI through failure avoidance, consumption control, and coordinated asset care across the facility or enterprise.

CASE APPLICATIONS AND FIELD EXAMPLES

Real-world deployments of IIoT-enabled lubrication systems demonstrate quantifiable value in uptime, compliance, and labor efficiency across high-impact industrial sectors. Each application leverages smart sensing, remote monitoring, and integrated control to overcome traditional lubrication limitations.

Mining: Autonomous Grease Systems Alerting on Blocked Lines

Surface mining operations deployed dual-line grease systems on draglines and shovels with integrated pressure transducers and cycle confirmation switches.

- System alerts were triggered when pressure exceeded setpoints at end-of-line blocks without corresponding metering piston pulses.
- Operators received fault codes via CAN bus interface in cab displays.
- Blocked-line conditions were resolved during scheduled maintenance, avoiding unplanned stoppage and component seizure.

This resulted in a 43% reduction in unplanned downtime due to lubrication faults and a 21% drop in critical bearing failures over 18 months.

Food Industry: Automated Compliance Records for H1 Lubrication Delivery

A dairy processing plant implemented a centralized oil lubrication system using NSF H1 food-grade fluids, integrated with OPC UA to the plant's MES (manufacturing execution system) and CMMS.

- Delivery cycles were logged automatically with time stamp, lubricant ID, and asset tag.
- Ultrasonic level sensors and metering confirmations provided traceable evidence of correct product and volume per lubrication point.
- Audit reports were auto-generated monthly to meet ISO 21469 and internal QA requirements.

As a result, the system reduced manual logbook errors to zero and enabled real-time quality assurance oversight without intrusive inspections, improving audit readiness and reducing recall risk.

Wind Turbines: Remote Lube System Monitoring in Inaccessible Assets

Offshore wind farm operators retrofitted nacelle-mounted gearboxes with progressive lubrication systems equipped with LoRaWAN-connected pressure and flow sensors.

- Alerts for low reservoir level, pressure loss, or cycle failure were transmitted to the control center onshore.
- Vibration and temperature sensors adjusted cycle frequency dynamically based on turbine load and ambient changes.
- Refill planning was based on real-time consumption trends, enabling supply boat dispatch only when needed.

Remote monitoring avoided unnecessary technician climbs, reduced helicopter access costs, and extended gearbox life in a location where failure response time often exceeds 24 hours.

These field examples underscore the operational gains of integrating lubrication systems with IIoT architecture in demanding, compliance-sensitive, or remote environments. Performance data validates investment and drives continuous improvement in lubrication strategy.

CHAPTER 9

DESIGN FOR MAINTAINABILITY: EQUIPMENT COMPATIBILITY CONSIDERATIONS

Maintenance efficiency is tightly coupled with how well lubrication equipment is designed into a system. Design for maintainability includes ergonomic access to fittings, compatibility with existing hardware, safe mounting points, and standardization of connectors and reservoirs.

Selection criteria should include:

- Ease of installation and retrofit
- Maintenance access without machine disassembly
- Compatibility with lubricant viscosity and chemical composition
- Modularity for component replacement

Equipment designers must anticipate challenges in refilling, venting, leak detection, and purging. Poorly placed fittings, incompatible materials, or incorrect thread types contribute to failure and service delays. Cross-functional collaboration between design engineers, maintenance teams, and lubricant suppliers ensures field-proven compatibility.

DESIGN FOR MAINTAINABILITY—EQUIPMENT COMPATIBILITY CONSIDERATIONS

Lubrication Accessibility and Downtime Reduction

Component accessibility directly influences mean time to repair and determines whether lubrication is applied consistently. Inaccessible lube points are either neglected or serviced under unsafe conditions. Grease zerks buried beneath guards or obstructed by structural frames require machine shutdowns or disassembly, increasing total downtime. Field data confirms that poor lube point placement correlates with skipped intervals, component failure, and reactive maintenance behavior.

Equipment Design and Lubrication Strategy Alignment

Original equipment design must align with the lubrication method. High-speed bearings without provision for oil mist or air/oil delivery create lubrication starvation at peak loads. Slideways designed for periodic oil injection must include channels for even distribution and sealing to retain film thickness. Components relying on splash lubrication must maintain constant fluid levels; lack of sight gauges or overcomplicated fill ports compromise reliability.

Integration between machine elements and lubrication systems must account for:

➤ Flow path geometry for uniform lubricant delivery
➤ Structural access for metering device installation
➤ Vibration isolation for sensor stability and accurate feedback

Design that ignores lubrication system footprints, routing, or service zones results in retrofits, pipe routing complications, and sensor fouling. Design must accommodate lubricant selection. Grease compatibility requires port orientation for purge flow, and high-viscosity oils require short vertical suction heads and vented fill points.

Maintenance Burdens from Poor Design or Integration Gaps

Field evidence shows that OEMs often prioritize mechanical performance over serviceability. Common failure modes linked to poor design include:

➤ Undersized reservoirs with no level indication, leading to pump cavitation
➤ Nonstandard fittings requiring special tools or adapters
➤ Disparate lubrication types on adjacent components, increasing the risk of cross-contamination
➤ Lack of feedback provisions; no flow sensors, pressure taps, or stroke monitors

Installations lacking feedback loops or designed without sensor provisions compromise predictive maintenance programs. Service personnel are forced into work-arounds, manual confirmation, or unnecessary disassembly to validate lubrication performance.

Design for maintainability requires early coordination between reliability engineers, lubrication specialists, and OEM design teams. Field service feedback must be incorporated into design revisions, especially in high-criticality, high-frequency lubrication zones. Compatibility with centralized systems, real-time monitoring, and maintenance task duration targets should be established during equipment specification.

LUBRICATION EQUIPMENT INTEGRATION PRINCIPLES

Ergonomic Access and Mounting

Grease fittings, oil fill ports, sight glasses, and reservoirs must be mounted for direct, unobstructed access during normal operating conditions. Components should be accessible without climbing, panel removal, or process shutdown. Zerk fittings positioned on the underside of motor housings, behind guarding, or above shoulder height without platforms introduce delay and safety risk. All lube points should be reachable from ground level or permanent walkways using standard service tools.

Lubrication interfaces must avoid collision paths with rotating shafts, belt drives, or pneumatic actuators. Reservoirs and filters must be mounted away from vibration nodes or thermal radiation zones. Mounting brackets should isolate from structural flexure and permit visual inspection without mirror tools or disassembly. Oil fill ports must be angled and elevated to allow complete filling without trapped air pockets. Drain ports should be at the lowest point of the sump and accessible for tool-free evacuation.

Standardization and Modularity

Thread sizes, couplings, and fittings must be standardized across lubrication system interfaces. NPT (National Pipe Thread), BSPP (British Standard Pipe Parallel), or ISO metric threads should be specified plantwide to eliminate adapter use and cross-thread risk. Common fitting sizes—$\frac{1}{8}$", $\frac{1}{4}$", $\frac{3}{8}$"—simplify inventory and reduce maintenance error.

Quick-disconnect reservoirs and filter housings reduce service time and contamination risk. Pump modules should support tool-less removal and backward compatibility with older systems. Progressive divider blocks and injector modules must use consistent port spacing and mounting hardware across machine models to enable interchangeability.

Standardization allows cross-training, consolidates spare part SKUs, and reduces downtime during maintenance. Lube system components should be modular by design; pump, controller, reservoir, and metering units must allow field replacement without disturbing adjacent systems or requiring system reprogramming. Each lubrication circuit should be designed as a serviceable module with isolated shutoff points and dedicated diagnostics.

Standardization and ergonomic integration are critical to ensuring lubrication systems remain functional, safe, and maintainable throughout the asset life cycle.

COMPATIBILITY WITH LUBRICANT PROPERTIES

Viscosity and Flow Considerations

Lubrication system components must be matched to the viscosity profile of the lubricant in use. High-viscosity greases (e.g., NLGI 2 or 3) require positive displacement pumps with high torque motors and short, large-diameter discharge lines to prevent shear degradation and start-up delays. Systems using oils with poor cold-flow properties (e.g., ISO VG 320 in cold ambient conditions) require heated reservoirs, jacketed lines, or viscosity-compensated metering to maintain flow rates.

Back pressure must be calculated for each circuit based on line length, elevation, bends, and injector resistance. Progressive divider blocks and single-point injectors have maximum pressure ratings; exceeding these causes cycle failures or internal bypass. NLGI 1 greases may function in systems rated for NLGI 2 if line and injector volumes are optimized, but assumptions based on standard conditions often fail in field deployments.

Pump selection must account for pressure drop at full-load temperature extremes, including start-up under ambient lows. Incorrect pairing leads to cycle timing errors, pump cavitation, or grease separation due to excessive mechanical stress.

Material Selection and Chemical Resistance

Lubricants containing synthetic base stocks (e.g., esters, PAGs, PAOs) or aggressive additive packages can degrade seals, hoses, and polymeric reservoir components. EP additives with high sulfur or phosphorous content can leach plasticizers from elastomers or cause embrittlement in improperly specified seals.

Seal materials must be matched to lubricant chemistry:

➤ **Viton (FKM).** Preferred for high-temp and synthetic compatibility.
➤ **Nitrile (Buna-N).** Acceptable for mineral oils, but limited resistance to esters and ozone.
➤ **EPDM.** Incompatible with most petroleum-based lubricants; avoid unless system is dedicated to glycol-based fluids.
➤ **PTFE (Teflon).** High chemical resistance, but prone to cold flow under mechanical load.

Hoses and soft lines must resist swelling, delamination, and hydrolysis. Polyurethane and PVC degrade in the presence of ester-based or water-laden lubricants. Braided stainless with PTFE liners or reinforced rubber compounds are preferred in high-temperature or chemically aggressive systems.

Oxidative stability of the lubricant must be matched with component exposure duration. Long-life synthetic lubricants may oxidize incompatible rubber compounds over extended intervals, leading to internal debris, flow restriction, and pump damage.

Chemical compatibility validation must be conducted during system specification and verified by field sampling after initial runtime to confirm no material degradation or lubricant polymerization has occurred.

REFILL, VENTING, AND PURGING DESIGN

Fill Port Positioning to Avoid Air Pockets and Prevent Overpressure

Fill ports must be located at the highest accessible point of each reservoir, tank, or sump to ensure air is displaced during filling. Ports positioned off-center or on vertical surfaces promote air entrapment, resulting in false fill levels, cavitation, and erratic pump performance. Funnel fittings or quick-connect fill adapters should be installed with straight, vertical access to the reservoir interior.

When pressurized fill guns or centralized bulk supply is used, internal pressure relief valves must be installed to prevent tank overpressure. Relief paths must discharge to a catch container or be routed to atmospheric venting. Overpressure events damage gaskets, distort reservoirs, and risk blowout of push-on fittings.

Clear Vent Paths to Avoid False Full Readings or Backflow

Each reservoir or fluid-holding component must have a dedicated vent port, positioned opposite the fill port. Vents must be fitted with desiccant breathers, sintered filters, or check valves as appropriate to prevent contaminant ingress. Breathers must have adequate flow capacity to equalize internal pressure during rapid fill or lubricant drawdown.

Trapped air in gearboxes or reservoir sumps leads to false high-level readings, overflow during thermal expansion, and foam formation. In high-elevation or high-temperature installations, thermal expansion of trapped air can create back pressure, impeding pump stroke return or causing lubricant to siphon back into feed lines.

Drain and Purge Options for Complete System Evacuation During Changeovers

Drain ports must be located at the system's low points, allowing gravity-assisted evacuation. Drain valves should be full bore, metal-seated, and positioned for tool-free access. Systems with long runs or multiple elevation changes require distributed drains to remove residual lubricant and contaminants. Drain points must not share return lines with active circuits to prevent backflow or cross-contamination.

Purge fittings or bypass manifolds should be incorporated into progressive or injector-based systems to enable complete line flushing during lubricant changeover, seasonal viscosity transitions, or contamination events. Purge valves must route flow to containment and allow for full-volume displacement without pressure spike.

Failure to design for proper refill, venting, and purging leads to inconsistent delivery, degraded lubricant performance, and unplanned system downtime during maintenance cycles. Refill systems should always be specified with operational environment, lubricant type, and reservoir geometry in mind.

RETROFIT AND INSTALLATION CONSTRAINTS

Compatibility with Existing Lubrication Strategies (Manual, Centralized, Automated)

Retrofits must accommodate legacy lubrication approaches already in use. Manual systems transitioning to automated delivery require clear identification of existing lube points, port thread types, and current lubricant types. Progressive or injector-based systems must match flow requirements and delivery intervals of the manual routine to avoid underlubrication or overlubrication.

For sites using basic centralized systems (single-line resistive or timer-based pump units), upgrade paths to feedback-enabled systems must consider existing metering blocks, controller compatibility, and routing integrity. Retrofitting automated systems into assets without prior lube architecture often requires drilling, tapping, or bracket fabrication; this must be accounted for during planning to avoid structural compromises or warranty violations.

Mounting Options for Tight Spaces or Mobile Equipment

Compact installations require remote mounting of reservoirs and pump units. Grease lines may be extended using high-pressure tubing with appropriate reinforcement and fittings. For mobile equipment (loaders, haulers, AGVs), vibration-resistant mounts, damped enclosures, and low-profile component designs reduce interference risk.

Modular pump assemblies and vertical reservoir configurations minimize footprint. Component layout must allow for unrestricted access to fill ports, breathers, and control interfaces. For moving assemblies (e.g., booms, arms, telescoping equipment), flexible lines must include strain reliefs and swivel couplings to prevent fatigue failure.

Articulating assets must have lubrication circuits zoned to match movement planes. Routing through pivot points requires protected channels and adequate slack to avoid tension during full extension or rotation.

Power Requirements and Integration with Existing Control Systems

Power source compatibility must be verified; most lubrication pumps operate on 12V DC, 24V DC, or 110/230V AC. Field power availability and load capacity must be confirmed. Where no electrical supply is present, solar-assisted battery systems or manual priming cycles may be used, though these limit feedback and automation capabilities.

Control integration with PLC, SCADA, or stand-alone HMIs must match existing communication protocols. Systems using Modbus RTU, CAN bus, or digital I/O must be matched with pump controller logic or via signal converters. Input/output mapping must include cycle complete, low-level alarm, pressure confirmation, and fault status.

Retrofitting without proper signal conditioning or grounding results in false alarms, system resets, or controller failure. Power conditioning devices and surge protection should be included in harsh industrial environments or mobile applications with fluctuating loads.

Retrofit success is dependent on detailed site surveys, compatibility mapping, and allowance for physical, electrical, and control-level constraints during system design.

FAILURE AND SERVICE DELAY MODES CAUSED BY DESIGN OVERSIGHT

Misaligned Fittings Leading to Wear or Leakage

Improper fitting alignment introduces mechanical stress at joints, especially in rigid tubing systems or at the interface between stationary and vibrating equipment. Misalignment causes seal distortion, thread fatigue, and eventual leakage. Pressurized systems are particularly sensitive; angled compression fittings or unsupported branch junctions develop hairline cracks under cyclic load. Grease couplers subjected to sideload during service wear unevenly, resulting in slow leaks and loss of volume delivery accuracy. Misalignment also complicates service access, increasing technician time and risk of improper reconnection.

Material Fatigue from Vibration or Poor Support

Unsupported lubrication lines or components mounted without vibration damping degrade rapidly in high-duty environments. Tubing and hose lines exposed to mechanical resonance, especially near engines, conveyors, or rotating assemblies, develop microfractures at bends, connectors, and ferrule junctions. Repeated flexing without proper loop geometry or strain relief causes fatigue failures, particularly in braided hoses and hardline aluminum tubing. Reservoirs and pump brackets mounted on thin panels or cantilevered supports experience crack propagation and mounting bolt loosening.

Over time, fatigue contributes to air ingress, lubricant leakage, or complete line separation under pressure. In mobile equipment or compact installations, fatigue failures often go undetected until performance alarms or dry-run damage occurs.

Thread Mismatches Causing Leaks, Contamination, or Line Breakage

Incorrect thread pairing—mixing BSPP with NPT or metric with SAE—results in incomplete thread engagement or galling during torque. Sealing tape or dope is often used as a work-around, but pressure cycling and thermal expansion cause joint separation and leakage. In worst-case scenarios, incompatible thread forms cut cross-threads into soft metal ports, compromising housing integrity and requiring full component replacement.

Thread mismatch is a leading cause of oil ingress into greased cavities, water ingress at outdoor lube points, and sudden line failures. Field retrofits using locally sourced fittings often overlook compatibility unless the thread type is verified by inspection or documentation. Mismatched threads in food-grade applications introduce the risk of lubricant contamination and foreign material ingress, violating compliance standards.

Thread selection must be standardized, verified during installation, and supported with torque specs, sealing practices, and inspection protocols to prevent systemic failures. Design errors at this level often cascade into higher-cost service delays and loss of lubrication control.

CROSS-FUNCTIONAL COLLABORATION IN EQUIPMENT DESIGN

Role of Reliability Engineering in System Layout

Reliability engineers define lubrication system architecture based on asset criticality, failure history, and performance targets. Their role is to ensure that lubrication supports availability goals and aligns with predicted failure modes. They identify lubrication-sensitive components, determine risk exposure from lube-related faults, and specify the need for condition monitoring, redundancy, and feedback loops. System layout is driven by RCM (reliability-centered maintenance) principles, matching lubrication type, the delivery system, and monitoring requirements to actual operating conditions.

Reliability engineers also interface with design teams to ensure LCC models include maintenance effort, lubricant consumption, and failure recovery. Their input validates whether component positioning supports MTBF targets and whether planned delivery methods are compatible with real-world process conditions.

Input from Maintenance Teams on Service History and Access Challenges

Maintenance technicians provide practical insight on service bottlenecks, point accessibility, tool clearance, and failure trends. Their feedback identifies where lubrication points are frequently missed, where incorrect products are used, and where time-consuming disassembly is required just to access reservoirs or fittings.

Historical maintenance logs are used to map recurring lubrication failures, seized bearings, degraded seals, broken lines, and guide design changes in new assets or retrofits. Maintenance teams highlight the need for better routing, ergonomic fill ports, and accessible drain points. Their involvement ensures that system designs reflect actual shop-floor behavior rather than theoretical maintenance intervals or idealized access diagrams.

Technician feedback also helps in validating sensor placement, port labeling, and alarm interpretation to reduce misdiagnosis during fault events.

Consultation with Lubrication Experts During Design Reviews

Lubrication specialists bring domain-specific knowledge of fluid dynamics, compatibility, delivery systems, and failure mechanisms. They select lubricant types based on speed, load, contamination risk, and operating temperature. Their input ensures correct sizing of metering elements, material compatibility, and purge/fill protocol design.

Lubrication experts validate whether the proposed delivery system, such as progressive, injector based, or circulating, is appropriate for the machine type and duty cycle. They also define flush intervals, reservoir sizing, sensor thresholds, and filtration requirements.

Their review of P&IDs (piping and instrumentation diagrams), component layout drawings, and BOMs helps catch incompatibilities before equipment commissioning. When included early in design, they prevent common pitfalls such as mismatched fittings, undersized pumps, unserviceable routing, and improper lubricant selection.

Cross-functional collaboration is essential to deliver systems that are not only engineered to spec, but maintainable, reliable, and resilient under actual plant conditions.

FIELD-PROVEN DESIGN PRACTICES

Use of Transparent Reservoirs and Low-Level Indicators

Transparent or semitransparent reservoirs allow immediate visual verification of lubricant quantity and condition. Integrated low-level indicators, such as float switches, capacitive sensors, or ultrasonic level detectors, enable both local alarms and remote alerts through PLC or SCADA. This prevents dry-run conditions, pump cavitation, and air ingress into metering systems.

Sight windows with graduated markings must be positioned for visibility during normal operation. Avoid placement on shaded or obstructed sides of the machine. Use of UV-resistant materials is essential to prevent clouding and degradation in outdoor or high-UV applications.

In-Line Filters and Pressure Checks Built into Pump Outputs

Filters installed directly after the pump outlet capture particulates before they reach metering valves or end-point bearings. Filter housings should include differential pressure gauges or switches to signal clogging and indicate service intervals. Field installations with no filtration consistently show higher injector fouling rates and premature wear in progressive metering blocks.

Check valves and pressure test ports should be built into the system header to allow quick diagnostics. Portable pressure gauges can be attached to these ports during troubleshooting or system validation. Systems without this capability rely on guesswork or invasive disassembly to locate flow restrictions or leaks.

Pump outputs should also include pressure relief valves set just below the system max pressure rating to protect components during cold starts or line blockages.

Designating Lube System Mounting Zones in Equipment Schematics

Lube system components must be included in mechanical layout drawings and equipment schematics from the start. Designating physical mounting zones avoids last-minute bolt-on placements that compromise accessibility, vibration isolation, or service clearance.

Zones should be located in low-vibration, thermally stable areas with room for access to fill ports, displays, and manual override functions. Routing corridors for tubing or hose runs should be defined in conjunction with structural and safety guard layouts to ensure no conflicts during installation.

Anchor points, cable trays, and mounting brackets must be specified during fabrication, not added later. Designating these zones ensures consistent placement across asset classes, reduces install time, and standardizes training and inspection procedures.

Field-proven practices reflect real operating conditions, streamline servicing, and enhance system longevity by addressing the typical points of failure observed in poorly integrated or ad hoc installations.

STEPS FOR AN IMPLEMENTATION PLAN: DESIGN FOR MAINTAINABILITY AND EQUIPMENT COMPATIBILITY CONSIDERATIONS

Steps for an implantation plan include:

1. **Initial Assessment and Requirements Gathering**
   - Perform a comprehensive asset audit to identify lubrication points, existing access challenges, and current lubrication methods.
   - Engage stakeholders including reliability engineers, maintenance personnel, lubrication specialists, and OEM representatives.
   - Define equipment compatibility criteria: thread standards, reservoir types, lubricant compatibility, and spatial constraints.

2. **Cross-Functional Design Collaboration**
   - Initiate early collaboration during equipment design between mechanical, electrical, and reliability engineering teams.
   - Integrate lubrication requirements into design documentation (3D models, schematics, and layout drawings).
   - Conduct joint design reviews to validate accessibility, modularity, and sensor integration.

3. **Ergonomic and Structural Integration**
   - Position lube points, fill ports, and sight glasses for ground-level access or safe reach zones.
   - Mount reservoirs and control modules in low-vibration, visible, and serviceable locations.
   - Avoid interference with rotating equipment, heat zones, or inaccessible frame areas.

4. **Lubricant and Component Compatibility Specification**
   - Select pumps, hoses, and seals based on lubricant viscosity, chemical stability, and temperature range.
   - Validate material compatibility using lubricant data sheets and supplier recommendations.
   - Standardize on fittings (thread type and size) and modular pump units across equipment types.

5. **System Layout and Routing Optimization**
   - Route tubing and lines with minimal bends, equal-length branches, and accessible junctions.

> Include structural supports, strain reliefs, and flexible coupling at movement points.
> Define purge, drain, and vent points in the schematic and installation instructions.

6. **Retrofit and Legacy System Accommodation**
   > Identify assets targeted for lubrication system retrofits.
   > Map existing lubrication infrastructure (manual, centralized, or automated).
   > Select components that align with retrofit space constraints and current lubricant practices.

7. **Sensor and Monitoring Integration**
   > Define points for flow, pressure, level, and temperature sensor installation.
   > Ensure compatibility with plant PLC, SCADA, or CMMS platforms.
   > Program alerts for low levels, pressure anomalies, or failed cycles.

8. **Installation Planning and Commissioning**
   > Develop installation protocols with attention to torque specs, sealant practices, and verification tests.
   > Train installers on alignment, component handling, and start-up procedures.
   > Commission systems with pressure and flow validation, purging, and cycle verification.

9. **Documentation and Maintenance Planning**
   > Produce comprehensive documentation: BOM, schematics, calibration charts, and refill schedules.
   > Define preventive maintenance intervals and service procedures.
   > Integrate lubrication system tasks into maintenance work orders and inspection checklists.

10. **Continuous Improvement and Feedback Integration**
    > Monitor performance via CMMS logs, operator feedback, and condition monitoring trends.
    > Record service challenges, component failures, or lubricant issues.
    > Feed insights back to OEMs or engineering teams for design refinement and standard updates.

Outcome: A maintainable lubrication system aligned with operational constraints, technician ergonomics, system compatibility, and life-cycle performance targets—supporting reliability, safety, and service efficiency.

MISAPPLICATION OF LUBRICATION EQUIPMENT: CASE STUDIES IN FAILURE

Misapplication of lubrication equipment can lead to catastrophic failure, reduced machine life, or process contamination. Case studies highlight the following errors:

- **Overpressurization in grease systems.** A progressive divider block was used where an injector system was needed, leading to blocked lines and failed bearings.
- **Inappropriate use of mist lubrication.** The lubrication was deployed in a high-dust environment, resulting in rapid airborne contamination and seal damage.
- **Incorrect pump selection.** A high-viscosity gear oil was fed through a pump rated for low-viscosity fluids, causing cavitation and erratic flow.

Each case underscores the need for precise alignment between equipment specifications and operational environment. Root cause analysis typically reveals gaps in training, failure to consult OEM requirements, or assumptions based on unrelated use cases.

Misapplication Risks

Incorrect selection or deployment of lubrication equipment results in functional failures, mechanical damage, and operational disruptions. Equipment that does not match the demands of the application introduces lubrication starvation, overpressurization, or incompatible delivery rates.

Misapplication leads to three dominant failure modes:

1. **Mechanical damage.** Bearings seize due to underlubrication from inadequate metering systems. Gear sets run dry or operate with aerated or degraded lubricants due to pump undersizing or faulty routing. Improper grease selection in high-speed spindles causes churning, elevated temperatures, and premature wear.

2. **Process contamination.** Incorrect reservoir materials leach into food-grade systems. Incompatible seal materials degrade and introduce particulates. Improper breathers or unfiltered fill ports allow dust, moisture, or chemical vapors to contaminate lubricant circuits.

3. **Operational downtime.** Failures in progressive systems from blocked lines or improperly configured dividers halt lubrication to entire banks of points. Inaccessible fill ports extend service intervals beyond recommended frequency. System alarms are ignored or bypassed due to unclear diagnostics.

Common Causes of Misapplication

➤ **Specification errors.** Equipment is sized without proper viscosity correction. Line lengths and flow rates are calculated under lab conditions, not field temperatures. OEM recommendations are applied generically across all duty cycles without correction for ambient load, cycle frequency, or contamination exposure.

➤ **Lack of training.** Technicians install components without understanding pressure limitations, priming requirements, or purge protocols. Maintenance personnel bypass system interlocks, use incorrect grease types, or rely on outdated manual routes post-automation.

➤ **Assumption-based decisions.** Design teams assume uniformity in point demand. Reservoir size and pump flow are guessed based on runtime rather than calculated lubricant consumption. Legacy equipment is retrofitted without considering structural vibration, clearance issues, or electrical integration capability.

Misapplication is often not recognized until after failure, and even then, root cause attribution is misdirected unless a structured review is performed. Preventing these failures requires exact matching of lubrication system design to machine function, environment, and operational profile. Field validation must be enforced at commissioning.

CASE STUDY 1
OVERPRESSURIZATION IN GREASE SYSTEMS

Situation

A progressive divider block system was installed to lubricate a series of large pillow block bearings on a conveyor head drive. The lubricant was an NLGI 2 grease delivered through rigid steel lines. The circuit included over 20 lubrication points spaced across a 40-foot span with elevation changes and multiple directional bends.

Failure Mode

Bearings at the end of the line exhibited signs of lubrication starvation, elevated temperatures, noise, and accelerated wear. After several weeks, a grease line ruptured midspan due to excessive internal pressure. The progressive divider block stalled without triggering a fault alarm. Upstream points received grease; downstream points were dry.

Root Cause

The lubrication system was designed with a progressive divider block, which requires consistent back pressure to operate correctly. However, the long line lengths and variable resistance among points created unequal flow demand. The divider block compounded pressure losses, and the absence of relief mechanisms or pressure feedback allowed a pressure spike to rupture the weakest line segment.

    The application required injector-based delivery, which meters lubricant independently at each point and compensates for differential resistance. The engineering team failed to account for grease flow behavior under high back pressure, especially with NLGI 2 grease in ambient temperatures below 40°F. No field validation or commissioning test with actual grease viscosity was performed.

Lessons Learned

Delivery type must be selected based on total line length, point spacing, resistance variability, and grease type. Progressive systems are inappropriate where flow demand is inconsistent or where line lengths exceed typical pressure propagation thresholds.

All high-point-count systems must be validated with pressure drop calculations, especially when using high-viscosity lubricants. Incorporating pressure test ports and flow indicators enables early detection of flow imbalance. During commissioning, actual field conditions, including temperature, back pressure, and line routing, must be used to confirm system viability.

CASE STUDY 2
INAPPROPRIATE MIST LUBRICATION
IN A DUSTY ENVIRONMENT

Situation

A mist lubrication system was installed on overhead chain drives in a pulp and paper facility. The system was chosen to reduce manual lubrication frequency and provide continuous delivery. The lubricant was a light mineral oil atomized via air-mist generators positioned near the chains. The ambient environment included high concentrations of airborne cellulose dust and fiber fines.

Failure Mode

Within weeks of commissioning, the chains exhibited abnormal wear patterns. Visual inspections revealed tacky buildup on the chain rollers and links. Seal degradation and roller pitting followed, causing misalignment and chain elongation. The atomized lubricant mist captured airborne dust, creating an abrasive slurry that embedded into moving components. Chain tension increased, and multiple segments failed prematurely.

Root Cause

The system design prioritized labor reduction without assessing airborne contamination risks. Mist lubrication, while effective for high-speed, clean environments, was misapplied in an environment with continuous particulate suspension. The mist acted as a carrier for cellulose dust, which adhered to chain surfaces and concentrated around lubricant entry points.

There was no enclosure, deflector shielding, or ventilation to isolate the mist-lubricated components from the dust-laden air. No preinstallation contamination risk assessment was conducted, and the system lacked monitoring to detect early wear or residue buildup.

Lessons Learned

Mist and spray systems are inappropriate in environments with high airborne particulate concentration unless components are fully enclosed or shielded. In open-process environments like pulp and paper mills, sealed systems or brush-fed drip lubrication is more suitable to prevent contaminant adhesion.

System design must account for environmental exposure as a primary parameter, not just application mechanics. Lubricant form, delivery method, and atomization behavior must be matched to the airborne contaminant profile, airflow direction, and thermal convection patterns around the application. Preinstallation site audits and airborne particulate measurements are required to validate compatibility.

CASE STUDY 3
INCORRECT PUMP SELECTION FOR HIGH-VISCOSITY FLUIDS

Situation

A centralized lubrication system was implemented to supply ISO VG 460 gear oil to multiple enclosed gearboxes on a bulk material handling line. The pump installed was originally specified for ISO VG 68 oil, carried over from a previous system used in a lighter-duty application. The system operated in a climate with seasonal ambient drops to below 40°F. Line lengths ranged from 20 to 50 feet with elevation changes.

Failure Mode

Within days of start-up, intermittent flow interruptions were reported. Pump noise increased, followed by erratic pressure readings and low discharge flow.

Cavitation symptoms developed at the pump inlet, causing vapor bubble collapse and mechanical damage to the impeller. Over time, the pump failed completely. Downstream gearboxes showed signs of lubrication starvation, rising temperature trends, elevated wear particle counts in oil samples, and audible gear mesh noise.

Root Cause

The pump was undersized for the viscosity of the ISO VG 460 oil, especially under cold start conditions. At lower ambient temperatures, the oil's high viscosity exceeded the suction capacity of the pump, resulting in cavitation. The inlet line was not heated or oversized to account for increased resistance. No viscosity correction was applied during pump selection. Cavitation indicators, such as noise, pressure fluctuations, and inconsistent delivery, were dismissed as start-up anomalies and not investigated until after failure.

No cold-flow testing was conducted during commissioning, and system verification was based on nameplate values, not dynamic performance under site-specific conditions.

Lessons Learned

Pump selection must include viscosity-corrected flow rate and pressure requirements. Pump inlet vacuum limitations must be matched to oil viscosity at minimum operating temperatures. Use of high-viscosity lubricants requires heated reservoirs, enlarged suction lines, and, potentially, gear or vane pumps designed for thick fluids.

Cold-flow validation is required during specification, particularly when using oils above ISO VG 320. Inlet pressure should be monitored, and cavitation alarms should be installed where ambient temperature fluctuations are common. System commissioning must simulate worst-case start-up conditions, not rely on room temperature testing or historical system designs.

SYSTEMIC CONTRIBUTORS TO MISAPPLICATION

Lack of Cross-Functional Design Review

Lubrication systems are often selected and installed without input from reliability engineers, maintenance personnel, or lubrication specialists. This siloed approach omits critical information about field conditions, maintenance workflows, and asset-specific failure history. Design decisions made in isolation overlook key variables such as operating environment, access limitations, and contamination risk. Misapplication frequently originates from assumptions that go unchallenged due to absence of collaborative review.

Incomplete Documentation or Outdated Spec Sheets

Relying on outdated OEM lubrication charts or legacy design data results in systems that are incompatible with current equipment configurations or lubricant formulations. Equipment upgrades, lubricant changes, and operating condition shifts are not reflected in older documentation. Without up-to-date drawings and specifications, systems are built on incorrect flow, pressure, or compatibility assumptions. Misaligned component selection and routing errors frequently stem from version control issues in technical documentation.

No Commissioning Validation or Performance Trial Phase

Commissioning is often reduced to power-up and leak checks, omitting system flow validation, pressure mapping, or cold-start testing. Lubrication delivery is assumed functional based on component installation, without verifying point-to-point lubricant arrival, proper metering, or delivery timing. Lack of a defined validation protocol prevents early detection of line restrictions, metering imbalance, or cavitation. The absence of trial runs with the actual lubricant under field conditions leaves performance unverified until failure occurs.

Training Gaps Among Specifiers and Installers

Technicians and engineers often lack formal training on lubricant behavior, system configuration, and component function. Installers may incorrectly torque fittings, use incompatible thread sealants, or substitute parts that degrade system performance. Specifiers may overlook pressure limitations, environmental constraints, or compatibility with control systems. These gaps introduce errors at every phase—design, installation, start-up, and operation—leading to widespread misapplication that propagates across the equipment base.

BEST PRACTICE PREVENTION STRATEGIES

Conduct Environment-Specific Risk Assessment Before System Selection

Evaluate ambient temperature range, airborne contaminants, washdown frequency, vibration exposure, and access constraints. Use particle count studies, airflow modeling, or thermal imaging if needed. Identify environmental conditions that could compromise lubricant integrity, delivery accuracy, or component function. Exclude delivery types that are incompatible with the site's exposure profile.

Always Verify Equipment Rating Against Lubricant Properties

Match pump, valve, injector, and line specifications with actual lubricant viscosity, temperature-dependent flow behavior, and additive compatibility. Confirm that seals, hoses, and fittings are compatible with the lubricant's base oil and additive package, especially with synthetic or food-grade formulations. Validate performance at the lowest ambient temperature expected.

Cross-Check Application with OEM and Third-Party Validation Tools

Use OEM configurators, pressure-loss calculators, and line sizing tools to verify design assumptions. Run simulations for flow rate, metering accuracy, and delivery time per cycle. Validate component compatibility with both OEM guidelines and third-party selection software to confirm robustness. Never rely solely on legacy charts or generic sizing rules.

Include Real-World Use Case Simulations in the Commissioning Phase

Commissioning must replicate operating conditions, including cold starts, full system load, and extended cycle times. Use test ports to verify pressure and flow at end points. Perform system start-up with the actual lubricant and verify delivery to each lube point.

Capture baseline data for pressure, temperature, and delivery intervals. Document all results and validate against design expectations. Use real-world scenarios to confirm system resilience, not just theoretical function.

DELIVERING GREASE INTELLIGENTLY: MODERN PUMP AND METERING TECHNOLOGY

Grease lubrication poses challenges due to its high viscosity and need for volumetric precision. Intelligent grease delivery systems use electronic metering valves, pulse feedback systems, and variable-speed electric pumps to control delivery.

New technologies include:

- ➤ **Smart lubricators.** Battery-powered or wireless-controlled devices with programmable output.
- ➤ **Feedback-enabled injectors.** Detect delivery success at each point.
- ➤ **Dual-line and progressive systems.** Designed for high-volume and large-scale machinery.

Proper design ensures even distribution, prevents overgreasing, and adapts to operating load and speed. Integration with CMMS allows data collection on consumption, delivery frequency, and point health.

These systems are especially effective in the mining, pulp and paper, and food industries where precise grease dosing can prevent contamination or mechanical failure. Selection depends on system scale, grease type, and required accuracy.

Grease Delivery as a Precision Task

Grease poses unique challenges compared with oil. Its high viscosity and non-Newtonian flow behavior create resistance during transport, especially in long or elevated lines. Proper delivery requires accounting for shear thinning, start-up torque, and temperature-dependent flow variability. NLGI grade directly affects pumpability and metering response, especially in colder environments.

Undergreasing leads to boundary lubrication, elevated operating temperatures, and metal-to-metal contact. Failures typically initiate at the rolling element/raceway interface, progressing through spalling, brinelling, or cage damage. Overgreasing generates pressure buildup within bearing housings, ruptures seals, displaces shields, and traps heat.

Thermal degradation of grease is accelerated, and purge pathways are blocked. Both conditions reduce bearing life and introduce secondary equipment failures.

Automated grease systems integrated with smart metering technology provide consistent volume delivery matched to application demand. These systems reduce reliance on manual cycles and eliminate variability in interval timing, point accessibility, and grease volume estimation. Modern reliability programs prioritize condition-based and load-sensitive grease application driven by data from vibration sensors, temperature probes, or motor current readings.

System architecture is transitioning from time-based delivery to real-time, feedback-controlled systems using pressure monitoring, delivery confirmation, and adaptive algorithms. Smart grease systems now form a critical component of Industry 4.0 reliability frameworks, ensuring lubricant is applied in precise amounts and at the right frequency, and is adjusted dynamically based on operating conditions.

CORE TECHNOLOGIES IN INTELLIGENT GREASE DELIVERY

Electronic Metering Valves

Electronic metering valves control grease delivery by regulating volume per cycle with high repeatability. Each valve is programmed to deliver a specific quantity per lubrication event, eliminating variation associated with manual application or spring-actuated mechanical systems. Shot size is digitally adjustable and can be fine-tuned during commissioning or runtime without mechanical recalibration. Programmable logic enables sequencing across multiple lubrication zones with distinct volume and frequency requirements. Electronic actuation provides faster response and consistent dosing across wide temperature ranges.

Pulse Feedback Systems

Pulse feedback devices confirm grease movement through system lines and validate delivery at the lubrication point. These systems detect pressure pulses or piston travel within metering blocks, signaling successful actuation. A lack of pulse indicates blockage, air lock, or failed discharge. Pulse data can be logged and linked to alarms in SCADA or PLC environments, providing traceability and compliance documentation. In systems servicing critical bearings or inaccessible points, pulse feedback ensures that lubrication is verified, not assumed.

Variable-Speed Electric Pumps

Variable-speed electric pumps adjust output flow based on system resistance, required discharge rate, or dynamic scheduling. These pumps deliver consistent pressure across multipoint systems and compensate for viscosity changes due to ambient temperature variation. Integrated sensors monitor motor current, discharge pressure, and line fill rates, enabling closed-loop control. Pump speed can be modulated based on point demand or programmed to deliver staggered volumes across a lubrication network. PLC integration allows coordination with metering valves, fault diagnostics, and load-responsive scheduling. These systems are essential in environments with varying duty cycles or complex equipment geometries.

ADVANCED GREASE APPLICATION PLATFORMS

Smart Lubricators

Smart lubricators are self-contained, battery-powered units equipped with programmable logic controllers and wireless communication modules. Designed for remote or hard-to-access assets, these units allow field personnel to configure delivery interval and grease volume per cycle via onboard controls or mobile applications. Internal pressure sensors and cycle counters track delivery status and remaining grease volume. Some models include Bluetooth or LoRaWAN modules for integration with plantwide monitoring systems. Alert functions notify users of depletion, blockage, or cycle failure through synced mobile or SCADA-based platforms. These units are suited for isolated motors, conveyors, or field-mounted drives where routing centralized systems is impractical.

Feedback-Enabled Injectors

Injectors with embedded sensors confirm the displacement of grease in real time, capturing each actuation event and verifying lubricant delivery at the point of use. Sensor feedback includes stroke completion, flow volume, and anomaly detection such as incomplete cycles or delayed response. Systems equipped with these injectors can trigger alarms upon detection of flow interruption or excessive back pressure. Self-diagnostic logic isolates the affected line and generates maintenance flags. These injectors are critical in applications where undetected lubrication failure can result in catastrophic equipment damage, such as rolling mills, crushers, or robotic joints.

Dual-Line and Progressive Systems

Dual-line systems operate via alternating pressure cycles between two supply lines, enabling grease delivery over long distances with high-volume output. Each lubrication cycle switches the pressure between lines, activating injectors that discharge metered quantities of grease. These systems tolerate pressure fluctuations and are well suited for large-scale applications with extensive distribution networks, such as steel production and paper mills.

Progressive systems use a series of metering blocks that distribute grease in a fixed sequence. Each piston in the block must complete its stroke before the next begins, ensuring continuous flow monitoring. Progressive systems provide lower total pressure but require balanced line resistance and minimal flow interruption. Design must account for uniform distribution line lengths and resistance balancing to avoid cascading failures. Progressive systems are ideal for mobile equipment and machinery with moderate point counts where visual confirmation and modular block servicing are required.

SYSTEM DESIGN CONSIDERATIONS

Matching Delivery Architecture to Application Scale

System architecture must match the operational footprint and lubrication complexity of the equipment. Small-batch operations or stand-alone assets may be best served by single-point smart lubricators or compact progressive systems. Large-scale continuous plants require centralized, multizone systems, often dual line or hybrid progressive, with remote pump stations, looped supply lines, and dedicated zone metering. NLGI grade directly influences line sizing and pump selection. High-viscosity greases (NLGI 2 and above) require larger-diameter lines, reinforced tubing, and higher-displacement pumps to overcome flow resistance and avoid pressure drop at line ends.

Volumetric Accuracy and Flow Resistance

Thixotropic behavior of grease leads to flow inconsistency during start-up and low-temperature conditions. Systems must account for the minimum shear rates needed to initiate flow and maintain volume accuracy. Cold starts introduce high resistance and transient back pressure; components must tolerate this without leakage or cavitation. Metering valves and injectors must be pressure-rated for expected peaks and deliver within ±10% volumetric accuracy to ensure proper lubrication without overfeeding. Line routing should minimize sharp bends and elevation changes to control pressure loss and promote uniform delivery.

Synchronization with Load and Duty Cycle

Static time-based lubrication intervals often result in underlubrication during high-load periods and waste during idle cycles. Intelligent grease systems should align lubrication delivery with actual mechanical demand. Load-responsive control modules adjust dosing frequency based on runtime, torque draw, or sensor-derived input (vibration, temperature, or motor current). Integration with PLCs allows synchronization of lubrication events with machine state changes, shift schedules, or predictive maintenance intervals. Dynamic scheduling ensures lubricants are applied precisely when boundary layer conditions begin to degrade, extending component life and eliminating excess application.

INTEGRATION WITH DIGITAL MAINTENANCE SYSTEMS

CMMS and Lubrication Module Connectivity

Modern grease delivery platforms interface directly with CMMSs through programmable controllers or API-based middleware. Each lubrication event, such as volume per point, time stamp, or cycle status, is logged and indexed by asset ID. This data enables closed-loop maintenance tracking and forms the foundation for lubrication KPIs. Consumption trends at the point level expose wear anomalies, blocked injectors, or overlubrication. Integration with maintenance history allows correlation of lubrication patterns with bearing replacements, motor failures, or rising vibration levels. Trigger thresholds can be established to auto-generate corrective work orders or inspection tasks when usage deviates from the baseline.

IoT and Remote Monitoring

Intelligent grease systems outfitted with IoT sensors transmit real-time operating data from pumps, metering valves, and lube points to central monitoring platforms.

Parameters such as line pressure, injector cycle count, reservoir level, and fault codes are pushed to cloud-hosted dashboards. Maintenance teams can access this data remotely via secure mobile apps or web portals. Cloud analytics engines flag deviations, calculate delivery efficiency, and project refill intervals. Remote monitoring allows support for geographically distributed equipment, reduces on-site inspection frequency, and enables fast response to delivery failures without physical access. This architecture supports predictive maintenance, benchmarking across facilities, and centralized reliability oversight.

INDUSTRY-SPECIFIC DEPLOYMENT

Mining

Mining applications involve extended lubrication line runs across haul trucks, crushers, shovels, and conveyors. Equipment operates under high mechanical load with long duty cycles and frequent shock loading. Grease delivery systems must withstand extreme vibration, high dust ingress, and severe ambient temperature variation. Dual-line systems are preferred for their high-pressure capacity and tolerance for uneven point resistance. Injectors must be rated for heavy NLGI grades and equipped with purge verification. Dust exclusion is critical, and breather filters, sealed enclosures, and high-integrity fittings are mandatory. Wireless-enabled smart pumps reduce technician exposure in remote or hazardous zones, allowing scheduled refills and diagnostics via telemetry.

Pulp and Paper

Paper machines demand reliable grease delivery to high-speed bearings, especially in drying and press sections where heat and chemical vapor exposure accelerate lubricant degradation. Progressive systems are commonly used due to their compact form factor and sequencing verification. Grease systems must deliver precise volumes to roll journals, doctor blade mechanisms, and felt guide systems at high frequency. Components must withstand steam, acidic by-products, and caustic washdown. Stainless steel metering blocks, PTFE-lined hoses, and automatic purge cycles are often standard. Integration with vibration and temperature monitoring enables adaptive scheduling based on load and thermal profile changes.

Food and Beverage

Lubrication systems must comply with NSF H1 standards and prevent cross-contamination with product zones. Greases must be food grade, inert, and nontoxic, with systems designed to avoid aerosol formation, drip, or overapplication. Airless grease delivery units, often battery-powered, are used in packaging lines, bottling equipment, and conveyors. Progressive systems with built-in cycle validation are preferred to meet audit and traceability requirements. Lubrication events are logged to CMMS with time-stamped verification. Cleanroom-compatible enclosures, stainless steel fittings, and color-coded lines are implemented for regulatory compliance and allergen control. Systems must support easy sanitation without compromising delivery accuracy.

SELECTION AND SPECIFICATION GUIDE

Pump-Type Selection

Match pump type to application scale, grease grade, and distribution architecture. Small systems (< 10 points) may use single-outlet electric pumps with onboard controls. Medium systems benefit from multi-outlet progressive-compatible pumps. High-point-count or long-run systems require high-output dual-line or injector-based pumps capable of overcoming back pressure. For mobile assets, battery-powered or hydraulic-driven pumps are suitable. Verify pump pressure range against system peak load conditions and ambient temperature extremes affecting grease flow.

Metering Strategy

Progressive metering is effective for medium-scale systems with uniform line lengths and moderate pressure variation. Choose injectors for systems requiring point-specific volume control, or where points differ significantly in back pressure. Dual-line systems are recommended where long distances, redundancy, and consistent pressure delivery are critical. Integrate electronic metering valves when dynamic scheduling, diagnostics, and precise volume control are needed.

Sensor Package Configuration

Select pressure sensors, pulse verification modules, and flow monitors based on system criticality and environmental exposure. Include reservoir level sensors for remote assets and dielectric sensors for grease degradation tracking. Install cycle counters and fault alert modules for critical points. For food-grade or hazardous zones, select sealed, corrosion-resistant sensor assemblies with wireless connectivity.

Tubing and Layout Design

Use steel or reinforced polymer tubing for high-pressure zones and long-distance runs. Avoid sharp bends and elevation shifts where grease may stagnate. Maintain equal lengths in progressive branches to balance flow. Include pressure test ports, isolation valves, and purge points in design. Apply color coding and tagging for traceability and service identification.

Flushing and Commissioning Protocols

Flush lines with compatible solvent or same-grade grease to remove debris. Use low-pressure prefill cycles to ensure all points are primed. Validate delivery volumes using manual stroke counts or pulse feedback. Run full-cycle simulations to confirm system performance under cold-start and max-load conditions.

Maintenance and Troubleshooting

Schedule inspection of all metering blocks, lines, and fittings for leaks, wear, or contamination. Monitor sensor output for delivery anomalies. Check injector response times and replace worn seals. Maintain spare pumps, cartridges, and sensor modules for fast swapout. Train technicians on system-specific fault codes and feedback logic. Review CMMS logs regularly to track abnormal consumption or skipped cycles.

CASE EXAMPLES IN INTELLIGENT GREASE DELIVERY

Downtime Reduction in Automated Bottling Line

A food and beverage facility implemented smart lubricators across its high-speed bottling line. Prior manual greasing during planned downtime was inconsistent due to accessibility and high cycle rates. NSF H1-grade grease was dispensed through battery-powered lubricators with programmable frequency and volume settings. Units transmitted status and fault alerts to mobile dashboards. After deployment, unplanned bearing failures dropped by 80%, and the facility eliminated scheduled shutdowns for lubrication. Lubricant consumption data was logged into the CMMS, enabling trend-based scheduling and early fault detection through deviations in usage patterns.

Life Extension of Conveyor Bearings in Mining Using Feedback Injectors

A surface mining operation upgraded its haulage conveyor systems with feedback-enabled grease injectors on all head pulley and idler bearings. Previously, overgreasing and skipped points led to frequent overheating and bearing degradation. New injectors provided real-time verification of each grease pulse and alerted for blocked lines. Integration with the site's SCADA system enabled live status tracking. Mean time between failures for conveyor bearings increased by 40%, and scheduled inspection intervals were extended. Maintenance labor hours were reallocated from daily greasing to system monitoring and proactive part replacement.

Grease Waste Reduction in Steel Mill Through Load-Based Variable Pumping

A hot strip steel mill transitioned from time-based central greasing to a load-responsive variable-speed system. High-temperature gearboxes and roll stand bearings previously received grease every hour regardless of process load. New logic-based controllers adjusted pump speed and injector actuation frequency in response to real-time load data and bearing housing temperatures. This adaptive approach reduced monthly grease use by 35% and eliminated purge contamination events on the shop floor. System alarms were tied to both underdelivery and overpressure, allowing instant isolation and correction without halting production.

THE ILLUSION OF CONTROL: RETHINKING THE 10 PPM STANDARD FOR H1 FOOD GRADE LUBRICANTS

Lubricants are widely used in food and beverage processing equipment under the premise that their accidental contact with food is safe up to a maximum of 10 parts per million (ppm). While this threshold appears to offer regulatory assurance, a closer inspection reveals that the 10 ppm standard is not a meaningful control point in real-world operations. In fact, the focus on ppm distracts from what truly ensures food safety: mechanical integrity, seal performance, and hygienic equipment design. There is flawed logic behind ppm-based risk control and advocates for a preventive engineering mindset.

What Is H1 Food Grade Lubricant?

H1 lubricants are specifically formulated for incidental contact with food. To be certified, they must be composed of ingredients listed on the FDA's Generally Recognized As Safe (GRAS) list and approved by NSF. However, H1 lubricants are not edible, they are only minimally toxic at low concentrations. Their inclusion in food products, even accidentally, must be limited to no more than 10 ppm.

Understanding 10 ppm in Context

Let's break down what 10 ppm actually looks like in practice:

- 1 gallon = 3.785 liters
- 55 gallons = 208.175 liters
- Assuming a water-like density, 55 gallons = ~208,175 grams of liquid
- 10 ppm = 10 parts per million = (10 / 1,000,000) × 208,175 grams = 2.08175 grams

This is roughly half a teaspoon of lubricant dispersed into an entire 55-gallon drum. Can you measure that on a single tortilla? In a batch of cereal? On a piece of sliced meat? No. You can't.

Why ppm Is Not Control

The 10 ppm allowance was never designed as a method of controlling contamination. It is a threshold to determine when action (like scraping product) is required. It assumes an already failed system. Yet we continue to see manufacturers and auditors obsess over ppm as if it were a real-time monitorable parameter. It's not.

You cannot measure 10 ppm of lubricant on most food surfaces with current quality control technologies. It's impractical, if not impossible, to detect such trace contamination in most processing lines. Therefore, the only real way to prevent contamination is to prevent lubricant escape altogether.

Engineering Food Safety: Real Controls

Instead of chasing ppm ghosts, food safety professionals and engineers should focus on:

➤ **Seal integrity.** Selecting and maintaining seals that withstand washdowns, CIP cycles, and thermal cycling.
➤ **Lubricant retention.** Using proper application methods, purge setups, and minimal-volume techniques to reduce lubricant migration.
➤ **Hygienic design.** Designing equipment that isolates lubrication points from food zones.
➤ **Condition monitoring.** Using tools like ultrasound, pressure differential, and vibration analysis to monitor lubricant migration and seal failure.
➤ **Safe tracers in oil.** How about an additive that is in the lubricant that could be measured but safe?

Focus on Performance, Not Allowance

The 10 ppm rule is not a protective barrier; it's a concession to failure. The real work of contamination prevention lies in proactive engineering: making sure lubricant never contacts food at all. That means higher standards for seals, smarter system design, and a commitment to true hygienic reliability.

Let's stop pretending we can manage half a teaspoon of grease across an entire processing line and start building systems where that scenario never happens in the first place.

FUTURE OUTLOOK

AI-Based Self-Optimizing Lubrication Loops

Machine learning models will increasingly drive lubrication delivery by continuously adjusting volume, frequency, and delivery paths based on evolving operating parameters. These systems will ingest data from vibration, thermal, and torque sensors, comparing it with historical wear and lubrication response patterns to optimize dosing in real time. Algorithms will self-correct for variables like seasonal temperature shifts, duty-cycle changes, and mechanical degradation signatures. AI-enabled controllers will dynamically reroute lubricant within multizone systems to prioritize critical points under transient load.

Embedded Condition Sensors in Grease Fittings

Next-generation grease fittings will integrate microsensors capable of measuring temperature, vibration, pressure, and contamination at the point of lubrication. These sensors will transmit data via low-power wireless protocols to edge computing nodes, enabling point-level diagnostics and autonomous alerting. Fit-for-service scoring per lubrication point will replace fixed-interval schedules. Predictive failure modeling will become hyperlocalized, flagging mechanical issues days before traditional methods detect anomalies.

Fully Autonomous Lubrication Microgrids Tied to Operational KPIs

Autonomous lubrication networks will function as closed-loop microgrids, interacting with equipment performance dashboards, energy usage metrics, and maintenance schedules. Lubrication delivery will be triggered by deviation from defined operational KPIs, such as increased torque draw, reduced line speed, or abnormal thermal output, rather than fixed cycles or basic sensor thresholds. These systems will self-regulate lubricant ordering, usage reporting, and compliance documentation. Integration with digital twins will allow simulation-based lubrication strategy optimization before physical implementation, reducing risk and accelerating deployment across fleet assets.

CHAPTER 10

SELECTION OF CONTAMINATION CONTROL PRODUCTS

BEYOND FILTERS: A TAXONOMY OF MODERN CONTAMINATION CONTROL SOLUTIONS

While filtration remains the cornerstone of contamination control, modern strategies encompass a much broader array of technologies aimed at exclusion, removal, and monitoring of particulate, moisture, and chemical contaminants. Today's taxonomy of solutions includes in-line filters, off-line kidney loops, magnetic separators, vacuum dehydrators, coalescers, electrostatic units, and advanced sensor-based diagnostics.

Each technology targets a specific contaminant profile. For example, electrostatic filters are ideal for removing submicron particles that pass through conventional filters, while coalescers are essential for removing free and emulsified water. Magnetic rod assemblies are increasingly common in gearboxes and circulating systems to trap ferrous debris at the source. The goal of contamination control is no longer just reducing particle count but managing the full fluid cleanliness profile to ISO 4406, NAS 1638, or SAE AS4059 standards. Effective programs use a layered defense strategy, combining passive and active control mechanisms with real-time diagnostics to anticipate fluid degradation and intervene proactively.

Filtration is only one element in a functional contamination control strategy. Comprehensive fluid cleanliness programs integrate multiple technologies, each selected based on contaminant type, fluid characteristics, system criticality, and operating conditions. Modern solutions fall into three operational categories: exclusion, removal, and detection.

Exclusion Technologies

Contamination exclusion prevents ingress before it occurs. Desiccant breathers are installed on reservoirs, gearboxes, and hydraulic tanks to eliminate moisture and airborne particulates from incoming air. These are critical in humid environments or systems with frequent thermal cycling. Closed-loop expansion tanks, sealed fill ports, and transfer filtration carts eliminate the most common contamination pathways introduced during lubricant top-offs or system breaching.

Removal Technologies

> **In-line filters.** Pressure and return-line filters remove circulating contaminants during operation. Selection criteria include filter β-ratio, collapse pressure, and dirt-holding capacity. Fine filtration (< 6 μm) is required for servo and proportional valve systems; coarser filters suffice for gear-driven hydraulics.

> **Off-line kidney loops.** These systems operate independently of the primary lubrication or hydraulic circuit. Flow rate, dwell time, and filter media selection are critical for controlling fluid conditioning without disrupting process dynamics. Kidney loops are effective for slow-acting contaminants such as varnish precursors and soft particulates.

> **Vacuum dehydrators.** Used for systems with water contamination from condensation, coolant ingress, or external washdowns. The systems pull fluid into a vacuum chamber, lowering pressure to vaporize water at low temperatures. They remove both free and dissolved water and help extend additive life.

> **Coalescers.** Ideal for fuel and turbine oil applications where water is finely dispersed or emulsified. Coalescing filters cause microscopic water droplets to merge and separate by gravity. Required for systems where emulsified water bypasses centrifugal or particulate filtration.

> **Electrostatic cleaners.** Target submicron particles and soft contaminants that evade mechanical filters. The systems use electrostatic charge to attract and trap particles on collector plates. Effective for varnish

removal in turbine and hydraulic systems where oxidation products remain suspended in the fluid.

➤ **Magnetic separators and rod assemblies.** Passive ferrous particle control installed in gearboxes, sumps, and reservoirs. High-strength rare-earth magnets attract steel and iron particles before they circulate. These devices require manual inspection and periodic cleaning. In circulating systems, magnetic filtration cartridges supplement conventional filtration to prevent gear and bearing damage.

Detection and Monitoring Tools

➤ **In-line particle counters.** Real-time measurement of ISO 4406 cleanliness codes. Required in mission-critical systems with tight fluid cleanliness tolerances. Optical or pore-blocking sensors are selected based on fluid opacity and application sensitivity.

➤ **Moisture sensors.** Capacitive and optical sensors monitor relative humidity or water saturation levels. Used for early detection in systems prone to water ingress. Sensor selection depends on fluid type, mineral oil, ester, or phosphate ester, as calibration varies.

➤ **Dielectric monitors.** Detect oxidation and additive depletion through changes in fluid permittivity. Enable prefailure detection of varnish formation or chemical contamination. Installed in-line or used for portable spot checks.

Cleanliness Standards

Cleanliness levels must align with component sensitivity. Hydraulic systems using servo valves typically target ISO 4406 15/13/10 or cleaner. Gearboxes may tolerate up to 18/16/13. Aerospace and defense systems refer to NAS 1638 or SAE AS4059 classifications. The specification of contamination control equipment must reflect the required cleanliness class, environmental exposure, and system duty cycle.

Integrated Strategy

A single filtration unit is insufficient in high-reliability systems. Layered contamination control integrates exclusion (sealed breathers, clean transfer procedures), removal (kidney loops, dehydrators, electrostatic units), and detection (real-time sensors). Selection must account for system operating pressure, fluid viscosity, temperature range, contaminant source, and response-time requirements. Real-time monitoring combined with predictive diagnostics shifts maintenance from reactive fluid replacement to active fluid conditioning.

BREATHERS, BARRIERS, AND BYPASS SYSTEMS: CHOOSING THE RIGHT MIX

Contaminants often enter lubrication systems through breathers, hatches, and imperfect seals. Effective exclusion strategies start at these points. Desiccant breathers, labyrinth breathers, and membrane barriers provide defense against moisture and airborne particles. Selection must be based on local ambient humidity, temperature fluctuations, and exposure to airborne contaminants.

Barriers such as expansion chambers, sealed reservoirs, or pressurized enclosures further reduce ingress risk, especially in marine and offshore environments. Bypass filtration systems, which continuously clean a portion of the fluid while equipment is in operation, offer an excellent supplement to in-line filters. These systems can operate with high-efficiency media and extended dwell times, capturing smaller particles that main filters may miss.

Blending these tools effectively requires an understanding of environmental severity, machine criticality, lubricant properties, and the economics of downtime. Combinations are often needed: desiccant breathers paired with vacuum dehydrators or bypass systems paired with magnetic filtration, depending on asset vulnerability.

Contamination exclusion must begin at known ingress points, breathers, fill ports, expansion zones, and mechanical seals. Ambient air is a primary vector for moisture and particulate contamination. The breather interface is a frontline defense.

Desiccant Breathers

Desiccant breathers are designed to remove both moisture and particulates from incoming air during thermal cycling. The breathers are ideal for systems with frequent temperature-driven expansion and contraction. High-humidity environments or systems operating outdoors require high-capacity silica gel or regenerable units. Flow rate ratings and saturation indicators must be matched to the system's breathing frequency and reservoir volume.

Labyrinth and Membrane Breathers

Labyrinth breathers use a tortuous path to trap particulates while allowing air exchange. They are best suited for gearboxes and enclosed systems where moisture ingress is less critical. Membrane breathers allow vapor to pass while blocking liquid water; they are used in mobile equipment exposed to splash and spray.

Barrier Systems

Expansion chambers and sealed reservoirs prevent external air from entering altogether. Pressurized enclosures create a slight internal overpressure using inert gas or dry air to prevent ingress. The systems are applied in offshore, marine, or caustic environments. Compatibility with the lubricant's vapor pressure and pressure rating of seals must be considered. Overpressure must not exceed bearing or shaft seal tolerances.

Bypass Filtration Systems

Bypass units divert a small portion of system fluid through fine-filtration media without disrupting operational flow. These systems allow for longer dwell time, enabling submicron and soft contaminant capture. Common media includes cellulose, depth type, and microglass rated at $\beta x(c) \geq 200$. The bypass rate typically ranges from 10–15% of the total system volume per hour. Systems must be sized to operate continuously without pressure imbalance.

Combined Strategies

Critical applications often require integration of multiple defenses. Gearboxes in dusty environments may pair desiccant breathers with magnetic filters and bypass filtration. Hydraulic power units with high ingress risk may combine membrane breathers, sealed fill ports, and vacuum dehydrators.

Selection Criteria

These criteria include:

> **Ambient humidity and temperature cycles.** Use desiccant or membrane-based exclusion.

➤ **Exposure to splash or dust.** Use labyrinth or sealed configurations.
➤ **Fluid sensitivity.** Select bypass systems for high-viscosity or oxidation-prone lubricants.
➤ **Criticality of asset.** Higher uptime requirements justify redundant exclusion and filtration.

System designers must account for the economic consequences of downtime, contamination source type, and lubricant vulnerability. The correct combination minimizes fluid degradation, reduces maintenance frequency, and prevents secondary equipment damage. Deployment should follow an exclusion-first principle: block entry before considering downstream removal.

SELECTION LOGIC FOR DESICCANT BREATHERS AND AIR MANAGEMENT DEVICES

Desiccant breathers are a frontline defense against moisture ingress, but their effectiveness hinges on proper sizing, media selection, and service practices. Choosing the right breather requires assessing the reservoir volume, breathing rate, operating environment, and system volatility.

Key selection criteria include:

➤ **Flow capacity** (measured in cubic feet per minute, or CFM). Must match or exceed the system's breathing requirements.
➤ **Desiccant material** (typically silica gel or molecular sieve). Tailored to moisture load and regeneration preferences.
➤ **Particle filtration capability.** Some breathers include HEPA-grade media.
➤ **Housing durability.** Polycarbonate, stainless steel, or hybrid housings, depending on chemical exposure and mechanical stress.

Air management devices go beyond breathers. Vacuum breathers, oil mist eliminators, and pressure-equalizing membranes control airflow and pressure changes while minimizing contaminant ingress. Air dryers and membrane separators are employed in high-moisture or temperature-sensitive environments.

Advanced breather systems include indicators for saturation, RFID tags for digital tracking, and modular replaceable cartridges. These features simplify inventory and maintenance while supporting contamination control programs with data-driven oversight.

Proper Selection of Desiccant Breathers and Air Management Devices

Desiccant breathers are specified based on system dynamics, environmental severity, and fluid volatility. Proper selection requires alignment of airflow capacity with reservoir breathing behavior, especially in systems subject to frequent thermal expansion or aggressive cycling.

Flow Capacity (CFM Rating)

The breather must match or exceed the maximum breathing rate of the system. Undersized breathers create vacuum conditions that draw contaminants past seals. Calculate the air exchange rate using fluid volume change per cycle and thermal expansion rates. Include a margin for rapid temperature shifts or aggressive refilling.

Desiccant Material Selection

Silica gel is standard for general industrial use. Molecular sieves are required for low dew point targets or high-temperature performance. Regenerability is critical for remote applications; select desiccants that can be oven-dried or supported by field-regeneration stations.

Particle Filtration Capabilities

Desiccant breathers often include integrated particulate filtration. Verify filtration efficiency (e.g., $\beta10 \geq 200$ or HEPA grade) for critical hydraulic or gearbox applications. Multistage breathers with prefilters and final filters extend desiccant life and improve particle control.

Housing Durability and Compatibility

Match housing construction to chemical and mechanical demands. Polycarbonate is suited to general-purpose use. Stainless steel is required for aggressive environments or high-pressure differential conditions. Hybrid designs combine visibility (clear media chambers) with reinforced base structures. Ensure UV resistance for outdoor installations.

Advanced Breather Features

Select models with desiccant saturation indicators for visual checks. RFID- or QR-enabled tags streamline digital tracking and preventive maintenance programs. Modular designs with replaceable cartridges reduce downtime and simplify field maintenance.

Air Management Device Integration

Devices include:

- ➤ **Vacuum breathers.** Allow pressure equalization without external air entry. Applied in closed systems with pressure fluctuation but low contaminant risk.
- ➤ **Oil mist eliminators.** Prevent lubricant carryover through breathers. Used in high-agitation reservoirs or systems with foaming risk.
- ➤ **Pressure-equalizing membranes.** Permit gas exchange while blocking liquid ingress. Suitable for washdown zones or high-humidity areas.
- ➤ **Membrane air dryers.** Deployed where ambient air has high moisture content. Provide ultra-dry air to reservoirs, increasing fluid longevity in hygroscopic fluids.

Application Considerations

Situations to be considered include:

- ➤ **High-moisture environments.** Use desiccant breathers with large media volume and air dryers.
- ➤ **Frequent cycling or large reservoirs.** Prioritize high CFM ratings and replaceable media.
- ➤ **Harsh chemical exposure.** Use stainless steel or chemically resistant housings.
- ➤ **Critical assets.** Employ sensor-integrated breathers and real-time tracking.

Breather systems must be maintained under condition-based schedules, not time-based intervals. Monitor saturation levels and breathing frequency to determine replacement intervals. Oversized breathers increase desiccant life and reduce service frequency. Undersized or mismatched breathers create pressure imbalance, desiccant bypass, and increased ingress risk.

PARTICLE EXCLUSION AT THE POINT OF INGRESS: DESIGN TO REALITY

Designing for particle exclusion involves anticipating ingress points and eliminating or controlling them during equipment specification, fabrication, and operation. This includes eliminating open-top reservoirs, replacing vented caps with sealed units, and relocating fill ports to reduce ambient exposure.

Seal selection is vital. Labyrinth, lip, and mechanical seals each have advantages based on speed, pressure, and environment. In high-contaminant zones, combinations of seal types, including air-purged or magnetic exclusion seals, offer superior protection. Material compatibility must be considered as well. Polymer seals may degrade in aggressive lubricants or elevated temperatures, releasing particulates internally. Likewise, poor design of gaskets and fasteners can allow pathways for contamination under dynamic pressure cycles.

Real-world execution often falters due to poor commissioning, improper installation, or retrofitting shortcuts. True particle exclusion requires cultural reinforcement, detailed procedural controls, and rigorous training in contamination control best practices.

Particle exclusion begins at the design stage. Open-top reservoirs, vented fill caps, exposed breather ports, and unsealed inspection hatches are direct entry points for airborne contamination. Eliminate open-fluid interfaces wherever possible. Replace vented caps with sealed, filtered breathers or pressure-equalizing membranes. Relocate fill ports away from splash zones and dust exposure areas; position them vertically or under protective shielding.

Seal System Design

Seals are critical containment interfaces. Selection must reflect shaft speed, housing pressure, fluid type, and contamination level.

- ➤ **Lip seals.** Suitable for moderate-speed, low-pressure environments. Prone to wear and ingress in dusty or wet zones unless backed by exclusion devices.
- ➤ **Labyrinth seals.** Noncontact designs that use tortuous flow paths. Effective in high-speed applications but require proper alignment and housing precision.
- ➤ **Mechanical seals.** Applied in pressurized systems or rotating equipment where zero leakage is critical. Material compatibility and flush plans must be validated.
- ➤ **Air-purged and magnetic exclusion seals.** Used in extreme contamination zones. Air purge maintains positive pressure; magnetic exclusion captures ferrous ingress.

Combinations, such as dual lip with labyrinth or mechanical with purge, are common in mining, pulp and paper, and food processing.

Material Compatibility

Seal and gasket materials must withstand fluid chemistry, operating temperature, and dynamic movement. Nitrile and EPDM degrade in ester- or PAO-based fluids under high thermal load. Incompatible materials swell, crack, or release particulates internally. Select Viton, PTFE, or fluoroelastomers where appropriate. Fasteners and housing materials must not corrode or flake into lubricants under thermal cycling or vibration.

Design Integrity Versus Field Reality

Clean design is often compromised during retrofits, unplanned maintenance, or budget constraints. Particle exclusion fails due to:

- ➤ O-ring reuse and improper torquing
- ➤ Misaligned fill couplings and open drain plugs
- ➤ Missing gaskets or incompatible seal replacements
- ➤ Bypassed breathers or missing filters post-servicing

Execution Requires Discipline

Effective exclusion is not just engineered; it is enforced. Commissioning protocols must include verification of all exclusion devices, seal alignments, and fill port integrity. Maintenance must adhere to clean handling procedures, filtered transfer containers, and sealed tools. Technicians require training in contamination pathways, seal inspection, and breather function.

Institutionalizing exclusion demands procedural ownership. Equipment drawings must include ingress points and corresponding exclusion devices. Job plans must verify seal type and orientation. Procurement must enforce material specification compliance. Exclusion isn't an accessory; it's part of system reliability.

FLUID CONDITIONING UNITS: WHEN, WHY, AND HOW TO DEPLOY

Fluid conditioning units (FCUs), also known as kidney loop filtration systems, are used to polish lubricants by removing fine particulate and water. These systems operate off-line from the primary flow path, allowing for extended dwell time and filtration efficiency.

FCUs are best deployed:

➤ During system flushing and commissioning
➤ For high-value or mission-critical assets
➤ In response to condition monitoring alerts
➤ As permanent installations for continuous filtration

Selection criteria include flow rate, filtration media type (cellulose, glass fiber, depth filter), water removal capability (absorptive versus coalescing), and system portability. Some FCUs include in-line heaters, moisture sensors, and real-time particle counters.

Advanced FCUs incorporate touch-screen diagnostics, variable frequency drives for flow control, and integration with SCADA or CMMS platforms. These units justify their cost by extending lubricant life, reducing wear, and enabling condition-based maintenance strategies.

FCUs operate independently of the main lubrication flow path. They provide off-line filtration, allowing for extended contact time with filter media. This improves contaminant capture efficiency and supports deeper fluid cleansing than in-line systems allow.

Deployment Scenarios

➤ **Commissioning and flushing.** New systems often contain residual manufacturing debris, seal material shavings, and metal fines. FCUs are essential for flushing new installations prior to operational start-up. Flow rate and filter efficiency should be selected based on reservoir volume and expected debris load.

➤ **High-value or mission-critical assets.** Equipment with tight tolerances, long overhaul cycles, or critical uptime requirements, such as turbines, gearboxes, and hydraulic systems, will benefit from continuous or scheduled FCU use. These assets often run at elevated temperatures and are sensitive to both particle and moisture contamination.

➤ **Condition monitoring triggers.** FCUs are deployed reactively when oil analysis shows rising ISO particle counts, moisture, varnish potential, or chemical degradation. Units with water removal features are critical when total water content exceeds saturation limits or emulsified water compromises film strength.

➤ **Permanent installations.** For systems that cannot tolerate process interruptions or are located in dusty, wet, or thermally unstable environments, FCUs may be integrated into the lubrication system infrastructure. Continuous off-loop filtration maintains fluid integrity even under varying load and environmental conditions.

Selection Parameters

- ➤ **Flow rate.** Must ensure complete reservoir turnover within a defined time frame, typically 5 to 10 times per day for mission-critical systems.
- ➤ **Media type:**
  - • **Cellulose.** Economical, suited for general particulate removal.
  - • **Glass fiber.** Higher dirt-holding capacity and beta ratio, better for fine filtration.
  - • **Depth filters.** Capture particles through the media matrix, not just the surface.
- ➤ **Water removal:**
  - • **Absorptive filters.** Use desiccant-style media to pull water from oil.
  - • **Coalescing units.** Separate free water via gravity or centrifugation. Required when dealing with high-water ingress or emulsified water.
- ➤ **Portability.** Mobile FCUs allow service to multiple reservoirs. Portable cart units are equipped with quick connects, strainers, and bypass protection.

Advanced System Features

- ➤ **In-line heaters.** Used to reduce viscosity for better filtration during cold starts or when working with high-viscosity oils (e.g., ISO VG 460+).
- ➤ **Sensors and feedback:**
  - • **Moisture sensors.** Alert for water saturation levels.
  - • **In-line particle counters.** Real-time verification of ISO 4406 improvements.
- ➤ **Controls and connectivity:**
- ➤ **Variable frequency drives (VFDs).** Allow flow control to match oil properties or system demand.
- ➤ **Touchscreen HMI.** For setup, diagnostics, and alarm logging.
- ➤ **SCADA/CMMS integration.** Enables remote monitoring, trend tracking, and automated service alerts.

Justification and ROI

The use of FCUs significantly extends lubricant service life, mitigates wear particle generation, and reduces the frequency of oil changes. In many cases, extending oil life by even

30% can offset the capital and operational expense of the unit. Reduction in unplanned downtime, elimination of reactive maintenance, and improved component service life contribute to measurable returns, especially in equipment with high repair or replacement costs.

THE ECONOMICS OF CLEAN: LIFE-CYCLE COST JUSTIFICATION MODELS

Contamination control is often undervalued because its benefits are preventive and nonlinear. However, data consistently shows that maintaining lubricant cleanliness can reduce component wear by 50% or more and double equipment life.

Justification models should include:

➤ **Downtime avoidance.** Quantify production losses from failures.
➤ **Component replacement.** Estimate cost of bearings, pumps, and valves.
➤ **Lubricant replacement.** Realize savings from extended fluid life.
➤ **Maintenance labor.** Reduce unplanned interventions.

Case studies from power generation, mining, and aerospace industries validate investments in desiccant breathers, FCUs, and bypass systems. Tools such as the life extension value analysis (LEVA) model and ROI calculators provided by major filter manufacturers help quantify the economic impact.

Cleanliness targets, such as ISO 17/15/12 or better for hydraulics, provide the benchmark for cost analysis. Performance metrics (MTBF, lubricant change intervals, equipment efficiency) serve as proof points for the value of clean lubrication.

Contamination control expenditures yield asymmetrical returns when evaluated over the full asset life cycle. While initial investment is front-loaded, payoff comes through reduced wear rates, extended component life, and minimized downtime events. Cleanliness efforts translate into tangible cost savings that accumulate over time across parts, labor, lubricant consumption, and productivity.

Downtime Avoidance

Lost production is the primary cost driver in high-capacity or continuous-duty operations. One hour of unscheduled downtime in power generation, steelmaking, or mining may result in losses exceeding $50,000. Maintaining target cleanliness levels reduces failure modes tied to fluid degradation, particularly abrasive wear, varnish formation, and seal erosion. Uptime preservation becomes a defensible capital line item when linked directly to avoided failures.

Component Replacement Costs

Hydraulic pumps, servo valves, actuators, and high-speed bearings are sensitive to particles smaller than 5 microns. Operating above ISO 18/16/13 triples the wear rate compared with systems maintained at ISO 14/12/10. The cost of premature failure for even one servo valve, often over $8,000, exceeds the cost of multiple desiccant breathers or FCU installations. Fleetwide reductions in component turnover present a measurable cost deferral.

Lubricant Replacement Extension

Fluid change intervals extend linearly with reduced contamination and moisture. In circulating systems, bypass filtration and dehydration can push lubricant life from 6 months to 3+ years. At $3–$10 per liter, high-volume systems (e.g., turbines, paper machines) justify kidney loop system investment on lubricant savings alone. Fewer changes also reduce waste disposal costs and regulatory exposure.

MAINTENANCE LABOR AND PLANNING

Scheduled maintenance reduces overtime, emergency labor rates, and the operational chaos of unplanned interventions. Fluid cleanliness enables predictable maintenance cycles and supports condition-based strategies. Lower demand on maintenance personnel improves bandwidth for precision tasks and reliability projects.

Analytical Models

- ➤ **LEVA** quantifies equipment life gain as a function of particle reduction, factoring in downtime, labor, and part cost.
- ➤ **ROI calculators** from filter OEMs estimate payback based on site-specific data: reservoir size, oil cost, contamination baseline, and failure history.

➤ **MTBF tracking** before and after contamination control improvements shows statistically valid reliability gains.

➤ **Lubrication cost per operating hour** provides a normalized metric across assets.

Benchmarks

Cleanliness targets serve as engineering goals tied to financial justification. For example:

➤ ISO 17/15/12 or better for hydraulics
➤ ISO 16/14/11 for turbine oils with servo controls
➤ ISO 19/17/14 for gear oils in moderate-duty gearboxes

Correlating fluid cleanliness to mean time between failures, asset health indexes, or scheduled oil analysis trends offers hard data for justification. Organizations that document improvements in pump efficiency, energy consumption, or filter lifespan can show line-item impact from implementing contamination control programs.

Long-term reliability metrics, not just initial procurement cost, must drive decisions around contamination control investment. Clean fluid is a multiplier across the asset life cycle, not a sunk cost.

CONTAMINATION CONTROL FOR MOBILE VERSUS STATIONARY SYSTEMS

Mobile equipment, such as construction vehicles or agricultural machinery, presents unique challenges in contamination control. These machines often operate in dusty, wet, or remote environments, and are subject to vibration, altitude change, and intermittent service.

For mobile systems:

➤ Use vibration-resistant breathers and seals.
➤ Incorporate onboard FCUs or filter carts.
➤ Design refill and service procedures to prevent ingress.
➤ Prioritize rugged, field-serviceable filtration components.

Stationary systems in controlled environments can benefit from permanent bypass filtration, advanced diagnostics, and higher-spec filters. However, assumptions about environmental cleanliness can lead to complacency. Indoor operations still suffer from construction dust, HVAC contamination, and operator-induced ingress.

Hybrid strategies are often needed. For example, mobile equipment in mining sites may dock at stationary filtration stations for scheduled fluid polishing. Likewise, mobile FCUs may be used to service both types of equipment on a route-based schedule.

Mobile systems are inherently more exposed to environmental variability and mechanical stress. Dust intrusion, vibration, shock loading, and intermittent usage cycles create high-risk conditions for lubricant contamination and degradation. Stationary systems offer better opportunities for engineered controls but are still vulnerable to ingress and operational contamination.

Mobile Equipment Contamination Control Strategies

- ➤ **Breather design.** Use desiccant breathers with vibration-resistant housings and reinforced mounting. Incorporate check valves or pressure-relief mechanisms to manage altitude-induced pressure fluctuations. Breathers must withstand repeated thermal cycling without media degradation.
- ➤ **Sealing systems.** Specify seals rated for dust exclusion under dynamic motion. Favor dual-lip, labyrinth, or magnetic exclusion designs for rotating components. Avoid vented caps and replace with closed-loop systems where possible.
- ➤ **Onboard filtration.** Equip systems with compact FCUs or portable filter carts stored onboard for route-based conditioning. Use quick-connect fittings with anti-drain-back valves for closed-circuit filtration during service intervals.
- ➤ **Field refill protocols.** Deploy closed refill containers with dry-break couplings. Train operators to perform lubricant top-ups using sealed systems only. Integrate contamination control into refueling trailers and field service units.
- ➤ **Component selection.** Use spin-on or cartridge filters with integrated bypass valves and visual clog indicators. Prioritize filter elements rated for off-road vibration profiles and temperature fluctuations. Design for tool-free filter changeouts where feasible.

Stationary System Contamination Control Strategies

➤ **Permanent bypass filtration.** Install dedicated kidney loop systems with high-efficiency elements and dehydration modules. Configure for continuous or demand-driven flow based on contamination monitoring.

➤ **Advanced monitoring.** Deploy in-line particle counters, moisture sensors, and varnish potential analyzers. Connect these to SCADA systems or CMMSs for real-time tracking and alarm generation.

➤ **Filter specification.** Use glass fiber or depth-type elements with beta ≥ 200 for fine particulate control. For synthetic fluids or ester-based lubricants, verify media compatibility to prevent swelling or breakdown.

➤ **Environmental controls.** Mitigate HVAC-induced contamination through positive pressure filtration, floor sealing, and equipment zoning. Identify high-traffic areas or shared airspace with fabrication zones as contamination hotspots.

Hybrid Models and Cross-Deployment

➤ **Docking stations.** Use permanent filter skids or flushing carts where mobile assets can dock at shift change or during maintenance windows. This allows scheduled fluid polishing without interrupting operations.

➤ **Mobile FCUs.** Assign trailer-mounted or cart-based FCUs to service both stationary and mobile systems based on route and condition priority. Configure with modular filters and interchangeable connections for multifluid support.

➤ **Fleetwide coordination.** Develop contamination control SOPs that span fixed and mobile assets. Align cleanliness targets across all fluid-handling operations, refill, sampling, storage, and disposal.

Effective programs acknowledge that mobility increases contamination exposure and complicates control. Solutions must be ruggedized, repeatable in the field, and supported by training and logistics. Conversely, fixed systems benefit from infrastructure but must resist assumptions of inherent cleanliness. Both environments demand engineered contamination control based on exposure profile, equipment criticality, and service constraints.

AIR AND WATER: THE PARADOX OF LIFE'S ESSENTIALS AS INDUSTRIAL LUBRICANT CONTAMINANTS

Air and water are fundamental to sustaining life, yet within the realm of industrial lubrication, they represent two of the most insidious contaminants. Their presence, though often invisible or underestimated, can compromise lubricant performance, accelerate equipment failure, and erode reliability gains painstakingly achieved through design and maintenance best practices.

Air Contamination: Forms, Issues, and Remedies

Dissolved Air

In lubricated systems, air enters in multiple forms. Dissolved air exists at the molecular level within the oil and under stable conditions remains harmless. However, when system pressures drop suddenly; such as at pump suction inlets. The air can come out of solution rapidly, leading to cavitation. Cavitation generates localized micro-jetting forces that pit metal surfaces, degrade bearings, and damage pump internals.

Entrained Air

Entrained air appears as microscopic bubbles dispersed throughout the lubricant, giving it a cloudy or milky appearance. It is commonly caused by turbulent return flows, suction line leaks, or low reservoir oil levels that induce vortexing. Entrained air reduces lubricant film strength, compromising hydrodynamic separation and leading to metal-to-metal contact. It also introduces compressibility into hydraulic fluids, creating erratic actuator performance and noisy pump operation. Furthermore, it increases oxidation rates by dramatically expanding the oil-air interface area.

Foam

Foam, a related issue, forms when air bubbles accumulate and coalesce at the oil surface. It typically results from mechanical agitation, defoamant additive depletion, or contamination with detergents and water. Foam obstructs level sensors, risks reservoir overflow, and can be ingested into pump suction lines, leading to air binding and hydraulic failure.

Free Air

Free air, while technically external to the oil phase, forms large voids in suction lines or reservoir headspaces. When ingested, it causes immediate cavitation, loss of pump prime, and potential equipment damage.

Effective control of air contamination demands a combination of mechanical and chemical strategies. Ensuring return lines discharge below oil surfaces away from suction zones, maintaining suction line integrity to prevent leaks, and using lubricants with robust air release properties are critical. For foam issues, selecting oils with effective defoamant packages and redesigning reservoir geometry to reduce agitation zones can eliminate persistent problems.

Water Contamination: Forms, Issues, and Remedies

Dissolved Water

Water contamination in lubricants also manifests in three forms. Dissolved water remains invisible but is corrosive and accelerates oxidation, leading to sludge formation and viscosity thickening.

Emulsified Water

Emulsified water occurs when water exceeds the oil's saturation limit, creating a stable oil-water emulsion with a cloudy or milky appearance. This condition compromises lubrication film strength, promotes micro-pitting, and reduces filterability, potentially clogging fine filtration systems.

Free Water

Free water is the most dangerous form, collecting at the bottom of reservoirs and sumps due to its higher density. Free water leads to direct corrosion of submerged metal surfaces, initiates rust formation, and creates an environment favorable to microbial growth in systems such as fuel storage tanks and hydraulic reservoirs. In hydraulic systems, free water can cause vaporous cavitation, leading to surface erosion and premature component failure.

Addressing water contamination requires eliminating ingress points and removing existing contamination. For dissolved and emulsified water, using lubricants with strong demulsifying properties enables effective separation and removal via settling or coalescing filtration. Free water must be drained routinely, and advanced water removal technologies such as vacuum dehydration or centrifugal separators deployed for critical systems. Preventive measures include using desiccant breathers to eliminate moisture ingress, sealing reservoir lids, and inspecting heat exchangers for leaks that can introduce water into oil circuits.

The Reliability Paradox

While air and water sustain all known life, in industrial lubrication they are relentless adversaries. Both compromise the very purpose of lubrication: to separate moving surfaces, control temperature, and protect against wear and corrosion. Their ingress into lubricating systems is inevitable without vigilant design, operational discipline, and maintenance practices.

We must treat air and water management as a core aspect of asset health strategies. Preventing their contamination is not merely a matter of oil life extension; it is a safeguard against unplanned downtime, catastrophic equipment failures, and compromised safety. Recognizing the paradox of these life-giving substances and controlling their destructive potential within lubrication systems remains one of the defining responsibilities in the pursuit of operational excellence and mechanical reliability.

WHEN GREASE MIXES WITH OTHER GREASES

Some greases thin down when mixed with other greases due to incompatibility between thickener systems or base oil formulations. Grease is not a homogeneous material; it's a structured system composed of base oil, thickener, and additives. When two incompatible greases are combined, the structure that holds the grease together can break down. The most common cause is thickener incompatibility. Different greases use different thickeners such as lithium, lithium complex, calcium sulfonate, polyurea, aluminum complex, or clay and these thickeners vary in chemical makeup, polarity, and structural behavior. When incompatible thickeners are blended, the soap fibers or gelling agents may dissolve, collapse, or repel one another, resulting in a breakdown of the grease's mechanical structure. This appears as softening, thinning, or even complete phase separation where oil and thickener separate.

In addition to thickener interactions, mismatched base oils can also destabilize grease. Some greases use mineral oils, others use synthetics such as esters, PAOs, or PFPEs. If the base oils in a mixture have different solvency properties, one may dilute or interfere with the other's thickener network. Additive packages can also interact in unexpected ways, although this is usually a secondary effect. When structural integrity is weakened by incompatibility, shear forces during operation can further accelerate thinning, transforming the grease into a runny, oil-rich slurry that leaks from seals or fails to stay in place.

The result is often poor lubrication, excessive leakage, and premature wear. Symptoms in the field include excessive purging, noisy or hot bearings, and failed grease injectors due to low resistance. To prevent these issues, it's critical to consult compatibility charts, minimize the number of grease types used across equipment, and thoroughly purge or clean bearings before switching to a new grease. In high-value applications, lab testing of mixed samples for consistency, oil separation, and drop point changes is a smart safeguard before transitioning.

Here's why thinning happens:

Mechanisms Behind Grease Thinning Upon Mixing

Thickener Incompatibility (Most Common Cause)

Different greases use different thickeners: lithium, lithium complex, polyurea, calcium sulfonate, aluminum complex, bentonite (clay), PTFE, etc. These thickeners have different polarities, soap structures, and stability conditions. When incompatible thickeners are mixed, the soap fibers or gelling agents can disperse, dissolve, or collapse, resulting in loss of mechanical structure which appears as softening or thinning.

Example: Mixing lithium-complex with clay-based grease often results in severe softening or phase separation.

Base Oil Mismatch

Greases may use mineral, PAO, ester, silicone, or PFPE base oils. When mixed, some base oils dilute or destabilize the thickener system of the other grease. In esters or PAOs, swelling or solvation effects can weaken fiber structure, leading to a drop in consistency.

Additive Interactions

Some additives (EP, tackifiers, anti-wear, thickeners, etc.) can neutralize or disrupt each other. This is less common as a primary thinning cause, but it can contribute when combined with base oil or thickener incompatibility.

Shear-Thinning Amplified by Incompatibility

If an incompatible mixture already weakens the structure, mechanical shear (e.g., in bearings or pumps) accelerates thinning and can turn grease into an oily slurry under dynamic load.

Field symptoms of grease thinning:

➤ Grease leaks from seals or purges excessively
➤ Bearing cavities fill with oil-rich soup
➤ Grease pumps and injectors lose pressure or fail
➤ Premature bearing wear from under-lubrication

Prevention and best practices:

➤ **Compatibility charts.** Use published compatibility tables (e.g., NLGI, OEMs), but validate with testing if critical.
➤ **Single grease standardization.** Minimize the number of grease types in a plant.
➤ **Purge old grease.** When switching, purge old grease thoroughly or clean components.
➤ **Lab testing.** When in doubt, test mixtures for penetration (ASTM D217), oil bleed (D6184), or drop point changes (D566).

Why Do Some Greases Thicken When Mixed With Other Greases

Cross-contamination is a fact of life in industrial lubrication, but it brings with it a poorly understood and costly problem: certain greases, especially PFPE/PTFE and silicone-based types, thicken dramatically when exposed to hydraulic oils, gear oils, or incompatible greases. This thickening impairs flowability, clogs lubrication paths, and accelerates failure. Understanding the mechanisms behind this phenomenon helps prevent downtime and protects critical assets.

Greases Are Systems, Not Single Substances

Greases are engineered systems consisting of three primary components: base oil, thickener, and additives. The base oil, be it mineral, synthetic hydrocarbon, ester, PFPE, silicone, or halocarbon, provides lubrication. The thickener—lithium complex, calcium sulfonate, polyurea, PTFE, barium complex, or aluminum complex gives structure and consistency. Additives enhance performance with EP, anti-wear, oxidation inhibitors, and tackifiers. When grease thickens after contamination, the cause is rarely chemical reaction. The real culprit is physical incompatibility between the grease's base oil and the contaminant, whether hydraulic fluid, gear oil, or another grease system.

PTFE-Based Greases and Hydrocarbon Contamination

PFPE/PTFE greases are widely used in vacuum and high-temperature applications because PTFE is chemically inert and PFPE base oils have excellent thermal stability. However, PFPE oils are immiscible with hydrocarbon-based fluids like hydraulic or gear oils. When contamination occurs, mutual insolubility displaces PFPE oil from the PTFE thickener matrix. The displaced PFPE oil migrates away, leaving a concentrated PTFE structure with less oil, resulting in dramatic thickening. Additionally, hydrocarbon contamination destabilizes PTFE particle dispersion, causing clumping and stiffness. The grease becomes too thick to flow through lubrication lines, leading to friction, heat generation, and early failure.

Silicone-Based Greases and Petroleum Contamination

Silicone-based greases experience similar problems. Silicone oils are highly immiscible with petroleum oils. When contaminated, the grease stiffens, seals swell, and lubrication films break down, compromising protection and accelerating wear.

Hydrocarbon-Based Greases: Better Compatibility

Hydrocarbon-based greases—lithium complex, calcium sulfonate complex, and polyurea—generally tolerate hydraulic and gear oil contamination better:

> ➤ **Lithium complex.** May soften or thicken slightly, depending on contaminant viscosity and additive interactions.
> ➤ **Calcium sulfonate complex.** Excellent stability, maintaining structure and EP performance.
> ➤ **Polyurea.** Generally stable with hydrocarbon fluids, though synthetic base oil polyureas require compatibility verification.
> ➤ **Barium complex.** Good compatibility but less common due to environmental restrictions.

Gear Oil Contamination Considerations

Gear oils often contain sulfur-phosphorus EP additives, which can soften or thin grease by dissolving or disrupting thickener structures. Alternatively, additive polarity can destabilize the grease matrix, increasing consistency and causing thickening. Both scenarios compromise lubrication performance.

Cross-Contamination With Dissimilar Greases

Cross-contamination between different grease types is an underappreciated threat. Incompatible base oils—such as mineral oil greases contaminated with silicone or ester-based greases—may separate, thicken, or soften unpredictably. PFPE or silicone greases contaminated with hydrocarbon greases typically thicken due to mutual insolubility.

Thickener incompatibility also poses risks. Polyurea and lithium complex greases often separate or suffer oil bleed and structural breakdown. Calcium sulfonate complex greases are broadly compatible with lithium greases, but compatibility with polyurea or barium complex greases should never be assumed without testing. Disparate additives can also react to form precipitates or destabilize the matrix, as when sulfur-phosphorus EP systems in gear oils interact with polyurea thickeners, causing structure loss or gelling.

Practical Implications

When thickening occurs from contamination, lubrication flow is restricted through lines and bearings, starving seals and bushings of lubrication. Friction increases, operating temperatures rise, and accelerated wear or catastrophic failures follow.

Halocarbon PFPE/PTFE Greases: Engineered Compatibility

Some halocarbon-based PFPE/PTFE greases are engineered to overcome typical incompatibility with hydrocarbons. These formulations remain chemically inert and nonflammable while maintaining compatibility with petroleum oils. They retain flowability in contaminated environments and tolerate temperatures up to 288°C and vacuum conditions. Field validation remains essential.

Engineering Recommendations

Reliability and lubrication engineers should:

➤ Understand system chemistry and evaluate base oil and thickener compatibility with potential contaminants.

➤ Perform contamination testing by mixing candidate greases with hydraulic oils, gear oils, and dissimilar greases at realistic ratios (10%, 25%, 50%), heating to operational temperatures, and observing for thickening, softening, or separation.

➤ Consult grease manufacturers for cross-contamination data and validation support.

➤ Enforce strict lubrication control programs to prevent unintentional mixing, with dedicated tools, clear labeling, and technician training.

Bottom Line

Grease thickening upon contamination is a physical phenomenon rooted in base oil compatibility and thickener interactions, not chemical reactivity. While PTFE itself remains inert, incompatibility with hydrocarbon fluids or dissimilar greases results in flow loss, asset failure, and unplanned downtime. Understanding these mechanisms equips engineers to make informed decisions that protect uptime, reliability, and safety.

Field Testing Protocol for Grease Compatibility

1. Identify potential contaminants (hydraulic oils, gear oils, other greases).
2. Mix with grease samples at realistic contamination levels (10%, 25%, 50%).
3. Heat to operating temperatures for appropriate dwell time.
4. Measure penetration (consistency), observe for separation, thickening, or softening.
5. Evaluate flowability through typical lubrication paths.

CHAPTER 11

MANAGEMENT OF LUBRICATION SUPPLIERS AND SERVICE PROVIDERS

STRATEGIC PARTNERING FOR TECHNICAL LUBRICANT SERVICES

Effective lubrication programs depend not only on quality products but also on the strength of supplier relationships. Strategic partnering goes beyond transactional purchasing to build a collaborative alliance between end users and lubricant providers. This includes joint problem-solving, shared technical training, co-development of performance targets, and alignment with the end user's reliability goals.

Key to this model is early engagement in specification and trial phases, where suppliers contribute expertise in base oil chemistry, application-specific additives, and system diagnostics. Long-term partnerships often include performance guarantees, exclusive support for high-criticality equipment, and integrated lubrication audits.

A strategic partner becomes a technical extension of the plant maintenance team, offering condition monitoring, laboratory services, contamination control guidance, and formulation recommendations tailored to actual operating conditions.

Strategic Partnering for Technical Lubricant Services

Effective lubrication management requires more than product procurement. Strategic supplier partnerships integrate technical service, application knowledge, and ongoing support into the lubrication program. The relationship should shift from vendor-client to a collaborative engineering function embedded in reliability improvement.

Early engagement during equipment specification, lubricant selection, and commissioning allows the supplier to provide formulation guidance based on base stock compatibility, additive behavior under thermal stress, and historical failure modes for similar assets. This front-end alignment prevents mismatched lubricants and simplifies qualification during start-up.

Joint development of KPIs, such as lubricant life extension, uptime improvement, and contamination control targets, establishes a framework for shared accountability. These partnerships often include:

➤ **Performance guarantees.** Suppliers commit to lubricant performance under defined operating parameters. Failure to meet metrics may trigger product replacement, root cause support, or credit mechanisms.

➤ **Exclusive support models.** Dedicated technical staff from the supplier assist with route audits, lube room setup, and implementation of condition-based maintenance strategies.

➤ **Field and laboratory diagnostics.** Regular oil analysis, wear debris interpretation, and varnish potential monitoring are managed by the supplier to ensure fluid integrity.

➤ **Customized formulations.** Suppliers may adjust additive packages or viscosity grades based on evolving conditions, load profiles, or regulatory requirements.

➤ **Training and knowledge transfer.** On-site education tailored to operator level, from lubrication fundamentals to advanced diagnostics, minimizes skill gaps.

➤ **Integrated contamination control.** Joint projects target ingress points, reservoir management, filtration upgrades, and breathers based on real-world particulate and water load profiles.

Strategic partners function as an extension of the reliability team. They contribute to root cause failure analysis, provide failure trend data from across industry, and act as advocates during OEM or warranty disputes. Selection should prioritize technical capability, response time, field service presence, and long-term alignment with plant reliability objectives, not just unit cost.

SETTING SLAs AND KPIs
FOR LUBRICANT SUPPLY CHAINS

To manage supplier performance with transparency and accountability, SLAs (service-level agreements) and KPIs are essential. SLAs should be clear, measurable, and tied to outcomes relevant to uptime, product quality, and operational continuity.

Examples of SLA metrics include:

- On-time delivery rates
- Emergency delivery turnaround time
- Stockout avoidance
- Technical support response time
- Lab report turnaround (e.g., oil analysis)

KPIs should be built into dashboards that track trends in lubricant consumption, waste generation, system cleanliness, and equipment uptime associated with lubrication tasks. These indicators help both parties align priorities and reinforce continuous improvement. SLAs should include escalation protocols, periodic performance reviews, and flexibility for system evolution. Collaborative review cycles can uncover root causes behind KPI shortfalls, leading to joint corrective actions and preventive planning.

Lubrication supply chains must be performance-driven. SLAs and KPIs define expectations, ensure accountability, and enable operational alignment between the end user and supplier.

SLA Structure and Metrics

SLAs must focus on operational continuity, responsiveness, and technical service delivery. Key SLA metrics include:

- **On-time delivery rate.** Minimum threshold typically > 95%. Should include lead time classification (standard versus urgent) and dock-to-shelf time.
- **Emergency order fulfillment.** Maximum allowable turnaround (e.g., < 24 hours) with supplier prepositioned stock or mobile dispatch capabilities.
- **Stockout incidence.** Count of stockouts per period. Zero tolerance for critical SKUs (e.g., turbine oil, H1 greases).

- ➤ **Technical support response time.** Response within 2 business hours for routine support; immediate escalation for critical failures.
- ➤ **Lab report turnaround.** Time from sample receipt to actionable report. Benchmark: ≤ 48 hours for routine oil analysis.

SLAs must also define escalation procedures, service penalties or incentives, and review cycles. Escalation protocols should trigger when delivery failures, missed diagnostics, or support delays affect asset availability or compromise critical lubrication points.

Lubrication-Specific KPIs

KPI dashboards track lubricant-related performance over time and provide data for trend analysis and process improvement. Key metrics include:

- ➤ **Lubricant consumption per asset.** Normalized against runtime or load to detect overuse or leakage.
- ➤ **Lubricant waste volumes.** Identify overdraining, contamination, or expired stock turnover.
- ➤ **Cleanliness compliance.** Percentage of samples meeting ISO 4406 or OEM particle targets.
- ➤ **Uptime correlation.** Link between lubrication task completion and equipment availability or MTBF.
- ➤ **Analysis compliance rate.** Percentage of scheduled samples received, tested, and reviewed.

Governance Model

Joint reviews, monthly for operational metrics, quarterly for strategic alignment, identify root causes of KPI drift. Typical causes include late replenishment, incorrect lubricant deployment, failure to interpret oil analysis, or procedural nonconformance.

Corrective actions may include route redesign, technician retraining, supplier stock repositioning, or lubricant spec adjustments. KPI ownership should be shared, with failure analyses jointly led by supplier and plant reliability staff.

SLAs and KPIs should remain dynamic. They should be updated as lubrication strategies shift from preventive to predictive models or as sensor-integrated delivery replaces manual routines. The goals are sustained reliability, minimized waste, and data-backed supply chain optimization.

QUALIFYING THIRD-PARTY SERVICE PROVIDERS: BEYOND THE BID

Selecting third-party lubrication contractors or service providers must go beyond price comparison. Qualification should assess technical proficiency, safety culture, field experience, regulatory compliance, and alignment with your facility's standards.

Key evaluation dimensions include:

➤ Technician certifications (e.g., ICML, STLE, NLGI)
➤ Insurance and liability coverage
➤ History of performance in similar industries
➤ Documentation practices and reporting systems
➤ Access to condition monitoring or mobile lab equipment

Site visits, reference checks, and trial engagements can provide deeper insight into a service provider's fit. A quality-driven provider will demonstrate a proactive stance on contamination control, safety procedures, and value-added reporting.

Formalizing expectations through a scope of work, reporting templates, and performance thresholds ensures consistency and mitigates risk.

Price should not be the primary selection criterion for lubrication service providers. Risk exposure, operational quality, and long-term reliability demand a deeper qualification process that verifies competency, safety discipline, and technical alignment with site standards.

Technical Credentialing

Service technicians must hold industry-recognized certifications relevant to the scope of lubrication work:

➤ ICML (MLT I, MLT II, MLA I, MLA II)
➤ STLE Certified Lubrication Specialist (CLS)
➤ NLGI Lubricating Grease Specialist (CLGS)

Credentialing ensures baseline understanding of lubricant compatibility, contamination control, system design, and inspection methods. Lack of certification is a red flag for potential misapplication, poor diagnostics, and unsafe handling.

Insurance, Compliance, and Legal Readiness

Providers must show current insurance coverage, including:

➤ General liability
➤ Workers' compensation
➤ Pollution liability if flushing, vacuum dehydration, or high-volume lubricant handling is involved

Confirm regulatory compliance with OSHA, MSHA, DOT, and environmental standards applicable to the operating region and process type.

Operational and Reporting Maturity

Effective service providers standardize execution through checklists, validated reporting formats, and mobile documentation platforms. Review their sample inspection reports, route logs, and condition monitoring records. Look for consistency, traceability, and actionable insights.

Ask to inspect or demo:

➤ Mobile filtration carts or dehydration skids
➤ Portable particle counters or viscosity meters
➤ Field software or digital CMMS integration tools

Performance Track Record

Require evidence of successful projects in similar environments, whether food-grade systems, corrosive chemical plants, remote mining operations, or continuous-duty rotating equipment. Solicit references, and confirm punctuality, quality of service, and communication during prior engagements.

Site Evaluation and Trial Work

Conduct in-person audits or supervised trial jobs. Evaluate:

➤ Safety briefings and PPE usage

➤ Equipment staging and cleanliness
➤ Job scope adherence
➤ Coordination with plant personnel
➤ Quality of post-job documentation

Contracting and Oversight

Define scope of work with precision:

➤ Frequency and task list (e.g., grease routes, filter replacements, oil top-offs, sampling)
➤ Reporting timelines and KPIs (e.g., missed points, fluid levels, equipment anomalies)
➤ Emergency support availability

Use formal templates for findings, service verification, and nonconformance identification. Include expectations for job walkdowns, root cause input, and joint review meetings.

A qualified third-party contractor extends the site's maintenance capabilities without introducing variability or compliance risk. Selection must be based on technical evidence, not bidding alone.

AUDIT-READY VENDOR EVALUATION CRITERIA

With increasing regulatory scrutiny and internal governance mandates, vendor evaluation processes must be audit-ready. Lubricant suppliers and contractors must be assessed against criteria that are traceable, documented, and aligned with corporate compliance systems.

Audit-ready criteria should cover:

➤ Documented training records and safety certifications
➤ SDS and regulatory disclosures
➤ Documented processes for waste handling, spill prevention, and chemical storage
➤ Quality assurance measures (e.g., batch testing, traceability)
➤ Environmental certifications or sustainability policies

Vendor evaluations must meet the requirements of both internal compliance systems and external regulatory audits. Lubricant suppliers and third-party contractors must be screened using objective, documented, and reproducible criteria. All records should be easily retrievable, auditable, and traceable to support corporate governance, ISO certification, and regulatory oversight.

Training and Safety Documentation

Vendors must supply documented training records for all personnel providing on-site service. This includes:

- Verified completion of OSHA-, HAZCOM-, or MSHA-required training
- Job-specific certifications (e.g., confined space entry, lockout/tagout)
- ICML, STLE, or equivalent lubrication technician credentials

Records should include expiration dates, issuing bodies, and renewal tracking. Safety training logs must be reviewed annually and updated for new hires or changes in regulations.

Regulatory Disclosures and Chemical Documentation

Suppliers are required to maintain up-to-date MSDS (SDS) for all products, with revision dates and GHS labeling conformity. Documents should be accessible on demand and included in initial qualification packages. Regulatory alignment must be validated against:

- REACH (EU)
- TSCA (USA)
- WHMIS (Canada)

Contractors handling bulk fluids, flushing chemicals, or used oil must submit documented procedures for:

- Spill response and containment
- Waste transport and disposal
- Secondary containment and chemical segregation

Quality Assurance Protocols

Lubricant manufacturers and blenders must provide evidence of:

- Batch quality control testing (viscosity, elemental content, particle count)
- Certificate of analysis (COA) availability on request
- Lot traceability across manufacturing and distribution

Field service providers should submit sample job checklists, route logs, and post-service inspection reports to confirm repeatable quality control. These documents must show adherence to best practices for contamination control, correct product application, and task completion.

Sustainability and Environmental Responsibility

Evaluate for environmental compliance and stewardship:

- ISO 14001 certification or equivalent policy documentation
- Use of re-refined base oils or biodegradable products where applicable
- Participation in lubricant recycling or recovery programs
- Low-emission, spill-resistant packaging and delivery systems

EVALUATION TOOLS AND DIGITAL RECORDKEEPING

Scoring rubrics should be created to ensure consistency across vendor types. Examples of categories are:

- Safety documentation (20 points)
- Regulatory and SDS compliance (20 points)
- Quality assurance (QA)/quality control (QC) traceability (20 points)
- Environmental policy and certifications (20 points)
- Technical capability and field experience (20 points)

Use digital vendor management platforms to centralize, filter, and retrieve all documentation during audit events. Include review time stamps, evaluator notes, and renewal cycles. Periodic requalification (annually or biannually) should be scheduled and logged.

Vendors maintaining ISO 9001 or ISO 14001 certifications simplify audit preparation and reduce administrative burden. Their quality management systems (QMSs) are typically aligned with the document control, continuous improvement, and corrective action expectations of industrial buyers.

BUILDING A MULTISUPPLIER RESILIENCE FRAMEWORK

Single-source dependency in lubrication supply chains exposes operations to substantial risk from disruptions due to transport, geopolitical shifts, or supplier-specific failures. A resilience framework involves qualifying multiple suppliers, establishing supply chain visibility, and creating substitution matrices for critical lubricants.

Strategies include:

➤ Dual-qualification of equivalent products from different vendors
➤ Cross-site supply coordination to leverage shared inventories
➤ Emergency stockpiles and alternative delivery routing
➤ Pre-vetted mobile service teams for contingency support

Reliability in lubrication supply chains demands mitigation of single-source risks through structured redundancy. A multisupplier resilience framework reduces operational exposure to disruptions caused by transportation delays, geopolitical instability, vendor insolvency, and material shortages.

Supplier Dual Qualification

Critical lubricants must be dual-qualified with performance-equivalent alternatives sourced from multiple approved vendors. Qualification includes:

➤ Matching viscosity grade, additive package, and base oil compatibility
➤ Cross-testing using ASTM or ISO performance methods (e.g., FZG, four-ball wear, RPVOT)
➤ Verification of seal, elastomer, and filtration compatibility
➤ OEM approvals for alternative products where required

Procurement must maintain current documentation of both primary and secondary options, including product codes, SDS, COAs, and shelf life.

Inventory Pooling and Cross-Site Coordination

Shared lubricant inventories across geographically distributed facilities reduce individual site risk. Establish:

- Centralized inventory databases with real-time visibility
- Transfer protocols and routing logistics
- Compatibility checks for dispensing equipment and storage containers across sites

High-turnover lubricants can be pooled and managed centrally, while site-specific SKUs remain localized but cross-referenced in the system.

Emergency Stockpiles and Logistics Contingency

Maintain critical lubricant stockpiles at strategic locations based on lead time and consumption rate. Define:

- Minimum on-hand thresholds based on run-rate and supply-delay estimates
- Periodic rotation to avoid expiry or degradation
- Secure warehousing and climate control for sensitive formulations

Establish alternative delivery providers or transport modes in case of disruption (e.g., rail fallback if truck routes are compromised). Routes and contacts must be documented and included in supply continuity plans.

Pre-Vetted Mobile Response Capability

Identify and pre-contract mobile lubrication service providers who are capable of emergency dispatch. These teams must:

- Be trained in your facility's specifications and documentation protocols.
- Carry interchangeable fittings and NSF-compliant products if needed.
- Have onboard filtration, flushing, and oil analysis tools.

Service-level expectations should be defined in advance, including response times, reporting format, and liability coverage.

Contractual Resilience Mechanisms

Supply contracts must include:

➤ **Force majeure clauses** with clear definitions of disruption triggers.
➤ **Minimum stockholding agreements** to ensure supplier buffer capacity.
➤ **Product change notification protocols** for additive reformulations, obsolescence, or discontinuation.
➤ **Dual-sourcing clauses** that mandate vendor support for alternative sourcing.

Performance bonds or financial penalties for critical delivery failures can be considered for high-risk assets.

Risk Modeling and Review Cadence

Run periodic risk assessments using supply disruption scenarios:

➤ Quantify impact in terms of MTTR, downtime cost, and safety implications.
➤ Rank SKUs by criticality, lead time, and availability of substitutes.
➤ Develop a substitution matrix with validated alternatives and usage guidelines.

Update plans annually or after any significant supply chain event. Review supplier solvency, geopolitical exposure, and transportation risk as part of enterprisewide risk management.

This framework embeds resilience by design, ensuring technical equivalence, operational continuity, and commercial readiness to handle disruptions without compromising lubrication performance or asset reliability.

SUPPLIER-INDUCED FAILURES AND CORRECTIVE ACTION PROTOCOLS

When supplier performance contributes to equipment failure or process deviation, clear corrective action protocols are essential. These must identify the root cause, quantify impact, and assign accountability.

Examples of supplier-induced failures include:

➤ Delivery of off-spec lubricant.
➤ Incompatible substitute product without approval.
➤ Improper field application or sampling by contractor.
➤ Late delivery causing lubrication starvation.

Corrective action should follow structured models such as 8D or root cause failure analysis (RCFA). Suppliers should be integrated into the investigation process and provide technical responses, including preventive strategies, such as reformulated product, retraining, or revised service protocols.

When equipment failures trace back to supplier actions, structured corrective protocols are necessary to protect operational integrity. Failures must be classified, investigated, and remediated with direct supplier involvement under formal accountability processes.

Failure Classifications

Examples of supplier-induced lubrication failures include:

➤ **Off-spec lubricant delivery.** Product fails viscosity, contamination, or additive spec upon receipt. Can lead to component scoring, foaming, or oxidation.
➤ **Unapproved product substitution.** Alternative lubricant supplied without cross-qualification, causing seal incompatibility, wear acceleration, or film breakdown.
➤ **Field service errors.** Contractor misapplication, overgreasing, cross-contamination, or improper sampling compromising root cause analysis.
➤ **Logistics failures.** Missed or delayed delivery windows leading to dry-run components, bypassed PMs, or forced substitution with nonconforming stock.

Corrective Action Framework

Corrective action must follow a structured, traceable format such as the 8D methodology or RCFA, initiated immediately upon event detection. Essential steps are:

1. **Immediate containment.** Isolate suspect lubricant, remove from active circulation, inspect at-risk equipment, and halt further contractor activities.
2. **Root cause identification:**
   - Test fluid samples (batch, in-system).
   - Review delivery documents, spec sheets, and service records.
   - Interview field personnel, and cross-verify procedures.
3. **Impact quantification:**
   - Equipment downtime
   - Maintenance hours
   - Component replacements
   - Lubricant loss or disposal costs
4. **Supplier engagement:**
   - Issue formal nonconformance report (NCR).
   - Require technical explanation, lab evidence, and proposed containment actions.
   - Demand immediate risk mitigation (e.g., revised process, batch quarantine, retraining).

Preventive Actions and Verification

Approved corrective actions may include:

- Reformulation or tighter product QC
- Respecification of delivery packaging or labeling
- Revised sampling procedures with dual verification
- Contractor retraining with competency assessments

Suppliers must submit documented countermeasures and undergo verification audits if recurrent events are identified.

Failure Tracking and Enforcement

All incidents are logged in a **shared quality tracking system**, linked to lubricant SKUs, delivery lot numbers, contractor IDs, and failure codes. Trend analysis supports:

- ➤ KPI tracking for supplier defect rate
- ➤ Escalation for repeated violations
- ➤ Performance scorecards for quarterly review

Penalty clauses may be triggered for high-severity failures, and continued supply is conditional on sustained corrective adherence.

This protocol ensures traceability, supplier accountability, and continuous improvement in lubrication supply chain performance.

INTEGRATING SUPPLIER DATA INTO YOUR LUBRICATION INTELLIGENCE SYSTEM

Supplier data integration with lubrication intelligence platforms enhances decision-making by consolidating real-time performance inputs, historical records, and compliance data into a single operational framework. This reduces manual reconciliation and strengthens failure prevention through traceable, condition-based insights.

Oil Analysis Integration

Automated ingestion of lab reports into CMMSs or EAM systems eliminates latency between sample testing and maintenance action. Reports must be standardized to XML, JSON, or CSV formats with structured fields including asset ID, sample point, test date, ISO code, elemental trends, and flags for water, oxidation, or fuel dilution. Systems must support API end points or secure file transfer for data push from supplier labs. Integration enables automatic alert generation, trending dashboards, and maintenance task creation based on defined thresholds.

SLA/KPI Performance Monitoring

Vendor logistics and service data should feed into dashboards tracking:

- Delivery timeliness by PO number
- Conformance with batch specifications
- Frequency of emergency response dispatches
- Oil analysis turnaround time

Deviation alerts tied to SLA contracts provide early visibility of performance risk. Data can drive supplier scorecards and inform contract renewal, escalation, or diversification decisions.

Lubricant Usage and Inventory Analytics

Integrating shipment records, consumption logs, and asset usage profiles enables:

- SKU-level burn rate calculations
- Forecasting for reorder and delivery scheduling
- Cross-site standardization of product usage
- Identification of excessive consumption or incompatible usage

Historical mapping of lubricant consumption against maintenance history improves root cause accuracy in equipment failure investigations.

Technical Documentation and Regulatory Updates

Suppliers must provide real-time updates to:

- Technical data sheets and SDS documents
- Regulatory notifications (e.g., REACH, GHS, NSF)
- Reformulation alerts and substitution advisories

These are indexed and linked to affected assets and lubricant SKUs, ensuring compliance and supporting automated risk assessments when formulations change.

Sustainability Tracking

Digital capture of supplier transport emissions, packaging reuse metrics, and lubricant reclamation volumes support Scope 3 emissions accounting and ISO 14001 initiatives. Sustainability KPIs include:

- Waste oil reduction
- Re-refined oil use percentage
- Delivery consolidation ratios

Integration supports ESG reporting and cost justification for eco-preferred product selection.

Data Standards and Connectivity Requirements

Suppliers must be required to support:

- Secure data formats (e.g., JSON over RESTful API)
- Compatibility with leading CMMS/EAM platforms (SAP, Maximo, Infor)
- Digital identifiers (batch codes, lab report keys, asset tags)

Audit trails are mandatory for all ingested data to validate traceability and meet quality assurance or regulatory demands.

By embedding supplier data into the lubrication intelligence layer, plants gain continuous operational context, reduce failure exposure, and enhance supplier accountability through performance-linked transparency.

CHAPTER 12

LUBRICATION PMS AND WORK ORDER MANAGEMENT

STRUCTURING PMS FOR LUBRICATION EFFECTIVENESS AND EFFICIENCY

P reventive maintenance tasks focused on lubrication must be designed not only to meet frequency guidelines but to ensure effectiveness in prolonging equipment life, minimizing downtime, and preventing failure. Structuring PMs begins with identifying critical assets, defining correct lubrication practices, and standardizing task instructions.

Key elements include:

➤ Lubrication method and volume
➤ Specified product (viscosity grade, additive type)
➤ Access and safety requirements
➤ Visual inspection criteria

Effectiveness is measured by lubricant film maintenance, contamination prevention, and minimization of human error. Efficiency, on the other hand, depends on route optimization, tool readiness, and technician training. PMs must be structured to balance thoroughness with time on task, supported by checklists and visual cues.

PM documentation should follow a controlled format, version tracking, and integration into centralized asset care plans. Consistency across PMs ensures reliability and comparability of task execution. Lubrication-specific PMs must be asset-specific, contamination-aware, and designed for repeatable execution. A properly structured PM includes unambiguous instructions, clearly defined lubrication parameters, and failure-preventive visual indicators. Frequency alone is not a reliability metric; quality of execution and environmental fit are.

Task Instruction Framework

Each PM must define:

- ➤ **Lubricant specification.** Product name, viscosity grade (e.g., ISO VG 150), NLGI grade for greases, additive package (AW, EP, H1), and OEM cross-approval
- ➤ **Volume and method.** Quantified delivery per point (cc, grams, shots), including the delivery mechanism (brush, nipple, spray, drip, or automated system)
- ➤ **Target point ID.** Asset hierarchy tag, lube point identifier, and physical location
- ➤ **Safety/accessibility requirements.** Lockout/tagout references, platform or lift use, confined space entry, or elevated access notes

Inspection Criteria

Visual inspection tasks must be embedded:

- ➤ Film presence and color
- ➤ Drip, leakage, or residue buildup
- ➤ Seal condition and lube pathway integrity
- ➤ Sight glass, desiccant breather, and in-line filter status

These must be binary (pass/fail or good/needs action) and recorded in checklists or digital forms. Photos or reference visuals improve repeatability.

Efficiency Controls

Route optimization tools must cluster PMs geographically and by system type to reduce travel and setup time. Job kitting must ensure that lubricants, tools (e.g., grease gun, meter, fittings), and PPE are ready. Preloading all PMs on mobile tablets or handheld devices eliminates paperwork lag.

Technician Qualification and Consistency

PMs must specify technician qualification requirements (e.g., ICML Level I), especially for assets with tight tolerances or food-grade lubricants. Documented training ensures execution consistency.

Version Control and Standardization

Lubrication PMs must be maintained under change control. Any updates to lube specifications, intervals, or equipment modifications trigger PM revision. Version numbers must appear on printed or digital work orders. Change logs should be audit-ready.

Alignment with Asset Criticality

Use RCM (reliability-centered maintenance) or FMEA output to prioritize PM detail. Critical assets require higher documentation granularity and closer inspection coupling. Noncritical or redundant assets may shift toward automated systems or reduced-touch PMs based on run-time behavior.

System Integration

PMs must reside in a CMMS with linkages to:

- Historical work orders
- Lubricant consumption logs
- Condition monitoring alerts (e.g., high particle count, viscosity shift)
- Manufacturer maintenance guidelines

Integrating PM execution with lubrication analytics enables failure prediction and task frequency adjustment. Missed PMs, delayed tasks, and repeated lube-related faults must trigger escalation workflows.

Work Order Feedback Loop

PMs must generate structured feedback opportunities:

- Consumption amount and deviation from baseline
- Lubricant condition at the point of use
- Observed contamination or abnormal wear
- Comments on accessibility or procedural deficiencies

These inputs must be closed loop and be reviewed by maintenance planning and reliability for continuous PM refinement.

Performance Metrics

Key lubrication PM metrics include:

> ➤ Compliance rate (% of PMs completed on time)
> ➤ Rework percentage from lube-related failures
> ➤ Mean time between replenishment (MTBR)
> ➤ Average lube usage per point over time
> ➤ PM execution time per asset class

These KPIs feed reliability reports and justify investment in automation, reskilling, or redesign. They also inform shift-level scorecards and department reviews.

Effective lubrication PMs are not calendar-driven templates but asset-integrated maintenance controls. Task design, technician execution, and data traceability determine program value, not task count.

PRECISION LUBRICATION ROUTES: FROM PAPER TO DIGITAL EXECUTION

Precision lubrication routes reduce waste, eliminate redundancy, and increase technician accountability. Transitioning from paper-based routes to digital platforms enables real-time tracking, automated updates, and integration with machine condition data.

Digitally executed routes often leverage:

> ➤ Mobile tablets with GIS or plant layout interfaces
> ➤ QR/barcode scanning at each lubrication point
> ➤ Route logic based on asset criticality and proximity
> ➤ Workflow prompts to reduce omissions or overapplication

These systems improve data accuracy, provide instant feedback, and support technician guidance. Moreover, route optimization algorithms can recalculate path efficiency as assets are added or modified.

Digitization also supports training by embedding SOPs, images, and video directly into route tasks. This significantly improves consistency, especially across multishift operations or temporary staffing. Precision lubrication routing improves time efficiency, lubricant control, and task accuracy. Manual systems rely on static paper checklists and memory-based sequencing, increasing the risk of point omission, misidentification, or incorrect lubrication volumes. Digital execution removes these constraints by guiding the technician through a validated route structure based on real-time data, asset layout, and lubrication history.

CORE DIGITAL ELEMENTS

Mobile tablets serve as route execution tools. Technicians receive task lists embedded with:

- ➤ GIS-based plant or site maps
- ➤ Equipment hierarchy and lube point metadata
- ➤ Route logic that optimizes proximity and criticality

Each lubrication point is identified using QR codes, barcodes, or RFID tags. Scanning confirms the correct location and logs time-stamped task verification. This prevents skipped points and eliminates manual data entry errors.

Real-Time Adjustments and Verification

Digital platforms allow for dynamic route optimization. As assets are reclassified, added, or removed, routes are auto-adjusted to maintain sequencing efficiency and workload balance. Sensor data or alerts can reprioritize points mid-route; for example, if an oil level sensor indicates low fill, the system triggers an unscheduled task.

Verification functions include:

- ➤ Mandatory completion prompts before proceeding
- ➤ Grease volume confirmation based on metering system or manual input
- ➤ Photo capture of completed points for audit purposes

Integrated SOPs and Multimedia

Each task can include embedded standard operating procedures, visual inspection guides, or application technique videos. This reduces technician variability, especially across shifts, and ensures uniform compliance with lubrication standards.

Feedback Capture and Analytics

Digital routes enable direct technician feedback at the point of work. Anomalies, such as inaccessible points, damaged fittings, or excessive contamination, can be logged immediately and routed to planners or reliability engineers.

Execution data feeds back into CMMS or EAM platforms, enabling:

➤ Route efficiency metrics (completion time, skipped points, delays)
➤ Lubricant consumption tracking
➤ Condition trends linked to lubrication quality

The benefits are:

➤ Improved accountability through scan-to-confirm systems
➤ Elimination of transcription errors and lost records
➤ Time reduction through automated path optimization
➤ Standardized application techniques with multimedia support
➤ Integration with asset condition and usage data for targeted scheduling

Digital route systems elevate lubrication tasks from routine rounds to traceable, condition-responsive maintenance activities.

CREATING TRIGGER-BASED LUBRICATION WORK ORDERS USING CONDITION DATA

Traditional lubrication PMs are often scheduled by calendar, regardless of actual machine condition. A more advanced strategy uses condition-based triggers such as temperature spikes, vibration changes, or contamination alerts to generate lubrication work orders dynamically.

Trigger-based work orders require integration of:

➤ Real-time sensor data (e.g., vibration, temperature, moisture)
➤ Oil analysis trends
➤ Alarm thresholds set by machine criticality and behavior

These work orders are more precise, allowing lubrication when it's truly needed, rather than as a routine placeholder. This method reduces overlubrication, cuts costs, and avoids labor redundancy.

Condition-based work orders also enable preventive diagnostics. For example, a sudden increase in bearing temperature may trigger not only a lubrication task but also an inspection for alignment or load imbalance. Trigger-based lubrication work orders shift the paradigm from interval-based PMs to event-driven precision tasks. Rather than relying on fixed schedules, these work orders are initiated when monitored parameters indicate that a lubrication event is warranted. This aligns with predictive maintenance strategies and supports asset-specific behavior profiles.

DATA INTEGRATION REQUIREMENTS

Effective implementation begins with integration between condition monitoring systems and the CMMS platform. Critical data sources include:

➤ **Vibration analysis.** An increase in acceleration or displacement near rolling elements can indicate boundary lubrication or loss of film.

➤ **Thermography or embedded RTDs** (resistance temperature detectors). A bearing temperature deviation from baseline triggers lubricant replenishment or investigation of root cause.

➤ **Moisture sensors and oil condition monitors.** The presence of water, oxidation, or particle load exceeding threshold levels prompts filtration, replenishment, or lubricant change.

➤ **Oil analysis reports.** Trends in viscosity shift, additive depletion, or wear-metal concentration flag degradation and initiate proactive tasks.

Thresholds must be asset-specific and based on historical performance, OEM limits, and operational criticality. Alarm conditions must be configured to avoid false positives and ensure that only actionable deviations generate work orders.

Dynamic Work Order Generation

Trigger-based work orders are generated automatically within the CMMS using preconfigured logic:

➤ Vibration rise > 20% from baseline → "Grease Point A, inspect for misalignment"

➤ ISO cleanliness code shift from 17/15/13 to 21/19/17 → "Polish fluid via FCU"

➤ Bearing temp > 180°F for > 30 minutes → "Lubricate and check housing integrity"

Tasks can include standard lubrication actions along with inspection steps to confirm root cause. Logic trees can differentiate between lubrication needs and mechanical faults (e.g., a lube task plus an alignment verification).

Benefits and Strategic Gains

➤ **Precision lubrication** reduces overlubrication risk, particularly in greased bearings that are sensitive to purge path obstruction.
➤ **Resource efficiency** is improved, as labor is only dispatched when needed, allowing redirection to higher-priority failures or inspections.
➤ **Lubricant usage optimization** lowers inventory demand and waste generation.
➤ **Preventive fault isolation** reduces cascading failures by combining lube action with adjacent diagnostics.

Implementation Considerations

➤ Sensor placement must ensure meaningful data capture, proximity to lube points, thermally representative zones, and sealed housings.
➤ Work order logic should be tested in sandbox environments to validate accuracy and response timing.
➤ Technician interfaces must differentiate trigger-based tasks from standard PMs for documentation and root cause capture.
➤ Historical data should be mined to refine threshold limits over time, using statistical process control or machine learning models if available.

Trigger-based lubrication elevates maintenance precision while reducing waste and redundancy. When configured correctly, it becomes a frontline defense against failure while conserving resources.

TASK INTERVALS VERSUS OPERATING CONTEXT: WHEN CALENDAR SCHEDULES FAIL

Lubrication PM intervals are commonly based on OEM guidelines or calendar time. However, real-world operating conditions often render these schedules inaccurate.

Factors like runtime, load variation, environmental conditions, and duty cycles all influence lubrication needs.

Context-driven intervals can be based on:

- Runtime hours or cycles
- Load severity ratings
- Temperature exposure profiles
- Historical failure data

CMMS and asset monitoring tools can be configured to adjust task intervals dynamically. Lubrication schedules for continuously operating equipment may differ from those of intermittently used machines, even if they are of the same make and model.

A contextual interval strategy improves resource allocation and reduces risk of lubrication starvation or waste. Fixed-interval lubrication PMs based solely on calendar schedules fail to account for actual equipment use and environmental severity. These rigid models often lead to underlubrication in high-demand assets or overlubrication in lightly loaded or intermittent duty cycles. Real-world variability in runtime, mechanical load, and ambient conditions directly influences lubricant degradation and consumption rates.

Context-Driven Interval Variables

1. **Runtime hours or cycles.** Use total operating hours or actuation counts to trigger lubrication tasks. Runtime-based triggers align lubricant delivery with wear-inducing motion, not elapsed calendar time. This is essential for assets operating on noncontinuous schedules, such as batch systems or seasonal equipment.
2. **Load severity ratings.** High dynamic loads accelerate lubricant shear and increase heat generation, requiring more frequent replenishment. Load multipliers, derived from torque sensors or process data, refine lubrication frequency. CMMS logic can incorporate severity factors to prioritize equipment under stress.
3. **Temperature exposure profiles.** Lubricants degrade faster at elevated temperatures. Thermally adjusted intervals account for operating heat rather than assuming ambient or design specs. Sensors or historical data can be analyzed to recalibrate PMs based on real-world thermal load.
4. **Historical failure data.** Recurring failure trends tied to lubrication timing, method, or volume offer valuable insights. Use RCFA output to adjust

intervals for similar asset classes or environments. Failures linked to lubricant starvation or contamination should prompt interval tightening or method revision.

SYSTEM CONFIGURATION

Modern CMMS platforms allow interval logic to be configured using conditional fields. For example:

- ➤ "Grease every 250 hours of runtime OR every 30 days, whichever occurs first."
- ➤ "Lubricate if torque > X for Y cumulative hours in 7-day window."
- ➤ "Adjust oil change frequency based on fluid life index derived from onboard sensors."

Fleet Variability

Assets of identical make and model may have divergent interval requirements based on usage context. Compressors in a high-humidity plant may need different lubrication strategies from identical units in a dry storage facility. PM standardization must be tempered with flexibility via context-aware task generation.

Operational Gains

Contextual task intervals reduce labor hours spent on unnecessary lubrication, extend lubricant life, and lower the risk of starvation-related failures. They support condition-based reliability programs and align maintenance frequency with actual risk. Properly configured, they transform PMs from generic routines into strategic interventions.

LUBRICATION PMS AS
A LEADING RELIABILITY INDICATOR

PM task adherence and effectiveness can serve as leading indicators of reliability health. Poorly executed or skipped lubrication PMs often precede mechanical failures, especially in bearings, gears, and hydraulic systems.

Metrics that reflect PM reliability value include:

- PM compliance rate (% on time, in full)
- Number of lubrication-related failures per month
- MTBLRF
- Oil analysis improvements post-PM

Monitoring these metrics enables root cause analysis of failures and supports the continuous improvement of lubrication practices. Trends can also reveal systemic training or procedural issues.

Lubrication PMs, when treated as data-rich activities rather than administrative obligations, provide valuable feedback loops for reliability programs. Lubrication PM execution provides direct insight into the reliability posture of rotating and hydraulic equipment. Missed or poorly performed lubrication tasks often precede critical asset failures. Consistent PM adherence, coupled with high-quality execution, can predict and prevent breakdowns in bearings, gearboxes, and hydraulic systems.

KEY RELIABILITY METRICS

1. **PM compliance rate (% on time, in full).** Tracks adherence to scheduled lubrication tasks, with "in full" defined as the correct method, volume, and product used. Drop-offs in compliance correlate with increased mechanical failure risk, especially in high-speed or high-load components.
2. **Lubrication-related failure count (monthly or rolling average).** Quantifies the number of equipment failures traced back to lubrication issues. This includes dry running, contamination ingress, improper product use, or overlubrication leading to seal failure. A rising trend signals systemic breakdowns in planning, training, or execution.
3. **Mean time between lubrication-related failures.** Isolates lubricant-specific failure intervals from total MTBF. A declining MTBLRF indicates that lubrication PMs are either inadequate in frequency or improperly executed. Trending this metric across asset classes helps identify weak points in the lubrication program.
4. **Post-PM oil analysis trends.** Tracks improvements (or deterioration) in fluid cleanliness, viscosity stability, additive retention, and wear particle count after lubrication PMs. The results of poor oil analysis post-PM suggest procedural failures, such as contamination during greasing or incorrect oil selection.

Analytical Use of PM Data

PM task data, especially when digitally captured, should be mined for deviations, delays, skipped entries, and inconsistencies. Systems such as CMMS or EAM can be configured to flag lubrication-related PMs completed late, without evidence of task performance, or without supporting inspection data.

Failure data from RCFA investigations should be linked back to the last lubrication PM execution to assess causality. This establishes a feedback loop where the reliability team can modify PM frequencies, improve route structures, or revise SOPs.

Organizational Impact

Lubrication PM effectiveness is often the earliest detectable sign of broader reliability discipline. Technicians who cut corners on greasing are likely to omit other PM tasks. Conversely, teams with high PM adherence and low lubrication-related failures typically demonstrate strong procedural discipline and technical knowledge.

Tracking lubrication PMs as leading indicators helps reliability leaders prioritize training, revise interval logic, and flag deteriorating practices before failures manifest. These metrics also support budget justification for automation upgrades or procedural reinforcements.

INTEGRATING LUBRICATION TASKS IN CMMS WORKFLOWS

A computerized maintenance management system is the backbone of modern maintenance planning. Integrating lubrication tasks into CMMS workflows enhances traceability, consistency, and response efficiency.

Integration best practices include:

- ➤ Creating task templates for each lubrication point or route
- ➤ Associating tasks with equipment hierarchies
- ➤ Including digital checklists, SOPs, and reference material
- ➤ Generating alerts for overdue or failed lubrication PMs

Technician feedback, lubricant consumption data, and sensor inputs can be logged directly in the CMMS, enabling comprehensive reporting. Tasks can also be linked to inventory levels and purchasing systems to trigger lubricant restocking or tool calibration. Such integration improves cross-functional communication among reliability, purchasing, safety, and operations.

A CMMS must treat lubrication not as a generic PM category but as a precision task with asset-specific execution requirements. Integration ensures traceable, standardized, and auditable lubrication activity.

Structured Task Templates

Each lubrication point requires a predefined task template specifying:

> ➤ Lubricant type and quantity
> ➤ Application method (manual, automated, brush, spray)
> ➤ Safety notes and PPE requirements
> ➤ Asset ID and location codes

Templates must support task cloning for route-based PMs and allow revisions as specifications change.

Hierarchical Equipment Association

Lubrication tasks should link to the equipment hierarchy. This allows visibility into lubricant demands per system, not just per asset. Criticality ratings can weight tasks during backlog prioritization and route scheduling.

Digital Documentation and Execution Aids

Each PM work order must embed:

> ➤ SOPs and OEM specifications
> ➤ Annotated diagrams or GIS references
> ➤ Visual pass/fail inspection checklists

Technicians access these via mobile devices or tablets to ensure procedural adherence and reduce knowledge gaps.

Exception Tracking and Escalation Triggers

The CMMS should flag:

- ➤ Incomplete lubrication PMs
- ➤ Missed tasks by technician or shift
- ➤ Overdue tasks beyond tolerance thresholds

ESCALATIONS ROUTE TO SUPERVISORS OR RELIABILITY ENGINEERS FOR INTERVENTION

Sensor and Field Input Integration

Where condition-based lubrication is deployed, sensor alerts (temperature, vibration, flow) auto-generate lubrication work orders. Lubricant usage, failure codes, and technician notes feed back into the asset history.

Inventory and Procurement Linkage

Lubrication tasks deplete tracked stock. CMMS should decrement lubricant inventory and generate reorder triggers. Integrated calibration tracking for torque tools, grease guns, or transfer systems ensures readiness for precise delivery.

Functional Outcomes

This integration drives:

- ➤ Cross-team alignment (reliability, maintenance, stores)
- ➤ Transparent lubricant consumption trends
- ➤ Root cause correlation of failures to lubrication history
- ➤ Audit-ready records for safety, ISO, or food-grade compliance

Proper CMMS integration enables lubrication PMs to function as a measurable, controllable input to asset reliability, not just a maintenance routine.

CLOSING THE LOOP: VERIFICATION AND AUDIT OF LUBRICATION WORK ORDERS

To ensure that lubrication PMs are completed accurately and deliver their intended value, organizations must implement verification and audit practices. This process involves:

- Field verification of lubrication points and volume
- Supervisor review of completed work orders
- Random audits using condition monitoring tools
- Feedback loops for continuous task refinement

Audit protocols may include photo documentation, lubricant labeling checks, and post-task asset condition measurements (e.g., temperature, ultrasound).

CMMS platforms can be configured to require completion verification steps, such as barcode scans or sensor readings. This enforces accountability and prevents pencil whipping. Verification and audit also support regulatory compliance, warranty validation, and certification programs like ISO 55000. They build confidence in the lubrication process and strengthen the reliability culture. Verification of lubrication tasks ensures that execution matches intent. Without field-level validation and structured audit procedures, lubrication PMs degrade into administrative compliance rather than reliability drivers.

FIELD VERIFICATION

Inspect lubrication points for residue, color, consistency, and cap seal integrity. Confirm that the lubricant applied matches the specified product (viscosity, additive profile). Grease fitting or port checks should ensure that access was possible and not bypassed.

Volume delivery must be validated against OEM specs or baseline logs. Grease output should match known stroke or cartridge displacement per task. For oil systems, dipstick or sight glass levels should reflect change and topping expectations.

SUPERVISOR REVIEW

Work order closure should include supervisor sign-off. Randomized post-task spot checks, especially on critical assets, help catch false completion or poor technique. Review includes visual confirmation, lubricant labeling match, and equipment access notes.

Random Audits Using Condition Monitoring

Deploy ultrasound, thermography, or vibration analysis to confirm the lubrication effect. Compare post-lube readings against pre-task baselines. Use acoustic emission to identify missed or underlubricated bearings. Conduct particle count or water content testing where applicable.

COMPLETION VERIFICATION MECHANISMS

Configure CMMS or mobile route tools to require:

- ➤ QR or barcode scan at point of application
- ➤ Confirmation of lubricant used via dropdown or photo
- ➤ Metered grease gun reading or pump cycle log
- ➤ Time-stamped digital signature

Require entry of field readings (e.g., temperature, visual inspection notes) before the work order is eligible for closure.

Documentation and Traceability

Photo evidence (e.g., grease purge at bearing, oil sight level) is stored with the work order. Labeling checks confirm correct product used, especially for NSF H1 or synthetic lubricants. Include tag photos or before or after gauges where feasible.

Audit Program Framework

Schedule routine lubrication PM audits per asset class or technician. Create scorecards to evaluate task execution quality. Assign audit findings to feedback loops, either corrective (technician retraining) or preventive (task rewrite, SOP update).

Compliance and Certification Alignment

Audit records support ISO 55000, food safety (FSMA, ISO 21469), and OEM warranty claims. Integrated verification systems build defensibility in external audits and internal reviews. When lubrication work is traceable, measurable, and auditable, it becomes a lever for asset longevity, not a liability.

CHAPTER 13

LUBRICANT HANDLING, STORAGE, CONSUMPTION, AND CONSERVATION

LUBRICANT LIFE CYCLE FROM DRUM TO DRAIN

The lubricant life cycle encompasses every phase from procurement to final disposal or recycling. Mismanagement at any point can compromise product integrity, damage assets, and inflate operational costs. Understanding and controlling the complete lubricant life cycle ensures consistent performance, reduces waste, and supports sustainability goals.

The cycle begins with product selection and supplier validation, followed by storage, internal transport, transfer, application, condition monitoring, and, finally, recovery or disposal. Each phase involves handling practices, material compatibility checks, contamination control, and documentation.

Poor handling or improper storage can introduce moisture, particulates, or thermal degradation before the lubricant even enters a machine. In-service practices, like topping off with incompatible fluids or using contaminated dispensing equipment, accelerate degradation and cause premature failures.

Post-use, lubricants must be drained, tested for residual value, and disposed of in accordance with regulatory frameworks. Re-refining or energy recovery can offset disposal costs, provided proper collection practices are in place.

DESIGNING TIERED STORAGE SYSTEMS FOR PRODUCT INTEGRITY

Proper lubricant storage minimizes degradation and contamination risks, extending the shelf life and preserving performance. Tiered storage systems segregated by product type, usage frequency, and environmental sensitivity support inventory accuracy, fluid cleanliness, and safe handling.

Key design considerations include:

- Dedicated, clearly labeled storage bays for each lubricant
- Color-coded containers and dispensing tools
- Secondary containment to manage spills
- Controlled temperature and humidity environments for sensitive fluids
- Fire protection and chemical compatibility safeguards

Bulk storage should include filtered breathers, bottom-drain tanks, and circulation systems to prevent stratification. Intermediate storage (e.g., transfer containers) must be sealed, durable, and cleaned regularly. Avoid storing lubricants near chemicals, solvents, or reactive substances.

Tiered storage aligns product rotation with FIFO (first in, first out) inventory practices, minimizing expiration losses and maintaining traceability.

HANDLING ERRORS THAT COST MILLIONS: CASE-BASED INSIGHTS

Improper handling is one of the most preventable causes of lubricant failure, often resulting in catastrophic equipment damage. Several case studies illustrate the high cost of mishandling:

Cross-contamination: A food processor experienced systemwide contamination due to accidental mixing of mineral and synthetic oils in a shared pump, leading to $2.5 million in downtime and cleanup.

Ingress during transfer: An offshore platform failed to cap a transfer hose between uses. Condensed salt air introduced moisture into a gearbox, ultimately seizing the drive.

Thermal degradation: Lubricants stored near a furnace lost viscosity and additive function. When applied, oil starvation led to bearing failure in a critical turbine.

Each case underscores the importance of sealed transfer equipment, temperature-controlled storage, and cross-contamination prevention through rigid procedural enforcement.

Training, signage, and mechanical safeguards (e.g., key-coded connectors) significantly reduce these risks.

CLOSED-LOOP TRANSFER SYSTEMS: BEST PRACTICES AND PITFALLS

Closed-loop systems eliminate exposure to air, moisture, and particulate contamination during lubricant transfer. A properly executed closed-loop approach maintains product purity from bulk storage to application.

Best practices include:

➤ Quick-connect, sealed fittings to prevent spillage and ingress
➤ Dedicated, color-coded containers for each lubricant
➤ In-line filtration during transfer (typically 3 to 5 micron filters)
➤ Gravity-fed or pump-assisted systems designed for fluid compatibility

Key pitfalls to avoid include:

➤ Using generic or multiuse containers without cleaning between products
➤ Neglecting pressure equalization, leading to container collapse or vacuum lock
➤ Overfilling or underfilling due to lack of calibrated transfer controls

Maintenance of transfer equipment is crucial. Hoses, nozzles, and couplings should be inspected for wear, cleaned regularly, and replaced on a schedule.

A robust closed-loop system not only protects the lubricant but serves as a visual and procedural reinforcement of contamination control culture.

CONSUMPTION BENCHMARKING: ESTABLISHING BASELINES AND DETECTING ANOMALIES

Tracking lubricant consumption is critical to identifying usage trends, operational anomalies, and potential sources of waste or leakage. Benchmarking begins by categorizing usage by machine, fluid type, and time interval.

Establish a baseline:

➤ Quantify expected usage rates by asset and operating conditions.
➤ Normalize data for run hours, duty cycle, and environmental variables.
➤ Compare against OEM guidelines and historical records.

Deviation from the baseline can indicate:

➤ Overlubrication or underlubrication
➤ External leaks or internal losses (e.g., seal degradation)
➤ Operator error or improper application methods
➤ Lubricant degradation requiring more frequent replacement

Benchmark data also supports cost forecasting, inventory planning, and supplier evaluation. Use of digital tracking tools ensures accurate data collection and allows for visual dashboards that reveal patterns. Lubricant consumption benchmarking begins by segmenting data across individual assets, lubricant types, and operational time frames. Each category must be tied to specific variables, such as run hours, load conditions, and ambient environment, to produce normalized usage metrics.

Baselines are established through direct measurement and historical data analysis. Expected usage is calculated by correlating application frequency and lubricant volume to OEM specifications, adjusted for real-world operating profiles. Where OEM benchmarks are unavailable or unrealistic, site-specific data from comparable assets under similar conditions should be used.

Unexpected increases in consumption often indicate overlubrication, technician error, or automatic system miscalibration. Sudden drops may suggest plugged lines, starved components, or skipped tasks. Losses due to external leaks—typically visible near seals, hoses, or couplings—and internal bypass due to worn components are also common causes of variance.

Degradation-driven overconsumption occurs when oxidation, contamination, or additive depletion forces premature lubricant replacement. In systems with no real-time monitoring, only tight benchmarking reveals these trends.

Consumption data should be integrated into digital platforms for automated collection and analysis. CMMS-linked dashboards enable threshold-based alerts and visualization of anomalies. Cross-asset comparisons can identify underperforming systems or incorrect practices by shift or technician.

Benchmarking supports predictive maintenance, inventory management, and supplier accountability. Trends over time validate improvements or expose systemic gaps in lubrication strategy.

CONSERVATION STRATEGIES IN HIGH-USE FACILITIES

Facilities with high lubricant throughput, such as power plants, steel mills, and manufacturing hubs, must employ conservation strategies to control costs and environmental impact.

Key conservation techniques include:

- Extended drain intervals based on oil analysis and condition monitoring
- Use of high-performance, longer-life lubricants
- On-site filtration and reconditioning of used lubricants
- Predictive maintenance practices to reduce avoidable consumption
- Minimization of flushing cycles through targeted clean-in-place solutions

Inventory control also plays a role. Centralized dispensing, locked storage, and technician authorization systems reduce unauthorized usage or waste. SOPs should include volume guidelines for each lubrication point to prevent overapplication.

Conservation must be framed as a reliability and cost initiative, not simply environmental compliance. Metrics should include cost per hour of lubrication, consumption per unit produced, and environmental impact factors (e.g., carbon or water footprint). High-throughput operations require precision strategies to manage lubricant use efficiently. In plants like steel mills, power generation facilities, and high-volume manufacturing, uncontrolled lubrication consumption directly impacts operating cost, waste generation, and environmental liabilities.

Start by extending drain intervals through oil analysis. Condition monitoring must confirm lubricant viability through viscosity, oxidation, TAN/TBN, water content, and wear metals. Deviation triggers replacement, not calendar schedules.

High-performance lubricants with thermal stability, oxidation resistance, and robust additive retention reduce top-off frequency and changeouts. Selection should be based on proven field longevity, not just lab test results.

Deploy on-site fluid reconditioning. Kidney loop filtration units with water removal, varnish mitigation, and particle count reduction can restore lubricants to within service limits. This minimizes waste disposal and new lubricant demand.

Predictive maintenance, integrating vibration, temperature, and acoustic emission data reduce secondary consumption caused by abnormal wear, overheating, or overcompensation.

Clean-in-place solutions reduce the need for high-volume flushing. Fluid-compatible detergents or temporary side-stream filtration cleans internal surfaces without complete fluid turnover.

Inventory control is nonnegotiable. Dispensing should be centralized and metered. Locked cabinets, RFID tagging, and control of technician access prevent unauthorized usage. CMMS-linked inventory software can tie usage to specific tasks and trigger reorder points based on actual consumption, not estimates.

Every lubrication SOP must include volume guidance. Eliminate the guesswork that leads to chronic overlubrication.

Track conservation effectiveness with metrics: liters per operating hour, volume per production unit, and lubricant cost per asset. Environmental indicators such as CO_2-equivalent emissions from production and disposal also support environmental, social, and governance (ESG) compliance frameworks.

DIGITAL TRACKING OF INVENTORY AND USAGE: FROM BARCODE TO BLOCKCHAIN

Digital systems provide transparency, traceability, and analytical power in managing lubricants. Barcode scanning, RFID tagging, and blockchain platforms enable real-time tracking of product movement, usage, and condition.

Barcode and QR systems allow:

➤ Real-time inventory updates
➤ Task validation and technician accountability
➤ Automated reordering and stock alerts

RFID adds benefits for:

➤ Non-line-of-sight identification
➤ Long-range reading and bulk asset tracking
➤ Integration with mobile dispensing units and condition sensors

Blockchain, though emerging, introduces tamper-proof transaction logs. In lubrication management, it can track batch integrity, supplier chain of custody, fluid condition history, and regulatory compliance.

Integrating these tools with CMMS, ERP, and EAM platforms delivers synchronized planning, streamlined procurement, and actionable intelligence for reliability professionals. Digital traceability also simplifies audits, reduces administrative overhead, and supports data-driven decision-making. Effective lubrication management requires real-time traceability from receipt to application. Digital systems replace manual logs and guesswork with precise, verifiable records. Barcode and QR code systems allow each lubricant container or dispensing point to be tagged, scanned, and logged at every transaction.

Barcode scanning enables real-time updates of stock levels and usage per work order. Technicians can validate lubrication tasks by scanning points, ensuring correct product use and volume. Integration with CMMS platforms ensures that usage data populates automatically into work orders, streamlining task closeout and replenishment.

RFID technology extends functionality beyond line of sight. Tags embedded in containers or dispensing tools can be read remotely, allowing tracking of mobile lubricant carts and field inventory. RFID-enabled systems also support condition tracking by pairing with temperature, vibration, or moisture sensors at the point of use.

Blockchain introduces immutable recordkeeping. Each transaction—procurement, storage, dispatch, or application—can be logged in a distributed ledger. This ensures batch integrity, verifies chain of custody, and supports traceability for contamination or warranty investigations. Blockchain can document lubricant age, fluid condition history, and alignment with regulatory or customer-specific compliance mandates.

Digital inventory tracking systems should interface with CMMS, ERP, and EAM platforms. This integration synchronizes planning, triggers auto-reorders, and correlates usage with maintenance activities. Dashboards can highlight anomalies, such as excess usage on a single asset or frequent reorders for a specific product, enabling proactive corrective actions.

These systems simplify auditing, reduce clerical error, and support continuous improvement by turning every fluid movement into actionable data.

CHAPTER 14

NEW MACHINERY SPECIFICATIONS AND COMMISSIONING

LUBRICATION REQUIREMENTS IN MACHINERY SPECIFICATION SHEETS

Lubrication specifications must be embedded early in machinery procurement documents to ensure optimal long-term performance. Machinery specification sheets should define lubrication requirements for each component, including lubricant type, viscosity grade, volume, intervals, and environmental conditions.

Effective specification includes:

➤ ISO viscosity grade or SAE equivalents
➤ Lubricant compatibility with seals, coatings, and materials
➤ Temperature and load-based performance requirements
➤ Minimum and maximum contamination thresholds

Critical systems such as gearboxes, hydraulics, and bearings must be clearly described with all lubrication touchpoints noted. Including lubrication parameters in specification sheets ensures alignment across engineering, procurement, operations, and maintenance. Lubrication parameters must be integrated into machinery specification sheets during procurement to prevent misapplication, underperformance, and warranty issues. Specifications must be developed per component, not generically, and reflect actual site operating conditions.

Each lubrication point should be listed with the following:

➤ **Lubricant type.** Mineral, synthetic, semi-synthetic; NSF H1 for food grade where applicable.

➤ **Viscosity grade.** ISO VG or SAE grade specified per OEM or load/speed calculations.

➤ **Additive requirements.** EP, AW, anti-foam, demulsifiers, or oxidation inhibitors, depending on duty.

➤ **Volume requirements.** Initial fill and normal operating level defined in mL or liters.

➤ **Lubrication interval.** Defined by time, run hours, or condition-based trigger.

➤ **Ambient and operating conditions.** Temperature range, humidity exposure, ingress risk.

➤ **Compatibility considerations.** Seal material, coating interactions, and elastomer tolerance.

➤ **Contamination limits.** ISO 4406 cleanliness targets, moisture content thresholds (ppm), or NAS/SAE codes.

System types requiring detailed lubrication criteria include:

➤ **Gear drives.** Include gear type, mesh frequency, and loading class to justify lubricant selection.

➤ **Bearings.** Rolling element or journal, load zone geometry, and orientation affect grease/thickener choice.

➤ **Hydraulics.** Detail fluid viscosity, foaming tendencies, and filtration needs.

➤ **Chains, guides, slides.** Specify spray or mist lubrication parameters if automated delivery is included.

All lubrication-related hardware, such as breathers, reservoirs, filters, drains, sight glasses, and sampling ports, should be included in the BOM and layout drawings. Equipment specifications must require suppliers to list approved lubricants, relubrication instructions, and service access provisions.

Cross-functional alignment among engineering, maintenance, and reliability teams during specification reviews reduces life-cycle friction and prevents procurement of machinery with inadequate or inaccessible lubrication infrastructure.

WRITING LUBRICATION CLAUSES FOR PROCUREMENT CONTRACTS

Lubrication-related clauses in procurement contracts protect the buyer from ambiguity and ensure that vendors deliver systems aligned with life-cycle reliability goals. These clauses should mandate specific deliverables and verifications.

Typical clauses include:

- Lubrication product details: brand, formulation, and approvals
- Requirement for clean fluid delivery (ISO 4406 targets)
- Factory fill documentation and sample results
- Lubrication point maps and service intervals
- Requirement for training and OEM support during commissioning

Clauses should also address warranty conditions related to lubricant compatibility and contamination and require that any lubricant changes during manufacture be disclosed.

Including these terms in the contract elevates lubrication from an afterthought to a core specification, giving maintenance teams a stronger voice in capital equipment decisions. Lubrication clauses embedded in procurement contracts provide enforceable standards for equipment suppliers and reduce life-cycle risk. These clauses ensure that lubrication performance, cleanliness, and support are built into the equipment from delivery through operation.

Contract language should include:

- **Lubricant specification requirements.** Mandate OEM-recommended or preapproved lubricant brand, formulation, and viscosity grade. Require confirmation of compatibility with seals, coatings, and system materials.
- **Factory fill verification.** Require submission of initial fill documentation, including batch numbers, volume per system, and sample oil analysis showing ISO 4406 cleanliness codes, water content (ppm), and viscosity index.
- **Cleanliness standards.** Define minimum cleanliness levels for delivered systems (e.g., ISO 17/15/12 or better for hydraulics). Include post-transport contamination checks prior to start-up.
- **Lubrication map submission.** Vendors must provide a lubrication schematic or point map detailing each lubrication location, volume, frequency, and product. Format should align with end user's CMMS for integration.

- ➤ **Service interval documentation.** Require documented relubrication schedules, compatible automated delivery options, and maintenance instructions tailored to actual plant operating conditions.
- ➤ **Disclosure of changes.** Mandate vendor disclosure of any lubricant substitutions during manufacturing, shipment, or installation. Require prior approval for any deviation from original factory fill.
- ➤ **Training and support.** Require OEM-provided training for technicians on lubrication systems at the time of commissioning. Include support obligations for calibration of automated systems or verification of delivery rates.
- ➤ **Warranty provisions.** Define warranty invalidation criteria related to off-spec lubrication, contamination ingress, or unapproved fluid use. Require written guidance for lubricant storage and handling to maintain compliance.

These clauses ensure that the procurement process includes reliability engineering input and creates a verifiable lubrication framework from day 1.

The following is a detailed example of lubrication clauses for procurement contracts, formatted as language suitable for inclusion in a capital equipment or systems purchase agreement.

Section X: Lubrication System Specifications and Compliance

X.1 Lubricant Specification Requirements

The Supplier shall use only OEM-recommended or Buyer-approved lubricants for all equipment provided under this contract. This includes brand, formulation, base oil type, viscosity grade (ISO VG or NLGI rating), and additive compatibility. The Supplier must confirm compatibility with seals, elastomers, coatings, and materials used in the supplied systems. No lubricant substitution is permitted without prior written consent from the Buyer.

X.2 Factory Fill Verification

The Supplier shall provide documentation of the initial factory lubricant fill, including:

- Lubricant name, formulation, and supplier
- Quantity used per system/component
- Batch or lot numbers
- Certificate of analysis showing:
  - ISO 4406 cleanliness rating
  - Water content (ppm)
  - Kinematic viscosity at 40°C and 100°C
  - Total Acid Number (TAN) or Total Base Number (TBN), as applicable

Samples of the factory fill lubricant shall be retained and made available to the Buyer upon request for verification testing.

X.3 Cleanliness Standards

All lubrication systems, reservoirs, and piping must meet or exceed ISO cleanliness code 17/15/12 prior to shipment. The Supplier shall flush, filter, and validate system cleanliness before factory acceptance. A post-transport sample must be taken and analyzed before start-up. Documentation shall be submitted to the Buyer with final delivery.

X.4 Lubrication Map and Point Identification

The Supplier shall provide a comprehensive lubrication point map, including:

- Component ID and location
- Lubricant type per point
- Volume per application
- Relubrication frequency
- Delivery method (manual, automated, circulating)

This information must be submitted in a format compatible with the Buyer's CMMS and asset database, including tagging aligned with the Buyer's standards.

X.5 Maintenance and Service Documentation

The Supplier shall provide lubrication service interval schedules tailored to operating load, temperature, and duty cycle, including:

- ➤ Recommended lubrication interval tables
- ➤ Compatible automatic lubrication options
- ➤ Filter service and replacement intervals
- ➤ Start-up, shutdown, and seasonal lubrication guidance

Instructions must reflect actual plant conditions, including temperature range, humidity, vibration, and exposure to contaminants.

X.6 Lubricant Change Disclosure

Any lubricant change made after design approval—including during factory assembly, storage, transport, or installation—must be disclosed in writing. The Buyer reserves the right to test substitute lubricants and may reject non-compliant fluids. No lubricant substitutions shall be made without written approval.

X.7 Training and Technical Support

At the time of commissioning, the Supplier shall provide hands-on training for the Buyer's maintenance and reliability staff, covering:

- ➤ Lubrication system architecture and function
- ➤ Lubricant storage, handling, and dispensing
- ➤ Inspection and verification procedures
- ➤ Fault detection and response protocols

The Supplier shall also support calibration of any automated lubrication components and provide on-site assistance in validating delivery volumes and cycles.

X.8 Warranty Provisions Related to Lubrication

Use of unapproved lubricants, failure to maintain lubrication cleanliness, or deviation from recommended lubrication intervals shall be grounds for warranty voidance. The Supplier must provide written procedures for lubricant storage, handling, and contamination control to maintain warranty eligibility.

FACTORY FILL STANDARDS AND FIELD-READY ADJUSTMENTS

Factory fills are often optimized for transport or short-term operation, not long-term service. Variability in fill procedures, cleanliness, or additive compatibility can undermine performance if not validated upon delivery.

Common issues with factory fills are:

➤ Residual machining debris and coolant contamination
➤ Inconsistent additive concentrations
➤ Incorrect volume or overfilling

Field-ready adjustments should include:

➤ Sampling and analysis of factory fill fluids on arrival
➤ Full drain and flush where contamination is present
➤ Refill with site-approved lubricants meeting cleanliness standards

Facilities should coordinate with OEMs to verify factory fill practices and establish whether initial lubricants meet local standards and application needs. Fluid compatibility and cleanliness targets should be confirmed before commissioning. Factory fills are typically optimized for equipment testing, preservation, or transport, not long-term operational reliability. Assumptions about quality, cleanliness, or compatibility often lead to early wear, fluid degradation, or warranty disputes if the fill is not verified before start-up.

Common deviations in factory fills include:

- **Machining residue.** Metal fines, sealant particles, and residual cutting fluids are frequently introduced during assembly and not fully removed prior to shipment. Without filtration or flushing, this debris accelerates wear.
- **Additive variability.** Fill batches may lack consistent additive concentrations or may include short-life break-in formulations unsuitable for full-duty operation. Overbased or underdosed additives can destabilize film strength or promote varnish.
- **Volume errors.** Overfilling or underfilling is not uncommon, especially when volume specifications are estimated or partially filled by multiple vendors. Overfilled reservoirs can cause foaming and aeration; underfilled systems risk starvation and cavitation.

Field acceptance procedures should include:

- **Initial sampling.** Collect representative samples of factory fill fluids upon arrival. Analyze for ISO particle count, water content, additive levels, and base oil identity.
- **Contamination response.** If contamination is found, perform a full drain, flush, and filter loop if necessary. Document findings, and notify OEM for warranty protection.
- **Approved refill.** Replace initial fill with site-approved lubricants that meet cleanliness standards and are compatible with facility-wide consolidation plans. Ensure that fluid is verified for material compatibility and has supporting lab certification.
- **OEM coordination.** Request detailed factory fill specs including lubricant brand, batch ID, additive package, and cleanliness level. Confirm whether the fluid is intended for extended service or needs to be replaced after run-in.

Field validation of factory fill eliminates latent risks that may not be immediately apparent during commissioning. It also aligns new equipment with plantwide lubrication standards and asset management protocols.

The following is an example of a contract clause section titled "Factory Fill Standards and Field-Ready Lubrication Adjustments" suitable for inclusion in a capital equipment or lubrication-related procurement agreement. This ensures lubrication integrity from delivery through commissioning and aligns new equipment with plantwide reliability and fluid standards.

Section Y: Factory Fill Standards and Field-Ready Lubrication Adjustments

Y.1 Factory Fill Lubricant Requirements

The Supplier shall ensure that all equipment is shipped with a complete initial lubricant fill that meets the following minimum requirements:

- OEM-approved or Buyer-specified lubricant brand, formulation, and viscosity
- Documentation confirming batch number, additive package, and intended service duration (e.g., break-in versus extended service)
- ISO cleanliness code $\leq$ 19/17/14 for gear oils and $\leq$ 17/15/12 for hydraulic and circulating oils
- Certificate of analysis from the fill batch showing:
  - Kinematic viscosity at 40°C and 100°C
  - Water content (ppm)
  - Additive content or elemental fingerprint (e.g., Ca, Zn, P, Mg, B)
  - TAN/TBN, as applicable

The Supplier shall confirm that all reservoirs are filled to specification volume. Overfilling or underfilling shall constitute nonconformance, and correction shall be the Supplier's responsibility.

Y.2 Contamination and Residue Prevention

The Supplier shall ensure that lubrication systems are free of:

- Residual machining debris, including metal shavings, grinding residue, and sealant particles
- Cutting fluid or assembly compound contamination
- Packaging lubricants or temporary corrosion preventives unless explicitly approved for in-service use

The Supplier must flush systems post-assembly if machining, welding, or sealing operations pose a contamination risk. Documentation of flushing (flush volume, filter used, and sample data) must be submitted upon request.

Y.3 Field Sampling and Initial Inspection

Upon delivery, the Buyer reserves the right to collect and analyze lubricant samples from each system prior to commissioning. Sample analysis will include:

- ISO 4406 particle count
- Karl Fischer water content
- FTIR or elemental analysis for additive verification
- Visual inspection for debris, emulsion, or phase separation

If contaminants or inconsistencies are found, the Supplier shall, at their expense:

- Authorize a complete system drain and flush
- Supply and install OEM-approved filters for flushing and final service
- Provide refill lubricants approved by the Buyer, compliant with site fluid specifications

Y.4 Refill and Commissioning Compatibility

Refill lubricants shall:

- Match or exceed factory fill specification
- Be confirmed to be compatible with system seals, gaskets, coatings, and elastomers
- Be certified by the Supplier as acceptable for warranty continuity
- Align with the Buyer's lubrication consolidation plan, fluid shelf-life controls, and inventory compatibility

The Supplier must coordinate with the Buyer to determine if the factory fill is break-in, transport-preservative, or long-term service fluid. If factory fill is not suitable for extended operation, it must be replaced before start-up.

Y.5 Documentation and OEM Declaration

The Supplier shall provide a Factory Fill Declaration including:

- Lubricant type, brand, and formulation
- Volume per system and system total
- Batch number(s) and date of fill
- Intended service life or scheduled changeout interval
- Statement of fluid compatibility with system components
- Cleanliness targets achieved at time of shipment

This declaration shall be submitted with the final shipment or as part of the pre-commissioning documentation package.

Y.6 Warranty and Dispute Resolution

If the factory fill is determined to have contributed to equipment wear, start-up failure, or early lubricant degradation, the Supplier shall:

- Cover all associated corrective actions including labor, downtime, and materials.
- Provide replacement parts or components affected by the fill.
- Accept lubricant change documentation and lab reports as evidence of nonconformance.

The Buyer reserves the right to reject any system where factory fill has not been validated or fails to meet agreed cleanliness or compatibility standards.

SOP EXAMPLE

The following is an example of a standard operating procedure for pre-start-up validation verifying OEM lubrication parameters, suitable for maintenance teams, commissioning engineers, or reliability professionals.

Standard Operating Procedure (SOP)

Title: Verification of OEM Lubrication Parameters—Pre-Start-Up
Document ID: LUBE-SOP-001
Revision: 1.0
Effective Date: [Insert date]
Prepared By: [Insert name/team]
Approved By: [Insert approval authority]

1. **Purpose**
   To verify that all lubrication parameters on OEM-supplied equipment meet operational and reliability requirements prior to start-up. This procedure ensures that correct lubricant type, volume, cleanliness, and component compatibility are established to prevent early failures and warranty issues.

2. **Scope**
   This SOP applies to all capital equipment delivered with prefilled lubrication systems, including gearboxes, hydraulic units, bearing housings, and centralized lubrication systems. It is required before initial energization or operation.

3. **Responsibilities**
   - **Reliability Engineer.** Oversees validation; ensures that documentation is filed in the CMMS.
   - **Maintenance Technician.** Performs physical checks and sampling.
   - **Lubrication Specialist.** Reviews compatibility and test results and prescribes corrective actions.
   - **OEM Representative (if required).** Provides spec sheets and clarifications.

4. **Materials and Equipment Needed**
   - OEM lubrication documentation (manuals, spec sheets)
   - Oil analysis kit (ISO 4406 particle count, Karl Fischer moisture, viscosity)
   - Fluid sampling pump and bottles
   - Sight glass or dipstick (as applicable)
   - Torque tools, flashlight, inspection mirror
   - Documentation checklist (see Appendix A)

5. **Procedure Steps**

5.1 Lubricant Specification Cross-Check

➤ Obtain OEM lubricant specification.

➤ Verify product brand, viscosity grade, and specification code (e.g., DIN, ISO, AGMA).

➤ Compare on-site lubricant (via label or analysis) to OEM recommendation.

➤ Confirm additive compatibility with metals and coatings (e.g., yellow metals, DLC coatings).

➤ *Document findings (Pass/Fail/Notes).*

5.2 Fill Volume Confirmation

➤ Inspect sight glass, dipstick, or reservoir marking.

➤ Confirm that volume matches OEM fill chart.

➤ Record measured level, and verify against start-up tolerances.

➤ Identify and document any overfill/underfill conditions.

➤ *Correct discrepancies before proceeding.*

5.3 Breather and Access Hardware Inspection

➤ Confirm that breather type (standard versus desiccant) matches environmental exposure.

➤ Check breather installation, seating, and cleanliness.

➤ Inspect fill ports, sample valves, and drain locations for:
  • Accessibility
  • Labeling
  • Cap tightness

➤ Flag missing or incorrect components for correction.

5.4 Fluid Cleanliness Validation

➤ Take representative lubricant sample from the lowest point or designated port.

➤ Analyze for:
  • **Particle count** (ISO 4406): Target ≤ 19/17/14 unless otherwise specified.
  • **Water content** (ppm): Target ≤ 500 ppm for mineral oils unless noted.
  • **Viscosity** and **base number** (as needed for compatibility).

➤ If out of spec:
  • Initiate a complete drain, flush, and refill procedure.
  • Notify OEM of nonconformance.

5.5 Filtration and Seal Element Verification

➤ Confirm part numbers and ratings of installed filters against BOM or OEM sheet.

➤ Verify that seals (shaft, port, static) are:
  • Installed
  • Of correct material (e.g., FKM, NBR)
  • Not shipping or storage seals.

➤ Check desiccant breather functionality (color indicator).

➤ Replace or reinstall components as required.

5.6 Lubricant and Material Compatibility Review

➤ Cross-check elastomer materials with lubricant chemistry.
  • Esters, synthetics, or EP additives may degrade incompatible rubbers (e.g., nitrile).

➤ Confirm that bearing and coupling grease thickeners are compatible if more than one type is used.

➤ *If incompatibilities are found, escalate to engineering for lubricant or component substitution.*

5.7 Final Review and Documentation

➤ Record all observations, test results, and corrective actions.

➤ File records in CMMS or commissioning dossier.

➤ Generate exception report for any deviations requiring OEM resolution.

➤ Obtain approval from reliability engineering before start-up is authorized.

6. Acceptance Criteria

PARAMETER	TARGET	ACCEPTANCE RANGE
Lubricant identity	Matches OEM spec	Must match
Fill volume	As per OEM chart	±5% unless otherwise stated
ISO particle count	≤ 19/17/14 (standard)	OEM-specific overrides allowed
Water content (ppm)	≤ 500 ppm (standard)	< 1,000 ppm max
Filter type and rating	Matches OEM BOM	Must match
Breather type	Environmental match	Desiccant for humid locations

7. **References**
   - ➤ OEM equipment manuals
   - ➤ ASTM D6304 (water via Karl Fischer)
   - ➤ ISO 4406 (particle count)
   - ➤ ASTM D445 (viscosity)

Pre-Start-Up Lubrication Verification Checklist

VERIFICATION TASK	STATUS (PASS/ FAIL/NA)	NOTES/ OBSERVATIONS	CORRECTIVE ACTION REQUIRED
Verify lubricant product identity against OEM specification			
Check viscosity grade and additive package compatibility			
Measure lubricant fill volume and compare with OEM chart			
Inspect breather type, and confirm installation			
Verify accessibility of fill, drain, and sample points			
Collect fluid sample and perform ISO 4406 particle count			
Analyze fluid for water content (ppm) using Karl Fischer			
Check installed filter element against BOM			
Inspect seal materials, and confirm compatibility			
Validate cleanliness and presence of sample/drain ports			
Record any signs of overfill, underfill, or leakage			
Verify presence and accuracy of level indicators			
Document all results and corrective actions			
Confirm that all findings meet acceptance criteria			
Upload checklist and analysis results to CMMS			

VERIFICATION OF OEM LUBRICATION PARAMETERS PRE-START

Prior to start-up, verification of lubrication parameters ensures the asset is configured according to reliability expectations. This includes confirming lubricant identity, fill volumes, system cleanliness, and compatibility with operational conditions.

Verification steps:

- Cross-check lubricant specifications against OEM documentation.
- Inspect fill level indicators, breather types, and access points.
- Conduct particle count and moisture analysis.
- Ensure installation of correct filtration and sealing components.

This process often reveals mismatches, such as incorrect filters, missing breather elements, or incompatible greases used during assembly. Early identification allows for correction without the cost of premature wear or failure.

Documenting these checks provides traceability and supports warranty claims if issues arise post-commissioning. Pre-start verification of lubrication parameters is critical for preventing early-life failures and aligning the system with site-specific reliability standards. Equipment delivered from the OEM often arrives with generic or transport-optimized lubrication setups that may not meet operational demands. Validation before start-up mitigates these risks.

Required verification tasks include:

- **Specification cross-check.** Compare lubricant product name, viscosity grade, and specification sheet data against the OEM's recommended fluids. Confirm additive compatibility, especially for multimetal systems or specialized coatings.
- **Volume confirmation.** Verify lubricant fill volumes using dipsticks, sight glasses, or reservoir markings. Cross-reference against OEM fill charts. Check for overfills, which may lead to foaming, or underfills, which pose starvation risks during start-up.
- **Breather and access inspection.** Confirm that breathers match environmental exposure levels (e.g., desiccant for high humidity). Ensure that they're installed and seated properly. Check that fill ports, drain valves, and sampling points are accessible and properly labeled.
- **Cleanliness validation.** Perform ISO 4406 particle count and Karl Fischer moisture testing on sampled fluids. Establish baseline cleanliness

and moisture levels before load is applied. Noncompliant fluids must be drained and replaced prior to operation.

➤ **Filtration and seal verification.** Confirm that correct filter elements are installed, especially if filter housings are reused across models. Verify that sealing elements meet OEM material specs and environmental ratings. Identify any temporary or shipping seals that must be replaced.

➤ **Component compatibility review.** Assess lubricant compatibility with system elastomers, coatings, and metallurgies. Pay attention to greases used in bearings and couplings; misapplied thickeners can cause base oil separation or caking.

Common findings include missing breathers, overfilled tanks, incorrect filters, and incompatible grease substitutions during assembly. Each poses a significant risk if undiscovered.

Document all findings, measurements, and corrective actions. Attach records to the CMMS or asset dossier for traceability. This documentation supports warranty enforcement and provides a defensible record in the event of early failure or audit.

STEP-BY-STEP PROCESS FOR VERIFICATION OF OEM LUBRICATION PARAMETERS PRE-START

Step 1: Review Documentation

➤ Obtain OEM lubrication specifications for the asset.
➤ Identify required lubricants by brand, viscosity grade, and additive compatibility.
➤ Confirm recommended fill volumes and lubrication schematics.

Step 2: Cross-Check Lubricant Identity

➤ Locate each lubrication point (e.g., reservoir, gearbox, bearing).
➤ Compare actual fluid labels or certificates with OEM specs.
➤ Check product name, viscosity grade (e.g., ISO VG, NLGI), and formulation (e.g., synthetic, mineral).
➤ Verify additive compatibility, particularly for systems with yellow metals, specialty coatings, or seals.

Step 3: Confirm Fill Volume Accuracy

➤ Use sight glasses, dipsticks, or level gauges to verify fluid levels.

➤ Cross-reference volume readings with OEM fill charts or service documentation.

➤ Flag overfilled systems (risk of aeration or foaming) and underfilled systems (risk of starvation).

Step 4: Inspect Breathability and Accessibility

➤ Verify that breather type (e.g., desiccant, sintered, filtered) matches environmental exposure risk.

➤ Ensure that breathers are installed, clean, and correctly seated.

➤ Check physical access to fill ports, sampling valves, and drains. Ensure that they are labeled and safely reachable.

Step 5: Validate Cleanliness and Moisture Control

➤ Take lubricant samples from reservoirs or circulating loops.

➤ Perform ISO 4406 particle count and Karl Fischer moisture testing.

➤ Compare results to OEM cleanliness requirements or site-specific standards.

➤ If contamination or water is present, plan for full drain, flush, and refill with clean product.

Step 6: Verify Filtration and Sealing

➤ Confirm that correct filter elements are installed (check part numbers, micron ratings, beta ratios).

➤ Inspect for any temporary filters or shipping plugs still in place.

➤ Check sealing materials (gaskets, O-rings) for compatibility with the lubricant's chemistry and temperature range.

Step 7: Assess Component Compatibility

➤ Review compatibility of lubricant with metals, elastomers, and coatings.

➤ Confirm that grease thickeners (e.g., lithium, calcium sulfonate, polyurea) are suitable for bearings or couplings.

➤ Ensure that there is no cross-contamination risk from incompatible legacy greases or oils.

Step 8: Document All Findings

➤ Record lubricant specifications, volumes, test results, and observed deviations.

➤ Note corrective actions taken: flushes, refills, filter swaps, or breather replacements.

➤ Attach findings to CMMS or asset file for traceability and warranty protection.

Optional Tools and Practices

➤ Use a standardized pre-start-up lubrication checklist to ensure that no steps are missed.

➤ Photograph each critical component during verification for future audits or warranty claims.

➤ Engage OEM support if there's uncertainty or conflicting specifications.

CLEAN START: COMMISSIONING FLUSH AND LUBRICANT VALIDATION

Commissioning flushes remove contaminants introduced during manufacture, storage, or transport. They are essential for high-speed, high-precision systems such as turbines, compressors, and hydraulic circuits.

Flush procedures vary but generally involve:

➤ High-velocity circulation with temporary filters
➤ Periodic sampling for cleanliness (ISO 4406 targets)
➤ Visual inspection for metallic debris, varnish, or fluid discoloration

Once flush targets are achieved, the system is filled with operational lubricants and retested to confirm cleanliness. Lubricant validation should include base number, viscosity, and additive concentration checks.

Skipping or shortening the flush process can negate warranty terms and reduce service life. A clean start builds a strong foundation for predictive maintenance and fluid reliability. Commissioning flushes are mandatory for systems where component precision, pressure sensitivity, and contamination susceptibility are high. This includes turbines, hydraulic units, compressors, and servo-controlled machinery. Residual assembly debris, weld slag, elastomer particles, and preservative oils must be removed before system oil is introduced.

Flush execution steps include:

- ➤ **Temporary loop configuration.** Set up a closed-loop flush circuit with temporary high-flow, high-efficiency filters. Flow rates should achieve Reynolds numbers > 4,000 to ensure turbulent flushing.
- ➤ **Filter media selection.** Use β > 2,000-rated elements with absolute filtration at 3–10 µm. Bypass filters are not acceptable during flush; full flow must be achieved.
- ➤ **Flush duration and sampling.** Continue circulation until samples consistently meet ISO 4406 cleanliness targets, often 18/16/13 or better for hydraulics. Sample at strategic locations including high spots, dead legs, and return lines.
- ➤ **Visual inspection.** Drain filter housings, and inspect for metallic swarf, varnish residue, or rubber particulates. Use white patch testing or magnetic plugs for additional confirmation.
- ➤ **Drain and fill.** Once flushing targets are confirmed, the system is drained hot and dry. The reservoir is inspected and cleaned manually if needed. Operational fluid is introduced through dedicated fill filters with < 10-µm absolute media.
- ➤ **Lubricant validation.** Analyze the fresh charge for viscosity at operating temperature, base number or acid number (depending on oil type), additive depletion (e.g., zinc, phosphorus), and water content. Confirm that product lot matches specification and delivery documentation.
- ➤ **Final cleanliness verification.** Post-fill sampling verifies cleanliness prior to start-up. Targets may be tighter for mission-critical assets, ISO 16/14/11 or better.

Omitting these steps risks early component wear, servo valve stiction, filter loading, or varnish formation. It may void warranty terms for hydraulic or rotating equipment. A validated clean start ensures system integrity, extends fluid life, and supports condition-based maintenance from day 1.

LUBRICATION TRAINING AND OVERSIGHT DURING COMMISSIONING

Commissioning presents a rare opportunity to train staff on system-specific lubrication practices, identify potential problems, and set expectations for long-term care. Involving lubrication experts during this phase enhances asset readiness.

Training topics should include:

➤ Overview of lube systems and components
➤ Product specifications and condition monitoring plans
➤ Access points and safety protocols
➤ Initial break-in and relubrication schedules

Oversight by certified lubrication professionals ensures that commissioning is performed to standard. This includes monitoring fill procedures, ensuring that tools and containers are clean, and validating procedures with OEM liaisons.

A documented commissioning checklist, cosigned by site and OEM representatives, should be created for each asset class. Commissioning provides the only full-system window before steady-state operation begins. It is the critical point to install best practices, identify vulnerabilities, and align expectations. Lubrication training during this window is most effective when hands-on and system-specific.

Training modules must include:

➤ **System configuration review.** Identification and walkthrough of all lubrication components, pumps, injectors, reservoirs, breathers, filters, sensors, and lubrication lines. Diagrams annotated with ID tags and flow paths are mandatory.
➤ **Product-specific orientation.** Review of lubricant type per component, viscosity index, base stock category, additive system, compatibility data, and handling requirements. Storage and transfer SOPs must be demonstrated using the actual products.
➤ **Access and safety protocols.** Technicians should be walked through physical access routes, isolation points, LOTO procedures, confined-space protocols (if applicable), and PPE requirements specific to lubricant handling.
➤ **Break-in strategy.** Initial lubrication intervals, relubrication cycles during load ramp-up, filtration behavior, and expected wear-metal trends must be communicated. Review of condition monitoring start-up baselines (particle count, viscosity, TAN/TBN, wear metals) reinforces accountability.

Oversight by lubrication engineers or certified tribology professionals:

➤ Ensures that fill and flush steps meet cleanliness targets
➤ Verifies fluid ID before fill using bottle tag, certificate of analysis, and pre-use sample
➤ Ensures that tools and containers are ISO 4406-compliant, sealed, and labeled

Commissioning oversight includes:

➤ Confirming breather type and orientation, filter installation torque, vent settings, and sampling port placement
➤ Supervising verification of reservoir fill levels and checking for leaks under static and dynamic pressure

A commissioning checklist is required per asset. The checklist must include lubrication sign-off points, date/time stamps, initial sample IDs, torque readings, and alignment with the OEM lubricant matrix. Signatures from both site maintenance leadership and OEM representatives validate compliance. The checklist remains part of the asset's permanent lubrication file.

OEM VERSUS SITE REALITIES AND BRIDGING THE SPECIFICATION GAP

OEM recommendations often reflect idealized conditions that differ from real-world plant environments. Factors such as higher ambient temperatures, airborne contamination, local regulations, or unique load profiles require site-specific adjustments.

Common gaps include:

➤ Lubricant brands specified by OEMs not available locally
➤ Recommended viscosity not suitable for temperature extremes
➤ Sealing or filtration systems not matching plant standards

Bridging this gap involves:

➤ Technical reviews of OEM specs by site reliability engineers
➤ Field trials or performance comparisons of alternative products

➤ Supplier consultations to ensure additive and base oil compatibility
➤ Documented deviations with justification and risk analysis

Establishing a formal review process for OEM-to-site adaptation ensures that lubrication strategies are tailored, documented, and defensible. Cross-functional alignment among procurement, engineering, and maintenance is key. OEM lubrication specifications are typically based on controlled lab conditions, standardized duty cycles, and global assumptions. Field conditions deviate due to thermal loads, contamination levels, duty severity, and climate. This creates a performance gap between OEM design intent and actual site needs.

Frequent disconnects include:

➤ **Unavailable products.** OEM-specified lubricants may not be stocked locally or may have long lead times. Equivalent products must be validated through data sheets, lab testing, or supplier matching tools.
➤ **Viscosity misfit.** OEM may call for ISO 68 in applications where high ambient temperatures push the need for ISO 100 or ISO 150 to maintain film strength. Cold starts may warrant lighter grades.
➤ **Sealing and filtration.** Supplied equipment may lack desiccant breathers, fine filtration, or purge seals common in the site's contamination control program. Retrofits may be required to meet plant reliability standards.

Engineering controls include:

➤ **Technical deviation reviews.** Reliability engineers review each spec for local fit. This includes lubricant suitability for operating temperature range, compatibility with site fluid handling protocols, and match with established supplier offerings.
➤ **Performance validation.** Alternative lubricants or filters undergo side-by-side trials or monitored pilot use. Lab oil analysis tracks wear metals, additive depletion, and contamination metrics to validate effectiveness.
➤ **Documented exception handling.** Deviations are logged with technical rationale, failure mode impact analysis (FMEA), and cross-reference to OEM risk exposure. These are stored within CMMSs or engineering change management systems.
➤ **Supplier collaboration.** Engage with formulators to confirm alternative lubricant chemical compatibility, especially with seal materials, paint finishes, or legacy residue from factory fill fluids.

Cross-functional reviews ensure that maintenance can execute lubrication consistently with available products, procurement aligns sourcing to technical needs, and engineering maintains compliance without compromising reliability. The result is a tailored lubrication plan grounded in field realities.

Table 14.1 is a structured presentation of OEM versus site realities, along with strategies for bridging the specification gap, and is designed to support alignment between OEM documentation and plant-specific lubrication realities.

TABLE 14.1   OEM Versus Site Realities: Bridging the Specification Gap

CATEGORY	OEM RECOMMENDATION	SITE REALITY	BRIDGING STRATEGY
Lubricant Brand Availability	Specific brands/ formulations (often under global supply contracts)	Local distributor does not carry the specified brand; long lead times or import issues	Use equivalent lubricant via technical cross-match; validate base oil and additive package compatibility
Viscosity Recommendations	Standard ISO grade based on moderate operating temperatures	Extreme ambient heat/cold; elevated loads; frequent cold starts	Adjust viscosity for thermal profile; validate film strength via lab testing under site conditions
Sealing and Elastomer Specs	Assumes standard compatible fluids such as Buna-N or FKM	Site uses specialty seal materials, or environmental exposure differs (e.g., ozone, chemicals)	Conduct seal-lubricant compatibility tests; consult OEM and seal supplier on deviations
Filtration and Cleanliness	Basic filtration (e.g., 25-micron filters), no desiccant breathers	Site requires finer filtration (e.g., 10-micron $\beta200$), breathers, or filter indicators	Retrofit filters/breathers to site spec; validate via particle count and pressure drop testing
Factory Fill Fluids	Short-life break-in oils, or preservation fluids for shipping	Site requires extended service fluids, or oil with ISO cleanliness certification before start-up	Drain and flush if needed; verify factory fill fluid lab specs; document and replace with site standard
Lubrication Interval Guidance	Based on duty assumptions (e.g., 8-hour operation, clean environment)	Real-world cycles are harsher: 24/7 ops, vibration, dust, chemical exposure	Recalculate relube intervals; validate with field inspections and oil analysis trend data
System Design Features	Limited feedback (e.g., no pressure sensors or flow indicators)	Site reliability standards mandate feedback loops for PM verification or control system tie-ins	Add sensors, indicators, or PLC inputs; integrate into CMMS or SCADA
Maintenance Access and Service	Assumes ideal ergonomics; access from test rigs or shop-floor assembly setup	Poor field access: guards, elevation, confined space, harsh weather	Reconfigure lube points for ergonomic access; use remote grease fittings or extension lines

Process for Bridging Strategy Workflow

Steps for bridging the strategy workflow include:

1. **Technical Deviation Review**
   - Reliability engineers audit each OEM spec.
   - Identify mismatches with site conditions, including lubricant chemistry, application method, filtration, and environmental exposure.

2. **Field Trial and Validation**
   - Test alternative lubricants or components side by side with OEM recommendations.
   - Track via lab oil analysis (ISO 4406, TAN/TBN, wear metals) and performance data.

3. **Supplier Consultation**
   - Coordinate with lubricant and seal vendors to confirm chemical compatibility.
   - Review SDS and technical data sheets to avoid adverse interactions or degradation.

4. **Documentation and Risk Analysis**
   - Log all deviations with rationale, technical justification, and risk mitigation steps.
   - Store records in CMMS, MOC (management of change), or engineering asset registry.

5. **Cross-Functional Alignment**
   - **Maintenance.** Ensures that field teams have access to compatible lubricants and service instructions
   - **Engineering.** Maintains system integrity and design compliance
   - **Procurement.** Sources technically acceptable alternatives with verified supply chains

Bridging the specification gap creates a field-proven lubrication strategy tailored to the actual operating environment. It improves reliability, simplifies maintenance, protects warranties, and ensures operational consistency across equipment platforms.

CHAPTER 15

LUBRICATION-RELATED SPECIFICATIONS FOR NEW MACHINERY

EMBEDDING CONTAMINATION CONTROL IN MACHINERY DESIGN SPECS

Contamination control requirements must be embedded into the original equipment specification, not addressed after delivery. The design phase is the point of maximum leverage to prevent particulate and moisture ingress and to ensure maintainability. Machinery vendors should be contractually obligated to include specific features that facilitate long-term reliability and reduce start-up and life-cycle contamination risk.

Specifications should mandate factory-installed breathers, preferably desiccant or membrane-type units, to manage humidity ingress and pressure equalization. Vented caps or open reservoirs are unacceptable in critical systems. The use of magnetic drain plugs and in-line strainers should be nonnegotiable, especially in gearboxes and circulating systems. These features capture debris and ferrous contamination at the source and reduce downstream filter loading.

Fill and drain systems must be closed loop by design. Open funnels or unsealed drain plugs should be prohibited. Specify quick-connect fittings compatible with on-site filtration carts. This allows clean transfer and handling practices to be executed without additional modifications.

Sampling infrastructure should be included in every asset with circulating or pressurized lubricant systems. Sample ports must be installed upstream of filters and downstream of critical components to ensure representative data collection. Vendors should not rely on drain samples or low points unless sample location can be verified by flow path schematics.

Required contamination control elements include:

➤ **Breathers.** Desiccant breathers rated for ambient moisture and particulate control, ideally with visual saturation indicators.

➤ **Filtration readiness.** Provisions for kidney loop filtration, including dedicated service ports and mounting surfaces for offline units.

➤ **Cleanliness specifications.** Delivered equipment must meet target ISO 4406 cleanliness levels. Typical benchmarks are 16/14/11 for hydraulics and 18/16/13 for gear oils before commissioning.

➤ **Fluid transfer hardware.** Color-coded, quick-connect fill ports with integrated filtration points to align with site fluid management protocols.

➤ **Materials compatibility.** All seals, hoses, and gaskets must be compatible with the specified lubricant chemistry and temperature range.

Procurement language should require preshipment cleanliness verification and documentation, including laboratory particle count results, photographic evidence of system internals, and confirmation of preinstalled contamination control hardware. Deviations or missing components must be addressed before acceptance or commissioning to prevent costly rework and early-life failures.

The following is an example of how to embed contamination control requirements into machinery design specifications for procurement or engineering documentation. This example can be adapted into RFQs, OEM technical requirement sheets, or capital project documentation.

Embedded Contamination Control Specification: Machinery Design Standard

Section 1: Purpose

To establish mandatory contamination control provisions in all lubricated mechanical equipment delivered to [company/facility name]. These provisions ensure system cleanliness, lubricant integrity, and extended component life from delivery through operation.

Section 2: Mandatory Design Features

2.1 Breather System

➤ All reservoirs, gearboxes, and hydraulic tanks shall be equipped with **factory-installed desiccant breathers** or equivalent moisture-control units.

➤ Breathers must include:
  - Particulate removal down to ≥ **3 microns**
  - **Moisture adsorption capacity** matching ambient RH levels
  - **Visual saturation indicator** or changeout schedule documentation

➤ Vented caps, open fill ports, or unsealed breathers are prohibited.

2.2 Magnetic Drain and Debris Capture

➤ Each gearbox, bearing housing, and oil sump shall include **magnetic drain plugs** to trap ferrous debris.

➤ **In-line strainers** or suction-side mesh screens (100–200-μm rating) shall be installed to protect downstream filters in circulating systems.

2.3 Closed-Loop Fluid Handling

➤ All fill and drain ports shall be **quick-connect style**, sealed, and color-coded per fluid type (per ISO 20443 or plant standards).

➤ **Open funnels, nonthreaded plugs, or dipstick-only fills are prohibited.**

➤ Fill connections must accommodate **filter carts or transfer systems** without spillage or exposure to ambient contaminants.

2.4 Sampling Infrastructure
➤ Equipment with oil reservoirs ≥ 5 liters or pressurized lube systems shall include:
- **Upstream and downstream sample ports** (Schrader or Minimess type) installed in representative flow paths.
- Sample ports must be metal, capped, and permanently labeled.
- Sample locations shall be indicated on supplied lubrication schematics.

Section 3: Cleanliness and Pre-commissioning Standards

3.1 Delivered Cleanliness
➤ Equipment shall meet **ISO 4406 cleanliness targets** upon delivery:
- **Hydraulic systems.** 16/14/11 or better
- **Gearboxes and circulating systems.** 18/16/13 or better
➤ Oil analysis reports (particle count, moisture ppm, viscosity) must accompany delivery documentation.

3.2 Internal Inspection
➤ Visual inspection photos of reservoirs, sumps, and filter housings must be provided with:
- Clean system interiors
- Verification of filter element installation
- No evidence of machining debris, rust, or shipping protectants

3.3 Filtration Provisions
➤ Provisions must be made for:
- **Offline filtration hookups** (kidney loop ports)
- Mounting brackets or skid clearance for filtration systems
- Installation of β ≥ **200-rated filter elements**, where applicable

Section 4: Materials and Compatibility

➤ All elastomers, seals, and hoses must be certified as compatible with the **specified lubricant(s)**, including base oil type (mineral/synthetic), additives, and temperature range.
➤ Vendor shall submit a **materials compatibility matrix** confirming chemical resistance of all wetted components.

Section 5: Acceptance Criteria and Documentation

Before commissioning or acceptance, vendor shall provide:

- ➤ Oil analysis report showing ISO cleanliness and water ppm
- ➤ Photographic documentation of internals and hardware installation
- ➤ Confirmation of:
  - Breather installation
  - Magnetic plug installation
  - Sample port functionality
  - Labeled, sealed fill and drain ports

Any missing elements or failure to meet documented standards may result in rejection or delayed commissioning at vendor's expense.

Section 6: Optional Enhancements (as specified by buyer)

- ➤ Online oil condition sensors for critical systems
- ➤ Pressure differential indicators on filters
- ➤ Integrated desiccant breather status monitoring via SCADA or PLC

This specification must be acknowledged and incorporated into final vendor submittals and OEM design sign-offs.

SPECIFYING LUBE POINTS, VOLUMES, AND ACCESSIBILITY AT PURCHASE

Maintenance efficiency is greatly influenced by the number, placement, and accessibility of lubrication points. Procurement specifications must require OEMs to provide detailed lubrication maps, including location, lubricant type, relubrication frequency, and volume.

Lubrication effectiveness is constrained by how accessible and well documented the lubrication points are on delivered equipment. Maintenance burden increases significantly when lubrication fittings are hidden, undocumented, or difficult to service under normal operating conditions. Procurement specifications must require OEMs to submit detailed lubrication maps that include all required data for asset integration into preventive maintenance programs.

At minimum, specifications must list the number and type of lubrication points. This includes grease fittings (zerk, button head, flush type) and oil fill, level, and drain ports. The total count per equipment type must be disclosed, along with the service interval and required lubricant volume per point. For rotating equipment, ensure that the spec addresses both stationary and rotating lubrication interfaces, such as couplings and pillow blocks.

Access requirements must be part of the design deliverables. All lubrication points should be clearly labeled with permanent, corrosion-resistant tags, laser-etched or engraved if possible. The tag should include the lubricant type, frequency, and volume. Points located in hazardous or physically restricted areas must include remote extension lines or be routed to a centralized lubrication manifold. Grease fitting extensions must be stainless steel tubing or hydraulic hose rated for the delivery pressure.

Digital integration is mandatory. OEMs must supply the lube point map in both PDF and CAD-compatible formats (e.g., DWG or DXF), with GPS or coordinate mapping for integration into CMMS route optimization tools. Annotated diagrams must correlate lube point IDs to the bill of materials and recommended PM task descriptions.

Required inclusions in the purchase specification are:

➤ Lubrication point count per component and system
➤ Service interval and volume per point in metric and imperial units
➤ Lube type and NLGI or ISO VG classification per point
➤ Access requirements including platform height, tools needed, or hot zone proximity
➤ Extension kits or manifolds where physical access is obstructed
➤ Digital lube map in editable format for CMMS integration
➤ Color-coded or symbol-based labeling for PM differentiation (e.g., weekly versus monthly)

Including these elements at the procurement stage prevents ambiguity, enables real-time planning, and reduces commissioning delays. It also ensures compatibility with existing lubrication tools, automation systems, and technician workflows.

The following is an example of how to formally specify lube points, volumes, and accessibility within a procurement or OEM equipment purchase specification. This can be embedded directly into an RFQ (request for quotation), purchase agreement, or design standard.

Lubrication Point Specification: Purchase and Procurement Requirements

Section 1: Purpose

To ensure maintainability, reliability, and integration of equipment into site lubrication programs, vendors must provide complete and accessible lubrication system documentation and infrastructure as part of the delivered asset.

Section 2: Minimum Documentation Requirements

2.1 Lubrication Map

➤ Vendor must provide a **comprehensive lubrication point map** in both:
  - **PDF format** for print and offline reference
  - **CAD-compatible formats** (.DWG or .DXF) for integration into CMMS or route optimization systems

➤ Map must include:
  - Unique **lube point ID numbers**
  - **Exact location coordinates** on the machine
  - Reference to **component BOM and function**

2.2 Lubrication Data Table

A tabulated list of all lubrication points must include:

LUBE POINT ID	COMPONENT	LUBRICANT TYPE	NLGI/ ISO VG	VOLUME (ML/OZ)	FREQUENCY	FITTING TYPE	ACCESS NOTES
EX123	Drive bearing	ISO VG 220 gear oil	ISO VG 220	150 ml 5 oz	Monthly	Drain plug	Ground-level access
GR456	Coupling zerk	Lithium complex grease	NLGI 2	20 g 0.7 oz	Weekly	Zerk (1/4") 6.35 mm	Requires extension

➤ Volume units must be provided in **both metric and imperial.**

➤ Lubricant type must include **manufacturer's name** (or approved equivalent).

➤ Frequency must be listed as **per calendar unit** (e.g., weekly, quarterly).

Section 3: Accessibility Requirements

3.1 Access Design
➤ All lubrication points must be:
- **Unobstructed during normal operations**
- **Reachable from floor level or permanent platform**
- **Serviceable with standard tools** (no disassembly required)

3.2 Extension and Remote Access Kits
➤ Any point requiring access above 2 meters, behind guarding, or near hot/moving zones must:
- Be extended via **stainless steel tubing or hydraulic hose**
- Terminate in an **accessible manifold or grouped service station**
- Include **mounting hardware and protective sleeving**

3.3 Labeling
➤ Each lubrication point must be permanently labeled with:
- **Laser-etched or engraved ID tag**
- Lubricant type, volume, and service frequency
- **Color-coded** indicators to differentiate PM tasks (e.g., red = weekly, blue = monthly)

Section 4: Digital Integration and Compliance

➤ Lube point data must be **machine readable** and CMMS compatible (CSV or XML export)
➤ All lube point identifiers must correspond with the machine's:
- P&IDs
- Maintenance manuals
- Operator training materials

Section 5: Acceptance Criteria

➤ Failure to provide complete lubrication maps, access accommodations, or digital data at time of FAT (factory acceptance test) or site delivery may result in:
- Commissioning delay
- Backcharge for corrective modifications
- Rejection of asset until compliant

Section 6: Optional Enhancements (as specified)
- Centralized auto-lubrication manifolds
- RFID/NFC-enabled smart tags for point verification
- QR-coded lube points linked to mobile PM instructions

By embedding this level of detail in your procurement spec, you ensure that OEMs design for maintainability, reduce technician burden, and enable fast, accurate PM execution from day 1.

REQUIRING COMPATIBILITY WITH CONDITION MONITORING TOOLS

Requiring compatibility with condition monitoring tools is nonnegotiable in reliability-focused operations. Equipment must be engineered with proactive diagnostics in mind. Lubrication-related components, such as pumps, reservoirs, piping, filters, sumps, and bearing housings, must allow seamless integration with oil analysis, infrared (IR) thermography, vibration analysis, and airborne/structure-borne ultrasound.

Sample ports must be engineered into the system in a way that does not disrupt laminar flow or introduce turbulence. Flow-disruptive sample points skew particle counts and lead to inconsistent wear-metal trending. Placement must follow ASTM D4057 or ISO 3170 practices, preferably on pressure lines or return lines where representative samples can be drawn at stable operating temperatures and flow rates. Sample valves must be compatible with vacuum sampling devices, and include check valves or purge systems to eliminate cross-contamination or stagnant fluid capture.

Bearing housings require clear access for ultrasonic probes. Grease-lubricated housings should include permanently mounted grease fittings positioned adjacent to the ultrasound contact point to enable real-time acoustic verification during relubrication. Housing geometry must not shield ultrasonic signal pathways. Castings or external guards that block sensor access must be redesigned or omitted. For thermography, unobstructed lines of sight to known heat transfer zones must be maintained. Paint types and surface finishes should not interfere with IR emissivity; anodized aluminum or high-reflectivity surfaces require corrective treatments or emissivity references to ensure thermal data accuracy.

Vibration sensor mount points must be free of protrusions, weld beads, and structural damping elements. Direct paths to the bearing load zone are critical. OEMs must avoid using ribbed or finned surfaces near mount locations. Structural resonance analysis should be performed to validate frequency transmission fidelity from the sensor location to the bearing assembly. Integrally cast sensor pads or machined flats should be incorporated into the design, ideally near vertical or horizontal shaft axes.

Thermal pathways must reflect actual bearing or fluid temperatures without influence from ambient interference. Heat sinks, shields, and enclosures must be evaluated for their impact on infrared transparency. Materials must not reflect or scatter IR readings. Components should be designed with test ports or windows compatible with calibrated thermal cameras.

OEMs must verify condition monitoring readiness with functional mock-ups or case study data from equivalent operating conditions. Standard submittals should include a condition monitoring integration report, including mounting locations, data trend access points, and prior examples of predictive diagnostics in operation. Designs requiring post-installation drilling, welding, or adaptation for monitoring device integration are unacceptable. Early alignment with site diagnostics architecture eliminates retrofit labor, accelerates predictive analytics onboarding, and enables full asset visibility from day 1.

The following is a step-by-step process for requiring compatibility with condition monitoring tools during the design, specification, and procurement phases of industrial equipment, ensuring seamless integration into reliability-centered maintenance programs.

Step by Step: Requiring Compatibility with Condition Monitoring Tools

1. Define Diagnostic Compatibility Requirements

➤ Identify which condition monitoring tools will be used on-site:
  - Oil analysis
  - Ultrasound (airborne and structure-borne)
  - Vibration analysis
  - Infrared thermography

➤ Communicate that these tools are part of standard reliability practices and must be supported by equipment design.

2. Specify Sample Port Design for Oil Analysis

➤ Require OEMs to install sample ports:
  - **Per ASTM D4057 or ISO 3170** for pressurized systems
  - **On pressure or return lines,** not at sump bottoms or dead zones

➤ Ports must:
  • Maintain **laminar flow**
  • Be **vacuum-pump compatible**
  • Include **check valves** or **purge valves** to eliminate cross-contamination

3. Engineer Bearing Housings for Ultrasound

➤ Mandate that bearing housings:
  • Include **flat, accessible probe contact points** adjacent to grease fittings
  • Be **free of interference** from shields or casting structures
  • Allow **real-time ultrasound feedback** during relubrication (grease-on detection)
➤ Specify that any grease fitting extensions must preserve probe alignment access.

4. Ensure Infrared Thermography Access

➤ Design for unobstructed **line of sight** to heat-critical zones (bearings, gearboxes, sumps)
➤ Avoid:
  • Surfaces with **low or variable emissivity** (e.g., polished aluminum)
  • Obstructions like shields or guards unless **IR-transparent windows** are included
➤ Require OEMs to apply **emissivity-correct coatings** or provide reference patches

5. Provide Vibration Sensor Mounting Features

➤ Sensor mount points must:
  • Be located **near the bearing load zone**
  • Be **machined flat** and **free of paint, ribs, or welds**
  • Offer **vertical/horizontal axis alignment** for orthogonal data collection

➤ Require OEMs to conduct **resonance path validation** from mount point to bearing
➤ Specify inclusion of **cast-in or machined vibration pads**

6. Require Integrated Test and Access Features

➤ For thermal and vibration measurements:
  - Include **inspection ports, sensor mounts,** and **nonobstructive covers**
➤ Fluid systems must include:
  - **Clean, turbulence-free sample ports**
  - **Filtration bypass ports** for trend sampling

7. Mandate OEM Submission of Diagnostic Readiness Documentation

➤ Require:
  - A **condition monitoring integration report** with marked diagrams of sensor/sample points
  - **CAD or 3D renderings** showing placement and accessibility
  - Examples of **prior use cases or pilot implementations** supporting diagnostics
➤ Include a **functional mock-up verification** if site-specific adaptation is required

8. Make Retrofit-Free Integration a Procurement Condition

➤ State explicitly in RFQs and contracts that:

"Any equipment requiring post-delivery drilling, welding, or field modifications to support standard condition monitoring practices will be deemed noncompliant."

9. Verify at Factory Acceptance Testing (FAT)

➤ Confirm presence, accessibility, and labeling of all required monitoring points

> ➤ Validate sample port function, sensor mount flatness, and thermal access during FAT
> ➤ Reject or document deviations before shipment

By embedding these steps in the specification and procurement process, you ensure that all assets are ready for condition monitoring on delivery, thereby reducing start-up delays, retrofit costs, and long-term reliability risk.

ESTABLISHING MINIMUM LUBRICANT FILM REQUIREMENTS BY APPLICATION

Minimum lubricant film thickness requirements must be explicitly defined in machinery procurement specifications based on actual operating conditions. Establishing a precise lubricant film thickness, quantified as a lambda ratio or directly measured in microns, ensures that components such as bearings, gears, and cams sustain appropriate separation during operation, avoiding metal-to-metal contact and subsequent wear or failure.

OEM submissions must include detailed geometric and load parameter data for critical lubricated components. Bearing specifications must detail internal clearances, contact angles, rolling element size, and raceway dimensions. Gear specifications require tooth geometry profiles, surface finishes, pitch line velocities, and load distribution across tooth faces. Precise definitions of anticipated operational regimes—boundary, mixed film, or hydrodynamic— are mandatory, directly correlated to real-world operating scenarios on-site.

OEMs must recommend lubricants capable of maintaining required film thickness under the most severe operational conditions. Lubricant recommendations must include base oil viscosity at operating temperatures, viscosity index, additive composition, and compatibility data to support robust lubricant selection. Specification of extreme pressure and anti-wear additives is necessary for applications prone to boundary lubrication conditions.

Film thickness data from OEMs should be validated through well-documented calculations or recognized software models (e.g., SKF bearing calculator, AGMA gear design standards). Calculated lambda ratios and film thickness measurements must reflect worst-case operational scenarios—minimum temperature, maximum load, and lowest relative speeds—to ensure adequate lubricant performance margins.

Inclusion of explicit minimum film thickness requirements moves specifications from lubricant-brand compliance toward quantifiable, performance-driven compliance. This clarity enables precise lubricant selection, optimal interval setting for lubrication tasks, and accurate risk assessments for each asset, thereby significantly enhancing reliability outcomes.

ENSURING SEAL AND GASKET COMPATIBILITY WITH PLANNED LUBRICANTS

Seal failure is one of the most common root causes of lubrication system degradation. Procurement specs must ensure that all elastomers, seals, gaskets, and adhesives used in new machinery are chemically and thermally compatible with planned lubricants.

Specification actions include:

> ➤ Requiring OEMs to provide seal material details and test data
> ➤ Mandating that all components be tested against the site's approved lubricant list
> ➤ Defining expected temperature and pressure ranges at each seal interface

Procurement specifications ensures that the selected lubricants won't cause hardening, swelling, or degradation of polymeric components. Compatibility testing, particularly for synthetic esters, PAOs, and PAG-based fluids, is especially critical in high-temperature or food-grade applications. Seal failure frequently contributes to lubrication system degradation, necessitating stringent procurement specifications that verify the chemical and thermal compatibility of elastomers, seals, gaskets, and adhesives with the planned lubricants. It is essential to require original equipment manufacturers to provide comprehensive details regarding seal materials and corresponding test data. This data should encompass extensive compatibility assessments with all lubricants listed as approved for use at the site. Additionally, the specifications must define the expected temperature and pressure ranges at each seal interface, which facilitates a clear understanding of the operational conditions that could impact seal performance.

The importance of these measures lies in their ability to prevent adverse reactions such as hardening, swelling, or degradation of polymeric components when exposed to lubricants. Compatibility testing becomes particularly crucial for fluids based on synthetic esters, paraffinic synthetic oils, or polyalkylene glycols, especially in applications that operate under high-temperature conditions or those requiring food-grade lubricants. These fluids can present unique challenges that necessitate careful evaluation to ensure reliability and longevity of seals under the specified operating conditions. Implementing these rigorous specifications is vital to maintaining system integrity and performance throughout the life cycle of the machinery.

The following is a structured breakdown for ensuring seal and gasket compatibility with planned lubricants, including identification methods and remedies as part of machinery specification and procurement practices.

Ensuring Seal and Gasket Compatibility with Planned Lubricants

1. Identification of Compatibility Risks

AREA	POTENTIAL FAILURE MODE	CONTRIBUTING FACTORS	DETECTION METHOD
Elastomer degradation	Cracking, embrittlement	Incompatibility with base oil or additives (e.g., PAO, PAG, esters)	OEM data sheet review; immersion testing
Swelling and softening	Excessive volumetric expansion	Lubricant absorption into seal material	ASTM D471 or ISO 1817 swelling tests
Loss of mechanical integrity	Compression set, extrusion	High temp/pressure cycling, poor material selection	Dynamic compression testing; tensile strength evaluation
Thermal breakdown	Hardening, loss of elasticity	Lubricant causes heat buildup or is exposed to extreme ambient temps	TGA and DSC (differential scanning calorimeter) tests
Chemical incompatibility with additives	Delamination, residue, leaks	EP/AW, food-grade or anti-oxidant additive reactions	Compatibility matrix analysis with actual formulation
Adhesive/ gasket attack	Separation or softening of bonding material	Solvent or polar additive incompatibility	Infrared spectroscopy (FTIR); visual inspection after soak

2. Specification and Procurement Requirements

SPECIFICATION ACTION	DETAIL
Seal material declaration	Require OEMs to list all elastomer types (e.g., NBR, FKM, EPDM, PTFE) used in seals, gaskets, and O-rings
Lubricant exposure matrix	Submit a matrix showing which seal contacts which lubricant (including factory fill and site fill types)
Operating envelope definition	Specify pressure, temperature, and dwell time for each lubrication zone where sealing occurs
Site lubricant compatibility testing	Mandate soak testing (e.g., 72-hour immersion at operating temp) of seal materials in site-approved lubricants
Food-grade application requirements	Specify NSF H1 or FDA CFR 21 compliance for seal materials in food and beverage applications
OEM warranty statement	Require a signed declaration from OEM confirming seal-lubricant compatibility under stated operating conditions

3. Remedies for Identified Incompatibilities

FAILURE MODE DETECTED	RECOMMENDED REMEDY
Seal swelling in immersion test	Switch to higher chemical-resistance elastomer (e.g., from NBR to FKM or HNBR)
Premature hardening after thermal cycle	Specify high-temp polymers (e.g., FFKM, PTFE), and require temperature derating
Additive incompatibility (e.g., ester reactivity)	Replace lubricant with nonreactive base oil, or specify chemically inert seal
Unavailable OEM test data	Conduct third-party lab compatibility testing using ASTM/ISO protocols
Undeclared gasket adhesives	Mandate adhesive specification disclosure, and cross-check with SDS of lubricants used
Seal cracking in high-frequency application	Use dynamic-rated seals with low-compression set materials (e.g., silicone blends, polyurethane)

4. Documentation and Control

REQUIRED DELIVERABLES	PURPOSE
Seal material compatibility report	Documents all lubricant interactions and test results for each seal component
Clean bill of material with material grades	Enables traceability of elastomeric components and future maintenance decisions
OEM sign-off letter	Enforces liability for material compatibility under defined site operating conditions
Engineering change request log	Tracks deviations from OEM specs if alternative lubricants or seal materials are approved on-site

DEFINING LUBRICATION EQUIPMENT INTERFACES IN RFPS

Interfaces between machinery and lubrication equipment, such as pumps, sensors, manifolds, and filters, must be standardized and clearly defined. This enables easier integration with automated lubrication systems, contamination control tools, and digital platforms.

RFPs (requests for proposal) should require:

➤ Standardized threads and ports (e.g., NPT, BSPT [British Standard Pipe Taper], and ISO)

> ➤ Defined torque and pressure ratings for connection points
> ➤ Physical space allowances for sensor and equipment retrofitting
> ➤ Electrical and control interface compatibility (e.g., 24VDC signal lines for sensors)

Properly defined interfaces prevent surprises during commissioning and improve the speed and reliability of lubrication system installation and maintenance. Standardization and clear definition of interfaces between machinery and lubrication equipment are crucial for seamless integration with automated lubrication systems, contamination control tools, and digital platforms. RFPs must mandate the use of standardized threads and ports, such as NPT, BSPT, and ISO, to ensure compatibility and ease of installation across various equipment. It is essential to specify torque and pressure ratings for all connection points to prevent mismatches that could compromise system integrity.

RFPs should also include requirements for adequate physical space allowances, enabling sensor and equipment retrofitting without significant modifications. Electrical and control interface compatibility must be clearly defined, with specifications such as 24VDC signal lines for sensors, to facilitate straightforward integration with existing systems. Properly defined interfaces eliminate unexpected issues during commissioning, enhancing the speed and reliability of lubrication system installation and maintenance. These measures ensure that equipment operates efficiently and effectively, reducing downtime and maintenance costs.

Defining interfaces in lubrication systems is a critical aspect of ensuring compatibility and functionality across various components. When interfaces are standardized, the integration between machinery and lubrication equipment, such as pumps, sensors, manifolds, and filters, becomes more streamlined. This standardization simplifies the process of connecting different components and reduces the risk of errors that might occur during installation or maintenance.

The use of standardized threads and ports, like NPT, BSPT, and ISO, allows for consistent connection points across different systems and manufacturers. This consistency is vital for ease of assembly and disassembly, reducing the potential for leaks and ensuring a secure fit.

Torque and pressure ratings must be precisely defined to ensure that connection points can withstand the operational demands of the system. By specifying these parameters in RFPs, manufacturers and suppliers can align their components with the expected requirements, preventing failures due to overtightening or pressure miscalculations.

Physical space allowances are equally important, particularly when retrofitting sensors and other equipment. Adequate space ensures that additional components can be added without the need for significant redesign or modification of existing systems.

This flexibility is crucial for adapting to technological advancements and integrating new functionalities.

Electrical and control interface compatibility, such as the use of 24VDC signal lines for sensors, ensures that all electronic components can communicate effectively within the system. This compatibility is essential for the operation of automated systems and the integration of digital platforms, allowing for real-time monitoring and control.

By clearly defining these interfaces in RFPs, the installation process becomes more predictable, reducing the likelihood of unexpected issues during commissioning. This clarity enhances the speed and reliability of installation and maintenance, leading to more efficient and effective lubrication systems. It also supports long-term operational reliability, minimizing downtime and maintenance costs.

Tables 15.1–15.4 define lubrication equipment interfaces in RFPs.

TABLE 15.1   Mechanical Interface Requirements

REQUIREMENT	DETAIL	PURPOSE
Standardized thread types	Specify all lubrication ports, fill/drain connections, and sensor mountings to use standardized threads: NPT (ANSI/ASME B1.20.1), BSPT (ISO 7-1), or ISO metric (ISO 6149)	Prevents adapter use; ensures cross-system compatibility
Port labeling and location	Require permanent markings for all lube-related ports: fill, drain, sampling, vent, and breather. Clearly identify function and media	Aids maintenance accuracy and reduces misapplication
Defined torque and pressure ratings	Include torque specs (Nm/ft-lbs) for each threaded port and max system pressure for manifolds and lines	Prevents undertightening/overtightening; ensures system integrity
Mounting flange and hole specs	For reservoirs, sensors, and pumps, specify bolt patterns and clearance zones per ISO 3019 (hydraulic components) or OEM compatibility charts	Enables easy component replacement or upgrade

TABLE 15.2 Spatial and Retrofit Design Provisions

REQUIREMENT	DETAIL	PURPOSE
Minimum clearance zone	Define required space (e.g., 150-mm radius) around lubrication ports for tool access or retrofitting equipment like flow sensors or breathers	Prevents rework during commissioning or upgrades
Sensor retrofitting accommodation	Include sensor "landing zones" with flat, machined surfaces, prethreaded bosses, or weldless pads for temperature, pressure, or ultrasonic sensors	Reduces need for field machining or redesign
Filter and reservoir access points	Specify open access for filter changes and reservoir level checks, including any required platforms or ladders if mounted above reach height	Supports routine maintenance and safety compliance

TABLE 15.3 Electrical and Control Interface Standardization

REQUIREMENT	DETAIL	PURPOSE
Sensor voltage/signal compatibility	Require all lubrication sensors (flow, pressure, temp, level) to use standard 24VDC supply and output (e.g., 4–20 mA, IO-Link, or Modbus RTU)	Enables integration with plant PLCs and IIoT systems
Cable and connector types	Specify connector types (e.g., M12 A coded, IP67 rated) and minimum cable shielding for EMI/EMC resistance	Supports plug-and-play installation, durability
Documentation of I/O mapping	Vendor must provide sensor I/O map, signal description, and controller interface sheet during submittals	Aids configuration of SCADA, DCS, or CMMS inputs

TABLE 15.4 Documentation and Quality Control Deliverables

REQUIREMENT	DETAIL	PURPOSE
CAD-compatible interface drawings	2D and 3D drawings (DWG, STEP) showing port locations, thread type, and torque spec	Accelerates system design and digital twin setup
Interface checklist at FAT	Factory acceptance test must include physical verification of all interfaces (mechanical, electrical, spatial) against drawings/specs	Ensures compliance before shipment
Deviation request protocol	Require vendors to submit written requests for any deviation from standardized interface definitions. Include justification and risk analysis	Maintains control over system uniformity

Implementation

➤ **Reduces cost and delay during commissioning** by eliminating interface mismatch surprises

➤ **Ensures seamless integration** of lubrication systems into plantwide predictive maintenance strategies

➤ **Simplifies upgrades and sensor retrofits,** supporting long-term asset life-cycle strategies

➤ **Improves safety and maintainability** through intentional access and port labeling

SPECIFYING LIFE-EXTENDING LUBRICATION TECHNOLOGY FROM THE START

Proactive lubrication engineering can extend the operating life of machinery significantly. RFPs and procurement specs should demand that OEMs integrate life-extending lubrication technologies wherever possible.

Examples include:

➤ Integrated oil filtration and recirculation systems
➤ Automated or centralized lubrication systems
➤ Synthetic or specialty lubricants specified for extended intervals
➤ Smart lubrication controllers with diagnostics
➤ Redundant sealing or isolator technologies

OEMs should also be required to disclose the expected lubricant life (in hours or cycles) and the basis for that estimate. Where extended intervals are not specified, vendors should justify the omission.

Including these requirements in procurement specifications aligns machinery design with long-term maintenance strategies, reduces total cost of ownership, and supports reliability-centered operations from day 1. Proactive lubrication engineering is essential for significantly extending the operational life of machinery. RFPs and procurement specifications must require OEMs to integrate life-extending lubrication technologies wherever feasible. This includes the incorporation of integrated oil filtration and recirculation systems, which play a critical role in maintaining oil cleanliness and extending the interval between oil changes. Automated or centralized lubrication systems should be specified to ensure consistent and precise application of lubricants, minimizing wear and tear on critical components.

The use of synthetic or specialty lubricants tailored for extended service intervals must be mandated, providing enhanced performance under varying conditions and reducing maintenance frequency. Smart lubrication controllers equipped with diagnostic capabilities should be included to monitor system performance in real time, allowing for timely interventions when deviations are detected. Redundant sealing or isolator technologies are also vital to prevent contamination ingress and ensure the integrity of lubrication systems.

OEMs must disclose the expected lubricant life, expressed in hours or cycles, and provide the basis for their estimates. This transparency is crucial for planning maintenance schedules and aligning them with operational requirements. In cases where extended intervals are not specified, vendors must provide a rationale for the omission, ensuring that all decisions are grounded in technical justifications.

Incorporating these requirements into procurement specifications aligns machinery design with long-term maintenance strategies, effectively reducing the total cost of ownership. This approach supports reliability-centered operations from the outset, ensuring that machinery not only meets immediate performance needs but also sustains high levels of reliability and efficiency over its life cycle.

Integrating life-extending lubrication technology from the start is a strategic approach that focuses on enhancing machinery longevity and reliability. By specifying advanced lubrication systems in RFPs and procurement specifications, organizations can ensure that their equipment is optimized for long-term performance. Integrated oil filtration and recirculation systems are crucial for maintaining lubricant purity, which directly impacts component wear and system efficiency. These systems help in extending the time between oil changes, reducing downtime and maintenance costs.

Automated or centralized lubrication systems are designed to deliver the right amount of lubricant at the right time, minimizing human error and ensuring consistent lubrication across all necessary points. This automation is particularly beneficial in complex machinery where manual lubrication might be challenging and time-consuming. The use of synthetic or specialty lubricants allows for longer intervals between servicing due to their superior thermal stability and resistance to oxidation. These lubricants are specifically designed to perform under demanding conditions, providing a protective film that reduces friction and wear.

Smart lubrication controllers enhance system monitoring by providing diagnostic feedback, which can be used to identify potential issues before they lead to failures. This real-time data enables maintenance teams to make informed decisions, optimizing the maintenance schedule and preventing unexpected breakdowns. Redundant sealing or isolator technologies further protect the lubrication system from contaminants, preserving the quality and effectiveness of the lubricant.

By requiring OEMs to provide detailed estimates of lubricant life and justifications for their choices, organizations can better plan maintenance activities and budget for operational expenses. This level of detail ensures that all stakeholders are aligned on expectations and performance metrics. Implementing these technologies supports a reliability-centered maintenance strategy, which focuses on maintaining system functionality and extending the useful life of equipment, ultimately reducing the total cost of ownership and enhancing operational efficiency.

The following is a structured example of how to specify life-extending lubrication technology in your RFPs and procurement specifications, formatted for inclusion in technical documents or supplier requirements.

Specifying Life-Extending Lubrication Technology in RFPs

Objective:

Integrate advanced lubrication engineering features into machinery design to maximize asset lifespan, reduce life-cycle costs, and align equipment performance with long-term reliability-centered maintenance strategies.

Mandatory Requirements for OEM Inclusion

TECHNOLOGY CATEGORY	SPECIFICATION REQUIREMENT	PURPOSE
Integrated oil filtration and recirculation	Equipment must include built-in filtration loops or return-line filters rated to achieve ISO 4406 cleanliness targets (e.g., 18/16/13 or better). Filtration must be serviceable without halting operation	Extends lubricant and component life by controlling wear debris and contamination
Automated/centralized lubrication systems	All multipoint grease or oil lubrication must be delivered via an OEM-installed automatic or centralized lubrication system. System must include pressure/flow verification and service alerts	Ensures consistent lubrication, reduces human error, and enables remote monitoring
Synthetic or long-life lubricants	Factory fill and recommended fluids must include synthetic or high-performance lubricants capable of extended change intervals. Must be listed by brand name and specification	Enhances thermal stability, oxidation resistance, and film retention for longer intervals

TECHNOLOGY CATEGORY	SPECIFICATION REQUIREMENT	PURPOSE
Smart lubrication controllers	Controllers must monitor cycle counts and delivery verification (pressure or flow) and issue fault codes. Interfaces must be compatible with plant SCADA/CMMS (e.g., Modbus, IO-Link)	Enables real-time diagnostics and predictive alerts, supporting proactive maintenance
Redundant sealing or isolator designs	Where applicable (e.g., gearboxes, rotating shafts), sealing systems must include double-lip, labyrinth, or isolator-type seals with ingress protection (IP65 or higher)	Prevents contamination ingress, protects lubricant quality, and extends system reliability

OEM Deliverables and Verification

DELIVERABLE	DETAIL	EVALUATION METHOD
Lubricant life expectancy	OEM must state expected lubricant service life (in hours or cycles) under nominal operating conditions. Must include test basis (ASTM D943, FZG, or OEM field trial)	Reviewed against duty cycle and application severity by reliability engineering
Justification of standard intervals	If standard lubricants or manual delivery are used, OEM must explain rationale and provide supporting FMEA or failure rate justification	Reviewed during technical bid evaluation or pre-award design review
Lubrication system diagram	Provide full lubrication schematic including product, flow path, metering method, service interval, volume, and access points. CAD and PDF formats required	Verified during design submittal and commissioning
Controller interface spec	Submit data sheet and I/O map for lubrication controller or sensor package. Must show power requirements, signal types, and communication protocol	Reviewed by instrumentation and controls team
Test and validation records	For equipment claiming extended lube intervals, include lab reports or customer case study results demonstrating performance under comparable conditions	Evaluated by reliability/tribology team prior to FAT

Procurement Contract Language (Example Clause)

"Vendor shall integrate life-extending lubrication technologies as standard equipment. This includes the use of synthetic lubricants, automated delivery systems, in-line filtration, and diagnostic-capable controllers. Expected lubricant life must be disclosed, with supporting test documentation. Failure to meet these specifications without preapproved variance may result in rejection of equipment at FAT."

Benefits of Early Specification

Early specification:

- ➤ Reduces unplanned downtime and early-life wear
- ➤ Aligns machinery with plantwide RCM and CMMS frameworks
- ➤ Decreases manual labor and operational oversight
- ➤ Lowers total cost of ownership through optimized lubrication intervals

CHAPTER 16

OUTAGE AND SHUTDOWN ACTIVITIES

PLANNING LUBRICATION ACTIVITIES DURING PLANT TURNAROUNDS

Turnarounds and planned outages present a critical opportunity to execute deferred lubrication tasks, upgrade systems, and conduct thorough inspections. A successful lubrication scope within a turnaround requires early planning, multidisciplinary coordination, and alignment with the overall critical path schedule.

Key planning actions include:

➤ Establishing a detailed lubrication work scope and budget
➤ Mapping lube system isolation points and drain zones
➤ Scheduling tasks in coordination with mechanical, electrical, and safety disciplines
➤ Defining resources: skilled personnel, equipment, materials, and PPE

A lubrication turnaround plan must be synchronized with master schedules and allow for float in case of delays in upstream work. It should also include checklists, contingency plans for equipment access restrictions, and temporary storage for recovered fluids.

Pre-outage inspections and walkthroughs help validate access and connection points and confirm the availability of correct lubricants. This stage is also the time to confirm that condition monitoring baselines are recorded before shutdown.

Turnarounds and planned outages are pivotal moments for executing deferred lubrication tasks, upgrading systems, and conducting thorough inspections. Success in these activities hinges on early planning, multidisciplinary coordination, and integration with the overall critical path schedule. Establishing a detailed lubrication work scope and budget is essential. This involves identifying all lubrication tasks that need to be performed during the turnaround and ensuring that they align with the financial constraints and goals of the project.

Mapping lubrication system isolation points and drain zones is critical for executing the work efficiently and safely. Understanding these zones allows for effective fluid management and prevents cross-contamination or accidental spills. Scheduling the lubrication tasks requires careful coordination with mechanical, electrical, and safety disciplines to ensure that all activities are synchronized and do not interfere with each other, maintaining the critical path of the turnaround.

Defining the required resources is fundamental. This includes identifying skilled personnel and the necessary equipment, materials, and personal protective equipment needed to execute the lubrication tasks safely and effectively. A comprehensive lubrication turnaround plan must be synchronized with the master schedules of the entire project, allowing for float in case of delays in upstream work. This plan should include checklists to ensure that all tasks are accounted for, that there are contingency plans placing restrictions on access to equipment, and that there are arrangements for temporary storage of recovered fluids.

Pre-outage inspections and walkthroughs are crucial. They validate access to necessary locations, confirm connection points, and ensure the availability of correct lubricants. This stage also involves recording condition monitoring baselines before shutdown, providing a reference point for assessing equipment condition post-turnaround. This preparation ensures that lubrication activities are executed smoothly, within budget, and contribute to the overall reliability and efficiency of the plant operations.

During plant turnarounds, the execution of lubrication-related tasks is a critical element that requires meticulous planning and coordination. These periods offer a unique opportunity to address deferred maintenance activities that cannot be performed during normal operations due to the need for equipment shutdown. The success of lubrication activities during turnarounds depends heavily on early and detailed planning.

Developing a comprehensive lubrication work scope is the first step. This scope must outline every lubrication task, from fluid changes and filter replacements to system upgrades and thorough inspections. Budgeting for these activities ensures that financial resources are allocated efficiently, avoiding any last-minute financial constraints that could impact the quality or extent of the work performed.

Mapping of lubrication system isolation points and drain zones is another essential planning action. These mappings allow teams to isolate sections of the system safely, facilitating maintenance without affecting other operations. This step also helps in planning the logistics of fluid removal and storage, ensuring compliance with environmental and safety regulations.

Coordination with the mechanical, electrical, and safety teams is vital. Each discipline must be aware of the lubrication schedule to prevent overlaps or conflicts that could delay the turnaround. This coordination ensures that all teams work harmoniously, adhering to the critical path schedule and meeting project deadlines.

Resource definition is crucial for ensuring that skilled personnel, appropriate equipment, materials, and PPE are available when needed. Identifying and scheduling these resources in advance prevents bottlenecks and ensures that all lubrication tasks are executed efficiently and safely.

The lubrication turnaround plan must be integrated with the master schedule of the entire project. This integration allows for flexibility and adaptability, providing float time to accommodate any delays in upstream activities without impacting downstream operations. Checklists and contingency plans for equipment access restrictions and temporary storage of recovered fluids are integral components of the plan, ensuring that all potential issues are anticipated and addressed.

Pre-outage inspections and walkthroughs are indispensable for validating access to equipment, confirming connection points, and ensuring that the correct lubricants are on hand. Recording condition monitoring baselines before shutdown provides a benchmark for post-turnaround assessments, ensuring that the equipment operates optimally once it is back online. These detailed preparations and coordinated efforts are executed with precision, minimizing the risk of cross-contamination, over- or under-lubrication, and missed service points. They also allow maintenance teams to quickly identify any deviations from expected performance upon restart, trace them to root causes, and respond with confidence. In short, successful lubrication during a turnaround begins long before the outage clock starts it's built on planning, verification, and the disciplined use of condition data to close the loop between inspection, execution, and post-startup reliability.

Table 16.1 is a step-by-step guide for effectively planning lubrication activities during plant turnarounds, structured for clarity and implementation within a broader outage management strategy.

TABLE 16.1 Lubrication Planning Process During Plant Turnarounds

STEP	ACTION	PURPOSE AND CONSIDERATIONS
1. Define lubrication scope	Identify all lubrication-related tasks (oil changes, filter replacements, regreasing, line flushing, component replacements, system upgrades)	Align with asset condition data, OEM recommendations, and deferred maintenance logs. Include inspection and verification tasks
2. Set budget and allocate funding	Estimate cost of lubricants, labor, filtration equipment, fluid handling systems, testing kits, disposal services, and contingencies	Build into overall turnaround budget. Include premium or synthetic fluids if reliability goals support extended intervals
3. Coordinate pre-inspection walkdowns	Conduct walkdowns to confirm equipment access, identify safety hazards, locate drain ports, and validate system schematics	Verify sample ports, ensure compatibility with fluid recovery tools, and check staging areas for drums and carts
4. Map isolation points and drain zones	Diagram system isolation valves, drain points, purge lines, and any necessary lockout/tagout zones	Ensure safe, complete fluid evacuation without cross-contamination. Plan for multisystem sequencing where necessary
5. Schedule with turnaround teams	Integrate lubrication tasks into the master turnaround schedule, coordinating with mechanical, electrical, scaffolding, and safety teams	Prevent task overlap, ensure access timing, and avoid critical path disruptions. Include time buffers for contingencies
6. Define resources	Identify required labor (in-house or contractors), lubricants (quantities and grades), filtration units, flushing rigs, PPE, and tooling	Ensure that resource availability matches task durations and access window
7. Confirm fluid handling and storage	Plan for temporary storage of drained fluids, designate filtration areas, ensure compliant labeling and segregation (used versus new), and prep disposal logistics	Prevent contamination and environmental violations. Use spill containment and SDS documentation
8. Establish QA/QC checks and checklists	Create pre-task checklists (fluid ID, cleanliness level, condition monitoring baseline), post-task checklists (volume recorded, sample taken, label applied), and deviation reporting forms	Standardize documentation, and reduce risk of missed steps. Attach checklists to CMMS work orders
9. Conduct baseline sampling and diagnostics	Sample oil from critical systems before shutdown to establish wear-metal and contamination baselines. Use vibration and ultrasound where needed	Allow post-turnaround comparison. Document in the CMMS

STEP	ACTION	PURPOSE AND CONSIDERATIONS
10. Train personnel, and review procedures	Review turnaround lubrication scope in toolbox meetings. Ensure that all technicians are trained on fluid handling, contamination control, and safety requirements	Avoid delays, and reduce errors due to unfamiliarity with turnaround-specific procedures
11. Execute work, and monitor progress	Perform lubrication tasks according to schedule, updating progress in the turnaround management software or CMMS	Escalate issues quickly. Use photos for documentation if systems are upgraded or deviations occur
12. Conduct post-start-up sampling and validation	Resample fluids after a set runtime (e.g., 8–24 hours) post-start-up. Compare with baseline to confirm successful execution	Detect residual contamination or improper fill. Enable early corrective action if needed

Turnaround Deliverables

- ➤ A complete lubrication task register
- ➤ Pre- and post-turnaround oil analysis reports
- ➤ A deviations log (with corrective actions)
- ➤ A labor and material cost report
- ➤ Lessons learned for continuous improvement

TEMPORARY LUBRICATION MEASURES FOR OFFLINE MACHINERY

Offline or idle machinery is vulnerable to corrosion, internal varnish formation, and elastomer degradation. Temporary lubrication measures help preserve these assets during the outage window.

Protective strategies include:

- ➤ Application of vapor phase corrosion inhibitors (VPCIs) in closed systems
- ➤ Hand-applied or spray-on rust preventives for exposed metal surfaces
- ➤ Inert gas purging or desiccant packs in reservoirs and sumps
- ➤ Manual turning of shafts to redistribute protective film

Gearboxes and bearing housings should be fully topped off or slightly overfilled (within OEM limits) to ensure lubricant coverage of critical components. Systems with open breathers should be sealed or replaced with desiccant breathers.

Documentation of all temporary protection measures should include type, date, personnel, and specific areas protected, ensuring traceability and accountability. Offline or idle machinery faces significant risks, including corrosion, internal varnish formation, and elastomer degradation. Implementing temporary lubrication measures is crucial for preserving these assets during periods when they are not in active use. One effective strategy is the application of VPCIs in closed systems. These inhibitors form a protective layer on metal surfaces, preventing corrosion even in the presence of moisture or other corrosive elements.

For exposed metal surfaces, applying rust prevention products by hand or using spray-on products is essential. These coatings create a barrier against environmental factors that can lead to rust and degradation. Inert gas purging or the use of desiccant packs in reservoirs and sumps can effectively control moisture levels, reducing the risk of corrosion and varnish formation. These methods help maintain a dry environment, which is critical for the longevity of idle machinery.

The manual turning of shafts is another important measure, as it redistributes the protective lubricant film across contact surfaces, ensuring even coverage and preventing localized corrosion or wear. For gearboxes and bearing housings, it is advisable to fully top off or slightly overfill them, staying within OEM limits. This ensures that all critical components remain submerged in lubricant, providing continuous protection.

Systems equipped with open breathers should be sealed or replaced with desiccant breathers. These breathers prevent moisture ingress, maintaining the integrity of the lubricant and protecting the internal components from contamination. Documentation of all temporary protection measures is vital. This documentation should detail the type of protection used, the date of application, the personnel involved, and the specific areas protected. Such records ensure traceability and accountability, providing a clear history of maintenance actions taken during the machinery's offline period. These steps collectively help in safeguarding machinery integrity and readiness for operation once the machinery is brought back online.

When machinery goes offline or becomes idle, it is exposed to various environmental elements that can lead to deterioration if not properly managed. Temporary lubrication measures are essential to mitigate these risks and ensure that the equipment remains in optimal condition during outages or extended periods of inactivity.

Vapor phase corrosion inhibitors are a critical component in protecting closed systems. By sublimating and forming a protective molecular layer on metal surfaces, VPCIs prevent oxidation and corrosion without the need for direct application, making them particularly effective in complex systems where access is limited.

For surfaces that are exposed to the environment, products that prevent rust can be applied either manually or via spray. These protective coatings shield metal surfaces from moisture and other corrosive agents, effectively preventing rust and extending the life of the components.

Inert gas purging involves the introduction of an inert gas, such as nitrogen, into reservoirs and sumps to displace oxygen and moisture. This creates an environment where corrosion is much less likely to occur. Similarly, desiccant packs absorb moisture within enclosed spaces, maintaining dry conditions that are less conducive to corrosion and varnish formation.

Turning shafts manually ensures that the protective lubricant film is evenly distributed across all components. This practice helps prevent the formation of flat spots or corrosion on areas that may not be submerged in lubricant during periods of inactivity.

For gearboxes and bearing housings, maintaining proper lubricant levels is crucial. Overfilling, within OEM-recommended limits, ensures that all internal components are adequately coated, preventing exposure to air and the associated risks of corrosion and wear.

Breathers play a vital role in protecting internal systems from the ingress of contaminants. Replacing open breathers with desiccant breathers allows for moisture control while still permitting necessary airflow. This switch is an effective way to maintain the integrity of lubricants and prevent the introduction of environmental contaminants.

Documenting all temporary protection measures is an essential part of this maintenance strategy. Detailed records provide a comprehensive overview of the actions taken, facilitating accountability and ensuring that protective measures are consistently applied. This documentation is invaluable for future reference and helps in assessing the effectiveness of the protection strategies employed during the offline period. By implementing these measures, the machinery is preserved in a state of readiness, minimizing the risk of damage and ensuring a smooth transition back to operational status when required.

Table 16.2 presents a step-by-step procedure for implementing temporary lubrication measures for off-line machinery, structured for operational use and long-term preservation.

TABLE 16.2 Temporary Lubrication Preservation Plan for Off-Line Equipment

STEP	ACTION	PURPOSE AND METHOD	DOCUMENTATION REQUIRED
1. Assess equipment vulnerability	Identify all machinery scheduled for downtime or idle state. Evaluate exposure risks (e.g., humidity, temperature, dust) and lubrication system type	Prioritize gearboxes, bearing housings, hydraulic reservoirs, and exposed rotating elements	List of equipment, idle duration, and environmental risk classification
2. Apply vapor phase corrosion inhibitors	Introduce VPCIs into closed-loop systems or reservoirs. Follow manufacturer dosage and distribution protocols	Provides noncontact corrosion protection to internal metallic surfaces. Ideal for enclosed systems like hydraulic loops and compressors	Record of chemical type, quantity, application method, and responsible technician
3. Apply surface rust preventives	Spray or hand-apply rust-preventive coatings to exposed shafts, fittings, or unpainted metal surfaces	Shields against atmospheric corrosion, especially for parts exposed to condensation or air exchange	Surface location, product type, date applied, inspection interval.
4. Seal or replace breathers	Remove open breathers and install desiccant breathers, or seal systems completely if safe to do so	Prevents moisture ingress and particle contamination during inactivity	Type and condition of breather, date of installation, system ID
5. Overfill gearboxes/ bearing housings (if OEM-approved)	Fill lubricant slightly above standard levels to immerse critical components fully	Ensures gears, shafts, and bearings stay submerged, preventing dry spots and oxidation	Document actual fluid volume added, target level, and fill date
6. Install desiccants or perform inert gas purge	For reservoirs and tanks, place desiccant packs or perform a nitrogen purge to reduce internal humidity	Maintains a dry environment to avoid water-related degradation (varnish, microbial growth, seal hardening)	Volume of nitrogen used or desiccant type and date installed
7. Manually rotate shafts	Turn shafts manually (weekly or per OEM guidance) to redistribute lubricant and prevent surface drying or brinelling	Evenly coats components with protective film; avoids flat spots and varnish formation	Rotation schedule log, personnel initials, and comments on any observed stiffness or noise
8. Inspect and reapply as needed	Schedule routine inspection during extended outages. Reapply protective products based on environmental exposure and product lifespan	Maintains ongoing protection; ensures no lapse in coverage due to degradation or consumption of temporary products	Inspection reports, replenishment records, and condition notes
9. Restore systems before recommissioning	Before restarting, drain overfilled fluids if necessary, remove desiccants/VPCIs per spec, and sample lubricants for contamination or degradation	Prevents start-up with compromised or inappropriate fluid levels or chemical carryover	Pre-start checklists, lubricant test reports, and recommissioning sign-off

Table 16.3 presents recommended products by function.

TABLE 16.3 Recommended Products by Function

PROTECTION NEED	TYPICAL PRODUCT TYPE	EXAMPLES
Internal corrosion protection	Vapor phase corrosion inhibitor	Zerust® ICT, Cortec® VpCl-329
External surface rust	Spray-on rust inhibitor	LPS® 3, CRC® SP-400
Moisture control	Desiccant breather/desiccant packs	Hy-Dry®, Des-Case® DC-BB
Shaft movement preservation	Manual shaft rotation	No product; mechanical action only
Inert gas purging	Dry nitrogen or argon	Site gas supplier

LUBE SYSTEM PRESERVATION DURING EXTENDED SHUTDOWNS

Extended outages ranging from several weeks to months require formal preservation plans for lubrication systems. Passive and active preservation methods are selected based on system complexity, fluid type, and environment.

Key elements include:

➤ Draining fluids with a short shelf life or marginal condition
➤ Circulating oil with filtration at defined intervals (loop flushes)
➤ Using nitrogen blankets or controlled humidity enclosures
➤ Applying dielectric grease or protectant to exposed fittings and sensor ports

Hydraulic and circulating systems may require bypassing sensitive components during preservation circulation. Sight glasses, sample valves, and pressure sensors should be inspected and cleaned or replaced.

Maintenance logs should document all preservation activities, test results, and next review points. Any preserved system must be tagged and locked to prevent unintended operation. Extended shutdowns lasting several weeks to months necessitate a formal preservation plan for lubrication systems to prevent degradation and ensure operational readiness upon restart. The preservation strategy must be tailored to the specific system based on its complexity, the type of fluid used, and the environmental conditions. Draining fluids that have a short shelf life or are in marginal condition is critical to prevent the deterioration of these fluids, which could compromise system integrity. Circulating oil with filtration at defined intervals, known as "loop flushes," helps in maintaining fluid cleanliness by removing particulates and moisture, thus preserving the quality of the lubricant.

Implementing nitrogen blankets or controlled humidity enclosures is crucial in environments where moisture control is necessary to prevent corrosion. These methods displace oxygen and maintain low-humidity conditions, reducing the risk of rust and oxidation on internal surfaces. Dielectric grease or protective coatings should be applied to exposed fittings and sensor ports to shield them from environmental contaminants and moisture ingress.

In hydraulic and circulating systems, it may be necessary to bypass sensitive components during preservation circulation to avoid damage or wear. This ensures that preservation activities do not inadvertently harm delicate parts. Regular inspection and maintenance of sight glasses, sample valves, and pressure sensors are required. These components should be checked for cleanliness and functionality, and replaced if necessary, to ensure accurate monitoring and sampling capabilities.

Comprehensive maintenance logs are essential for documenting all preservation activities, test results, and scheduled review points. These records enable traceability and accountability, providing a detailed account of the preservation measures implemented. Each preserved system must be clearly tagged and locked to prevent any unintended operation during the shutdown period. This tagging and locking procedure ensures that systems remain in their intended state of preservation and are not accidentally activated, which could undo preservation efforts and lead to potential system damage. Implementing a well-structured preservation plan maintains the integrity and readiness of lubrication systems throughout extended shutdowns, facilitating a seamless transition back to full operation when required.

Table 16.4 presents a step-by-step plan for lube system preservation during extended shutdowns, formatted for field execution, system integrity assurance, and audit traceability.

TABLE 16.4 Preservation Plan for an Extended Shutdown of a Lubrication System

STEP	TASK	PURPOSE/NOTES	DOCUMENTATION REQUIRED
1	Assess system type and conditions	Identify system type (gearbox, hydraulic, circulating oil) and environmental exposure (humidity, temperature, contaminants)	Asset ID, lubricant type, system schematic, ambient risks
2	Drain marginal or short-life fluids	Remove aged, degraded, or low-shelf-life fluids (e.g., vegetable esters, heavily oxidized oils)	Record volume drained, fluid condition, disposal method
3	Perform loop flushes (where applicable)	Circulate oil through fine filtration to remove particulates and water. May use portable kidney loop carts. Repeat every 2–4 weeks	Record loop duration, filter condition, ISO cleanliness code

STEP	TASK	PURPOSE/NOTES	DOCUMENTATION REQUIRED
4	Isolate or bypass sensitive components	Temporarily isolate sensors, servo valves, proportional valves, and instrumentation that may be damaged by repeated flushing or pressure cycling	Annotated diagram of bypass points and locked valves
5	Apply nitrogen blankets or desiccants	For tanks and reservoirs, apply a nitrogen purge or install desiccant breathers/packs to minimize moisture and oxidation	Pressure/vacuum relief logs, nitrogen fill records, desiccant install dates
6	Coat external ports and fittings	Apply dielectric grease or protective wax coating to sensor ports, breathers, and open fittings to shield against ingress	List of locations treated and product type used
7	Inspect and clean accessory components	Clean/replace sight glasses, level indicators, pressure sensors, and sample valves. Ensure visibility and sampling accuracy post-preservation	Before/after photos, replacement part records, inspection checklist
8	Tag, lock, and isolate systems	Physically tag preserved systems: "Do Not Operate—Preserved for Extended Shutdown." Lock isolation valves and remove power if applicable	LOTO log, tag IDs, authorized personnel signatures.
9	Establish monitoring and review schedule	Define intervals (e.g., monthly) for reflushing, nitrogen top-off, desiccant inspection, or visual checks	Calendar entry, task owner assignment, baseline data comparison
10	Pre-restart evaluation and recommissioning	Before restart, test fluid quality (ISO code, water content, TAN), confirm component cleanliness, restore bypassed sections, and remove protective coatings	Restart checklist, lab results, reactivation log

Table 16.5 provides examples of preservation materials and equipment.

TABLE 16.5 Preservation Materials and Equipment Examples

COMPONENT	PRESERVATION MATERIAL OR METHOD	EXAMPLES
Reservoirs and tanks	Nitrogen blanket, desiccant breathers	Numatics® N2 Kit, Des-Case® DC-BB
Hydraulic/recirculating oil	Loop filtration (off-line), drain and store	Pall®, Donaldson®, OilSafe® drums
Sensor ports and fittings	Dielectric grease, waxy barrier coating	Dow Corning® 4, CRC® Heavy Duty Corrosion Inhibitor
Exposed metal surfaces	Rust preventive	LPS® 3, Boeshield® T-9
Elastomers and seals	Compatibility check before idle	Reference seal compatibility matrix

COORDINATION WITH OTHER TRADES TO PREVENT CROSS-CONTAMINATION

Multidisciplinary activity during outages creates high potential for cross-contamination of lubricants, especially when trades are working near open systems. Risks include ingress of debris, cleaning fluids, weld slag, paint, and water.

Preventive steps include:

➤ Locking or capping fill, drain, and sample ports
➤ Assigning lubrication stewards to monitor open systems
➤ Installing signage to alert others of lubricant-sensitive areas
➤ Requiring pre-task coordination meetings involving lubrication personnel

Designated buffer zones should be established around lubricant storage and transfer areas. Ventilation patterns and cleaning agent usage must also be managed to prevent airborne ingress.

Contractor briefings should include lubrication-specific contamination risks and controls. Any breach must be documented, investigated, and addressed immediately. During outages, the simultaneous activities of multiple trades significantly increase the risk of cross-contamination of lubricants, particularly when work occurs near open systems. This contamination can arise from debris, cleaning fluids, weld slag, paint, and water entering the lubrication systems. To mitigate these risks, several preventive measures must be implemented. Fill, drain, and sample ports should be locked or capped to prevent unintended entry points for contaminants. Assigning dedicated lubrication stewards to monitor open systems is crucial. These stewards ensure that all openings are properly sealed, and they oversee activities in lubricant-sensitive areas.

Installing clear signage around lubricant-sensitive areas serves as an effective warning to other trades about the potential risks of contamination. These signs help maintain awareness and caution among personnel working in proximity to open lubrication systems. Pre-task coordination meetings involving lubrication personnel are essential. These meetings facilitate communication and planning among different trades, ensuring that all parties understand the critical importance of maintaining lubricant purity and the specific measures needed to achieve this.

Establishing designated buffer zones around lubricant storage and transfer areas helps protect these critical points from cross-contamination. These zones act as barriers, minimizing the risk of contaminants reaching sensitive lubrication equipment. Additionally, managing ventilation patterns and the usage of cleaning agents is necessary to prevent airborne contaminants from entering open systems. Proper ventilation helps direct potential contaminants away from lubricant areas, while careful selection and application of cleaning agents reduce the risk of chemical ingress.

Contractor briefings must include detailed information on lubrication-specific contamination risks and the controls in place to prevent such incidents. This ensures that all personnel, including external contractors, are aware of the protocols and their roles in maintaining system integrity. Any breach of these protocols must be documented, promptly investigated, and addressed to prevent recurrence. By implementing these strategies, the risk of cross-contamination is minimized, ensuring the integrity and performance of lubrication systems during and after outages.

Table 16.6 offers an example of a coordination plan to prevent lubrication system cross-contamination during outages, formatted in checklist and planning form for use during project briefings and outage execution:

TABLE 16.6 Cross-Contamination Prevention Plan for a Lubrication System

STEP	ACTION ITEM	PURPOSE/NOTES	RESPONSIBLE PARTY	DOCUMENTATION/ TOOLS
1	Lock or cap all lubrication system openings	Prevent ingress of particulates, liquids, and fumes into fill, drain, and sample points	Lube team/ stewards	Lockout tags, port caps, tamper-evident seals
2	Assign lubrication stewards during outage	Monitor lubrication-critical assets; enforce protection protocols in field	Maintenance supervisor	Steward assignment log; area checklist
3	Install clear signage around lubricant-sensitive equipment	Notify other trades (welding, painting, blasting) of high-sensitivity zones	Lube team/safety	"Lube Critical Area" signs, floor tape, physical barriers
4	Conduct pre-task coordination meetings involving lube personnel	Ensure lubrication team is aware of scope, sequencing, and trade activities	Outage planner/ all trades	Meeting minutes, task alignment matrix
5	Establish buffer zones around lube storage and transfer areas	Prevent intrusion, foot traffic, and airborne debris near drums, carts, or tanks	Logistics/facilities	Floor marking plan, access restriction signage
6	Manage ventilation and cleaning agent usage	Prevent vapors, mists, and overspray from reaching open or vented systems	EHS/ housekeeping lead	Vent maps, SDS controls, cleaning SOPs

(continued on next page)

TABLE 16.6 Cross-Contamination Prevention Plan for a Lubrication System (*continued*)

STEP	ACTION ITEM	PURPOSE/NOTES	RESPONSIBLE PARTY	DOCUMENTATION/ TOOLS
7	Include lube risks and controls in contractor briefings	Educate external trades on how their work could affect lubricated systems	Contractor manager/QA	Contractor onboarding checklist, safety orientation deck
8	Do real-time monitoring for breach or intrusion	Identify and mitigate any breach (e.g., slag entry, paint overspray) as it occurs	Lubrication steward/QA	Breach logbook, incident report template
9	Document and respond immediately to any breach	Conduct root cause analysis, apply corrective measures, and document for traceability	Reliability engineer/QA	Root cause analysis form, photographic evidence, system flush records
10	Do a final inspection and system revalidation before start-up	Confirm that lubricant systems remain clean, filled, and uncontaminated	Maintenance and operations	ISO 4406 particle count, Karl Fischer moisture test, start-up clearance checklist

Table 16.7 shows examples of guidelines for buffer zones.

TABLE 16.7 Buffer Zone Guidelines (Examples)

AREA TYPE	BUFFER ZONE RADIUS	ACCESS RESTRICTIONS	VENTILATION REQUIREMENTS
Lubricant drums or totes	2 meters (6 feet)	No hot work, no liquids, no chemical use	Positive pressure ventilation if indoors
Open sumps or reservoirs	3 meters (10 feet)	Barrier plus signage, no foot traffic	Downdraft protection, vapor isolation
Portable filtration carts in use	1.5 meters (5 feet)	Tools-only access, PPE required	Controlled airflow if enclosed

The following are examples of signage:

➤ LUBRICATION-SENSITIVE EQUIPMENT—NO SPRAYING, GRINDING, OR WASHDOWN IN THIS AREA
➤ SEALED SYSTEM—DO NOT OPEN WITHOUT AUTHORIZATION
➤ CONTAMINATION CONTROL ZONE—ENTRY REQUIRES APPROVAL

FLUSHING PROTOCOLS
AND VERIFICATION TECHNIQUES

Flushing during outages helps remove contaminants accumulated during operation and introduces new lubricants under controlled conditions. The type of flush depends on system criticality, fluid condition, and OEM requirements.

Common flushing techniques include:

➤ **Drain and refill.** Simple systems with low contamination risk
➤ **Recirculation flush.** Extended circulation through external filtration until ISO target is met
➤ **High-velocity flush.** For hydraulic systems with complex flow paths

Verification involves:

➤ Particle count testing (ISO 4406, NAS 1638)
➤ Water content (Karl Fischer titration)
➤ Visual inspection for varnish, foam, or discoloration
➤ Conductivity or dielectric tests (for insulating oils)

Flushing must be accompanied by system tagging, controlled disposal of used fluids, and post-flush certification by lubrication or reliability personnel. Recordkeeping is critical for warranty compliance and quality assurance.

Flushing protocols during outages are essential for removing contaminants that accumulate during regular operation and for introducing new lubricants under controlled conditions. The choice of flushing technique is guided by the system's criticality, the condition of the fluid, and specific OEM requirements. Simple systems with a low contamination risk often use the drain and refill method, where old lubricant is drained and replaced with fresh fluid. This straightforward approach is effective when the risk of residual contamination is minimal.

For systems requiring more thorough cleaning, a recirculation flush is employed. This involves extended circulation of the lubricant through external filtration until the cleanliness level reaches the ISO target. This method is particularly beneficial for systems where contaminants may be more deeply embedded. High-velocity flushing is used for hydraulic systems with complex flow paths. This technique involves circulating the fluid at high speeds to dislodge and remove contaminants from intricate system components.

Verification of the flushing process is crucial to ensure the effectiveness of the cleaning. Particle count testing according to standards like ISO 4406 or NAS 1638 provides a quantitative measure of cleanliness. Water content is assessed using Karl Fischer titration to detect and quantify water levels in the lubricant, which is critical for preventing corrosion and degradation. Visual inspections help identify the presence of varnish, foam, or discoloration, indicating potential issues with the lubricant or system condition. Conductivity or dielectric tests are performed for insulating oils, ensuring that the fluid maintains its electrical insulating properties.

Flushing procedures must include system tagging to clearly indicate that the process has been completed, along with the controlled disposal of used fluids in accordance with environmental regulations. Post-flush certification by lubrication or reliability personnel is essential to verify that the system meets required standards and is ready for operation. Meticulous recordkeeping of all flushing activities is vital for warranty compliance and quality assurance, providing a documented history of maintenance actions and their outcomes. These records support future maintenance planning and help ensure the long-term reliability and performance of the lubrication system.

The choice of flushing techniques in lubrication systems is influenced by several key factors:

1. **System criticality.** Systems that are more critical, where failure could lead to significant operational downtime or safety hazards, often require more thorough and rigorous flushing methods to ensure the complete removal of contaminants.

2. **Fluid condition.** The current condition of the lubricant, including its level of contamination, oxidation, and degradation, determines the intensity and type of flushing required. Severely contaminated fluids may necessitate more aggressive or extended flushing processes.

3. **System complexity.** Systems with complex flow paths, such as those found in hydraulic systems, may require high-velocity flushing to effectively clean all components. Simpler systems might be adequately serviced with a straightforward drain and refill method.

4. **OEM requirements.** OEM specifications and recommendations play a significant role in determining the appropriate flushing technique. Adhering to these guidelines ensures compliance and maintains warranty coverage.

5. **Contaminant type.** The nature of the contaminants present, whether particulate, water, varnish, or chemical, influences the choice of flushing technique. Different contaminants may require specific methods for **effective removal.**

6. **Environmental considerations.** Environmental regulations and the need for controlled fluid disposal may impact the choice of technique, favoring methods that minimize waste and environmental impact.

7. **Operational constraints.** Time constraints, availability of resources, and operational schedules can also influence the choice, as certain flushing methods may require more time and specialized equipment than others.

By evaluating these factors, maintenance teams can select the most appropriate and effective flushing technique to ensure optimal system cleanliness and performance.

The following is an example of a standard operating procedure for lubrication system flushing protocols and verification techniques, designed for use during turnarounds or major maintenance events.

Standard Operating Procedure (SOP): Flushing Protocols and Verification Techniques

Document No.: LUBE-SOP-004
Effective Date: [Insert date]
Approved By: [Insert name/title]
Reviewed Annually

1. Purpose
To define standardized flushing procedures and verification methods to ensure lubricant systems are cleaned of contaminants and prepared for fresh lubricant introduction under controlled, validated conditions.

2. Scope
Applies to all plant equipment with circulating, hydraulic, or enclosed oil systems scheduled for maintenance or recommissioning following shutdown, outage, or contamination event.

3. Responsibilities

ROLE	RESPONSIBILITIES
Lubrication Technician	Execute flushing procedures; perform fluid handling, tagging, and basic inspection
Reliability Engineer	Select flushing type, verify completion, interpret test results, and authorize system release
Maintenance Planner	Schedule resources, coordinate with other trades; ensure system access and documentation
HSE Coordinator	Approve disposal methods for waste fluids; oversee environmental compliance

- ➤ **Lubrication Technician:** Executes flushing procedures and performs fluid handling, tagging, and basic inspection
- ➤ **Reliability Engineer:** Selects flushing type, verifies completion, interprets test results, and authorizes system release
- ➤ **Maintenance Planner:** Schedules resources, coordinates with other trades, and ensures system access and documentation
- ➤ **HSE Coordinator:** Approves disposal methods for waste fluids and oversees environmental compliance

4. Flushing Types and Procedures

FLUSHING TYPE	USE CASE	PROCEDURE SUMMARY
Drain and refill	Clean systems with low contamination risk or light-duty use	- Drain system to waste - Inspect reservoir - Refill with OEM-approved lubricant - Label and log
Recirculation flush	Systems with moderate to high contamination or varnish risk	- Fill system with flushing fluid or clean base oil - Circulate through high-efficiency external filtration - Run until ISO target achieved - Drain and refill with operational fluid
High-velocity flush	Complex hydraulic systems, directional valves, servo circuits	- Connect external pump skid - Install jumpers across components to create loop - Circulate fluid at 2x to 5x nominal flow rate - Monitor filter ΔP and particle count until clear

5. Verification and Acceptance Criteria

TEST	STANDARD/METHOD	PURPOSE	PASS CRITERIA
Particle count	ISO 4406/NAS 1638	Cleanliness level	Hydraulics: ≤ 16/14/11 Gears: ≤ 18/16/13
Water content	Karl Fischer titration	Detect moisture	≤ 500 ppm (unless OEM states otherwise)
Visual inspection	Direct examination	Identify varnish, foam, discoloration	Clear, bright, no odor, no residue
Dielectric strength (insulating oils)	ASTM D877/D1816	Confirm electrical integrity	≥ 30 kV, depending on OEM spec
Conductivity	< 200 pS/m (insulating oils)	Electrical reliability	As per OEM for EHC or turbine systems

6. Procedure Steps (Generalized)
Pre-Flushing Preparation
- Review OEM specs and previous oil analysis history.
- Isolate system electrically and mechanically.
- Identify drain, fill, and bypass points.
- Prepare flushing rig, hoses, filters, and sampling kit.
- Place spill containment, and acquire waste drums.

Flushing Execution
- Conduct selected flushing method as per Section 4.
- Collect in-process samples for visual and lab testing.
- Monitor and record system temperature, pressure, and ΔP.
- When target cleanliness is achieved, shut down and drain flush fluid.
- Refill with fresh lubricant and label.

Verification and Documentation
- Collect post-flush sample for lab testing.
- Inspect reservoirs, sight glasses, and sample valves.
- Tag system as "Flushed & Charged."
- Complete Flushing Certification Report.
- Update CMMS with test results and next review interval.

7. Waste Management
- All flushed fluids must be disposed of via a **certified waste oil contractor**.
- Flush rig filters must be tagged and bagged for HSE disposal.
- Waste volumes and disposal receipts must be logged with maintenance records.

8. Safety and Environmental Controls
- Use full PPE: gloves, goggles, flame-resistant coveralls, and face shield.
- Ventilation is required in enclosed or low-flow areas.
- Use drip trays and absorbent mats under all hose couplings.
- Lockout/tagout must remain in place until the flush is verified complete.

9. Documentation
- Flushing Certification Report (Form LUBE-04A)
- Post-Flush Lab Test Reports (attach to CMMS)
- Change-Out and Disposal Log (Form HSE-WASTE-001)

➤ Lubrication System Cleanliness Tag (green = pass; red = retest required)

10. References
➤ ISO 4406 Cleanliness Codes
➤ ASTM D4057 Sampling Practices
➤ OEM Equipment Manuals
➤ Plant Environmental Waste Policy

RECOMMISSIONING LUBRICATION SYSTEMS POST-SHUTDOWN

Bringing systems back online requires systematic recommissioning of lubrication circuits to avoid start-up failures. Recommissioning activities include:

➤ Verifying oil and grease types, volumes, and cleanliness
➤ Ensuring that reservoirs are full, and filtration systems are intact
➤ Replacing temporary protectants with operational lubricants
➤ Priming pumps and establishing flow in circulation systems
➤ Bleeding air from hydraulic lines

Technicians should monitor key indicators (pressure, temperature, flow) during the first hours of operation. Oil samples should be taken within 24 hours to assess fluid stability post-start.

Recommissioning lubrication systems after a shutdown involves a series of systematic steps to ensure that the systems are ready for safe and reliable operation. The process begins with verifying the presence of the correct types and volumes of oil and grease and confirming the cleanliness of these lubricants. Ensuring that reservoirs are completely filled and that filtration systems are functioning properly is critical to prevent contamination and maintain fluid integrity.

Temporary protectants that were applied during the shutdown must be removed and replaced with the appropriate operational lubricants before the systems are brought back online. This ensures compatibility and performance as per the original system specifications. Priming the pumps is essential to establish proper flow within the circulation systems, preventing dry starts that could cause damage to the components. Air must be bled from hydraulic lines to avert issues such as spongy operation or cavitation, which can impair system performance and lead to failures.

During the initial hours of operation, the technicians should closely monitor key indicators such as pressure, temperature, and flow. These parameters provide early warnings of potential issues and help ascertain that the systems are operating within their designed parameters. Oil samples should be taken within the first 24 hours post-start to assess fluid stability and detect any early signs of degradation or contamination.

Any anomalies detected during this initial phase should prompt immediate inspection and corrective action, particularly in machinery that is high-speed or sensitive to thermal variations. Secondary checks using post-start vibration analysis, thermography, and ultrasound can provide additional confirmation that the lubrication systems are functioning correctly and that all components are receiving adequate lubrication. These diagnostic tools can help identify hidden issues that might not be immediately apparent through basic monitoring alone, ensuring comprehensive validation of system readiness.

The following is a standard operating procedure for recommissioning lubrication systems post-shutdown, suitable for integration into plant reliability protocols and turnaround plans.

Standard Operating Procedure (SOP): Recommissioning Lubrication Systems Post-Shutdown

Document ID: LUBE-SOP-007
Effective Date: [Insert date]
Owner: Maintenance & Reliability Manager
Review Cycle: Annual

1. Purpose
To establish a structured process for safely recommissioning lubrication systems after a planned shutdown, ensuring that all circuits are clean, full, and functional and are protected against start-up-induced failures.

2. Scope
Applicable to all circulating oil systems, hydraulic systems, grease-fed components, and centralized lubrication networks affected during plant outages, shutdowns, or lay-ups.

3. Responsibilities

ROLE	RESPONSIBILITIES
Lubrication Technician	Execute lubricant changes, air bleeding, tagging, and sample collection
Maintenance Supervisor	Oversee compliance with recommissioning checklist and safety procedures
Reliability Engineer	Validate operational indicators, review start-up oil analysis, and authorize release
HSE Officer	Verify environmental controls and waste fluid disposal

➤ **Lubrication Technician:** Executes lubricant changes, air bleeding, tagging, and sample collection
➤ **Maintenance Supervisor:** Oversees compliance with recommissioning checklist and safety procedures
➤ **Reliability Engineer:** Validates operational indicators, reviews start-up oil analysis, and authorizes release
➤ **HSE Officer:** Verifies environmental controls and waste fluid disposal

4. Recommissioning Checklist

Pre-Start Activities

➤ Verify correct lubricant (type, volume, and cleanliness) against OEM specification.
➤ Confirm removal of vapor phase corrosion inhibitors (VPCIs), protectants, and desiccant breathers.
➤ Top off reservoirs to full marks with filtered lubricant.
➤ Replace all used filters; check that filtration systems are seated and intact.
➤ Remove any locks/tags applied during shutdown (with authorization).
➤ Ensure that sample ports, drain ports, and breathers are closed or fitted correctly.

Circulation System Preparation

➤ Prime oil pumps or central lubrication pumps to prevent dry start.
➤ Establish lubricant flow in all circuits (visual inspection or flowmeters).
➤ Bleed air from hydraulic lines via bleed screws or vented cycling.
➤ Check accumulator pre-charge pressure if present.

Initial Operation Monitoring
- ➤ Run equipment at no-load or low-load where possible.
- ➤ Monitor:
  - Lubrication pressure (against OEM specs)
  - Flow (if metered)
  - Reservoir and return oil temperature
  - Filter differential pressure
- ➤ Record all values at 15-minute intervals for the first hour.

Post-Start Testing
- ➤ Collect oil samples within the first 4–24 hours for:
  - Particle count (ISO 4406)
  - Water content (Karl Fischer or crackle test)
  - FTIR or TAN baseline if condition-based monitoring is used.
- ➤ Document observations: foaming, aeration, leaks, or odor.

5. Optional Diagnostic Validation

To ensure deeper verification on high-value assets, perform:
- ➤ **Thermal imaging** of bearings, gearboxes, and hydraulic circuits to verify thermal balance and lubrication film effectiveness
- ➤ **Ultrasound monitoring** to detect friction or lack of lubrication
- ➤ **Vibration analysis** for early bearing lubrication faults or resonance changes due to fluid dynamics

6. Documentation

RECORD	STORAGE SYSTEM
Completed Recommissioning Checklist	CMMS or reliability system
Start-Up Oil Sample Report	Lubricant analysis database
Diagnostic Reports (if used)	Asset reliability folder
Equipment Release Sign-Off	Signed by Reliability Engineer

7. Safety and Environmental Notes

- ➤ Dispose of used protectants and temporary lubricants per site waste management procedures.
- ➤ Use PPE—gloves, safety glasses, oil-resistant footwear, and face shield—during sampling or fluid handling.
- ➤ Never operate systems with open sample ports or uncapped vents.

8. References
- ➤ OEM lubrication manuals
- ➤ ASTM D4057 (sampling)
- ➤ ISO 4406 (cleanliness coding)
- ➤ Plant Oil Analysis Specification Guide

OUTAGE LESSONS LEARNED: BUILDING A CONTINUOUS IMPROVEMENT LOOP

Post-outage reviews should capture lessons related to lubrication planning, execution, and performance outcomes. These insights support future outages and drive procedural improvement.

Key review components include:

- ➤ Task completion rates and compliance with scope
- ➤ Incidents of contamination or missed lubrication steps
- ➤ Unplanned corrective work tied to lubrication deficiencies
- ➤ Fluid analysis comparisons (pre-shutdown versus post-start)
- ➤ Resource utilization and labor effectiveness

Documented lessons learned should feed into updated SOPs, training curricula, and vendor qualification protocols. They also inform capital expenditure (CAPEX) planning for system upgrades like automated lubricators or improved filtration.

A root cause approach to analyzing any lubrication-related failures allows organizations to identify gaps in planning, execution, or equipment design, leading to more resilient lubrication programs in future outages.

Post-outage reviews play a critical role in capturing valuable lessons related to lubrication planning, execution, and performance outcomes. These reviews are essential for supporting future outages and driving procedural improvements. Evaluating task completion rates and compliance with the set scope provides insights into the efficiency and effectiveness of the lubrication activities performed during the outage. Identifying any incidents of contamination or missed lubrication steps is crucial for understanding potential weaknesses in the process and for preventing recurrence.

Analyzing instances of unplanned corrective work that are tied to lubrication deficiencies helps pinpoint areas where lubrication practices may have fallen short. Comparing fluid analysis results from pre-shutdown and post-start periods allows for an assessment of the lubrication system's stability and performance, highlighting any deterioration or improvements.

Resource utilization and labor effectiveness are evaluated to ensure that the right skills and manpower were deployed efficiently. This assessment helps in refining future resource allocation and optimizing labor effectiveness for upcoming outages. Documented lessons learned from these reviews should be integrated into updated standard operating procedures, training programs, and vendor qualification protocols, ensuring that these insights lead to tangible improvements in processes and practices.

The insights gained also inform decisions related to CAPEX planning, such as investments in system upgrades like automated lubricators or enhanced filtration systems. These improvements can significantly enhance lubrication reliability and performance.

Adopting a root cause analysis approach for any lubrication-related failures allows organizations to systematically identify gaps in planning, execution, or equipment design. This approach leads to the development of more resilient lubrication programs, ensuring that lessons learned are effectively applied to enhance system reliability and performance in future outages.

POST-OUTAGE REVIEW CHECKLIST

1. **Task Completion:**
   - ➤ Verify the completion of all lubrication tasks against the planned scope.
   - ➤ Check for adherence to timelines and any deviations from the schedule.
2. **Contamination Incidents:**
   - ➤ Document any occurrences of contamination.
   - ➤ Identify the sources and types of contaminants involved.
3. **Missed Lubrication Steps:**
   - ➤ Record any lubrication steps that were missed or improperly executed.
   - ➤ Analyze the impact of missed steps on system performance.
4. **Unplanned Corrective Work:**
   - ➤ List any unplanned work that was required due to lubrication deficiencies.
   - ➤ Determine the root causes of lubrication-related failures.

5. **Fluid Analysis:**
   ➤ Compare the pre-shutdown and post-start fluid analysis results.
   ➤ Identify the changes in fluid condition and potential causes.
6. **Resource Utilization:**
   ➤ Evaluate the effectiveness of resource allocation and labor.
   ➤ Assess whether the skills and manpower were adequate for the tasks.
7. **System Performance:**
   ➤ Monitor the key performance indicators (pressure, temperature, flow) post-start.
   ➤ Record any anomalies or deviations from expected performance.
8. **Compliance and Documentation:**
   ➤ Ensure that all procedures and protocols were followed as per guidelines.
   ➤ Maintain comprehensive documentation of all activities and findings.
9. **Lessons Learned:**
   ➤ Compile insights and recommendations for process improvements.
   ➤ Discuss findings with relevant teams and stakeholders.
10. **SOP and Training Updates:**
    ➤ Integrate lessons learned into updated SOPs and training curricula.
    ➤ Review and adjust the vendor qualification protocols if necessary.
11. **CAPEX Planning:**
    ➤ Identify potential system upgrades or investments needed.
    ➤ Plan for enhanced equipment, such as automated lubricators or improved filtration systems.
12. **Root Cause Analysis:**
    ➤ Conduct root cause analysis for any failures related to lubrication.
    ➤ Develop action plans to address identified gaps and prevent future issues.

Using this checklist helps ensure a thorough evaluation of lubrication systems post-outage, fostering continuous improvement and enhanced reliability in future operations.

CHAPTER 17

WARRANTY AND REGULATORY COMPLIANCE MANAGEMENT

LUBRICATION RECORDS AS LEGAL EVIDENCE IN WARRANTY CLAIMS

I n warranty disputes, detailed lubrication records are indispensable as legal evidence. They prove adherence to OEM guidelines and proper asset maintenance, often determining the outcome of warranty disputes. These records provide crucial documentation, such as lubrication task completion, including dates, times, and the responsible technician. They must also specify the product used, its volume, and the application method, ensuring alignment with the manufacturer's specifications.

Oil and grease sample results are pivotal, detailing ISO cleanliness levels, viscosity, and additive data, reflecting the fluid's condition throughout its service life. Logs that document equipment operating conditions during service intervals add another layer of evidence, demonstrating that the machinery was operated within its design parameters. These records must be stored in systems that are accessible, time-stamped, and tamper-resistant to ensure integrity. Electronic maintenance management platforms, such as CMMSs or EAM, are integral, offering traceability and seamless integration with other maintenance logs.

In industries with high stakes, such as power generation or aerospace, these logs are routinely scrutinized by legal and compliance officers. Ensuring that these records are meticulously maintained and readily accessible is critical, as they serve as the backbone of any warranty claim process and compliance verification. The precision and accuracy of these records can significantly influence the decision-making process in legal and regulatory contexts, highlighting their importance in maintaining operational credibility and legal standing.

Critical documentation includes:

➤ Lubrication task completion records (date, time, technician)
➤ Product used, volume applied, and method of application
➤ Oil and grease sample results (with ISO cleanliness, viscosity, and additive data)
➤ Logs showing equipment operating conditions during the service interval

Records must be stored in accessible, time-stamped, and tamper-resistant systems. Electronic maintenance management platforms (e.g., CMMS, EAM) provide traceability and integration with other maintenance logs.

Common Causes of Warranty Disputes in Machinery

Common causes of warranty disputes include:

➤ **Improper maintenance.** Failure to adhere to the recommended maintenance schedules or procedures often leads to disputes. If a component fails, and maintenance records do not show compliance with OEM guidelines, warranty claims may be denied.
➤ **Use of non-OEM parts or lubricants.** Utilizing parts or lubricants not specified or approved by the original equipment manufacturer can void warranties. Disputes arise when such substitutions are identified as contributing to failures.
➤ **Operational misuse.** Operating machinery outside its designed parameters, such as overloading or inappropriate use, can lead to failures and subsequent disputes over warranty coverage.
➤ **Inadequate documentation.** Lack of comprehensive maintenance records or discrepancies in documentation can hinder the verification of proper care, leading to denied claims.
➤ **Installation errors.** Incorrect installation or setup of machinery components can result in premature failures. If installation is not performed by authorized personnel or according to OEM specifications, warranty claims may be challenged.

➤ **Defective manufacture.** Disputes can also arise from manufacturing defects. However, proving a defect existed at the time of manufacture often requires detailed inspection and evidence.

➤ **Environmental factors.** Exposure to extreme environmental conditions outside the specified operating range can cause damage not covered by warranty, leading to disputes.

➤ **Delayed reporting.** Failing to report issues promptly after they occur can complicate warranty claims, as timely notification is often a condition of warranty coverage.

Addressing these issues proactively through proper maintenance practices, thorough documentation, and adherence to OEM guidelines can help minimize the risk of warranty disputes.

Effective strategies for maintaining machinery to prevent issues include:

➤ **Scheduled preventive maintenance.** Establish regular maintenance intervals based on OEM recommendations and the specific operating conditions of the machinery. This helps to identify potential issues before they lead to failures.

➤ **Condition monitoring.** Implement technologies such as vibration analysis, thermography, and oil analysis to continuously monitor machinery health. These tools provide early warnings of wear or malfunction.

➤ **Proper lubrication management.** Use the correct lubricants as specified by the manufacturer, and ensure that they are applied in the right quantities and at the right intervals. Regularly check and replace lubricants as needed.

➤ **Training and skill development.** Ensure that maintenance personnel are well trained and knowledgeable about the specific machinery they are servicing. Regular training updates can help technicians stay current with the latest maintenance techniques.

➤ **Detailed recordkeeping.** Maintain comprehensive maintenance records, including dates, tasks performed, parts replaced, and any anomalies observed. This documentation is crucial for both troubleshooting and warranty claims.

➤ **Use of OEM parts.** Always use the original equipment manufacturer's parts for replacements to ensure compatibility and maintain warranty coverage.

➤ **Regular inspections.** Conduct routine inspections to detect signs of wear, corrosion, or misalignment. Addressing these issues early can prevent more significant problems.

➤ **Calibration and alignment.** Regularly calibrate and align machinery components to ensure that they operate efficiently and reduce undue stress on parts.

➤ **Environmental control.** Maintain a clean and controlled environment, free from dust, moisture, and extreme temperatures, which can accelerate wear and cause damage.

➤ **Feedback loop for continuous improvement.** Use maintenance data to identify trends and areas for improvement, refining maintenance strategies over time to enhance machinery reliability.

Implementing these strategies helps ensure that the machinery operates efficiently, reducing downtime and extending the lifespan of the equipment.

NAVIGATING REGULATORY FRAMEWORKS: LUBRICANTS AND ENVIRONMENTAL LAW

Lubricants are subject to a variety of environmental, health, and safety regulations. These frameworks govern production, labeling, application, disposal, and chemical content.
Key regulatory bodies and guidelines include:

➤ **REACH (EU)**—Registration, Evaluation, Authorisation and Restriction of Chemicals

➤ **EPA (USA)**—Environmental Protection Agency. Clean Water Act (CWA), Resource Conservation and Recovery Act (RCRA)

➤ **OSHA (USA)**—Occupational Safety and Health Administration. Hazard Communication Standard (HCS)

➤ **NSF International** (formerly the National Sanitation Foundation). Standards for food-grade lubricants (H1, H2, H3)

Companies must ensure that lubricant formulations meet regional requirements for toxicity, volatility, and biodegradability. Safety Data Sheets (SDSs), proper labeling, and spill prevention plans are essential.

Navigating regulatory frameworks for lubricants involves understanding and complying with a complex array of environmental, health, and safety regulations. These regulations dictate the production, labeling, application, disposal, and chemical content of lubricants. Key regulatory bodies and guidelines include REACH in the European Union, which oversees the registration, evaluation, authorization, and restriction of chemicals. In the United States, the EPA enforces regulations such as the CWA and the RCRA, focusing on water pollution and waste management, respectively.

OSHA mandates compliance with the HCS, ensuring that hazards related to lubricants are clearly communicated to workers. NSF International provides standards for food-grade lubricants, with classifications such as H1, H2, and H3, to ensure safety in environments where lubricants may come into contact with food products.

Companies must ensure that their lubricant formulations adhere to regional requirements concerning toxicity, volatility, and biodegradability. This involves maintaining accurate and up-to-date SDSs, ensuring proper labeling, and implementing spill prevention plans to mitigate environmental impact. For cross-border operations, it is crucial to monitor differences in permissible additives and disposal methods, as these can vary significantly among jurisdictions.

Noncompliance with these regulations can result in severe consequences, including fines, operational restrictions, and damage to a company's reputation. Therefore, a thorough understanding of, and adherence to, these regulatory frameworks is essential to ensure legal compliance and protect both the environment and public health.

COMPLIANCE-DRIVEN LUBRICANT SELECTION FOR FOOD, PHARMA, AND CLEAN INDUSTRY

Industries with heightened hygiene or contamination concerns require lubricants that comply with strict safety standards. Selection must balance performance with regulatory acceptance.

For food and beverage, pharmaceutical, and clean manufacturing environments, selection criteria include:

- ➤ NSF H1 (incidental contact), H2 (no contact), and ISO 21469 certifications
- ➤ Resistance to washout and chemical cleaners
- ➤ Use of base oils and additives recognized as safe (e.g., white oils, synthetic esters)
- ➤ Absence of heavy metals, allergens, and prohibited substances

OEMs serving these industries often provide certified lubricant lists. Substitutions must be validated with documentation and, where necessary, third-party certification. Failure to use compliant lubricants can lead to product recalls, contamination incidents, or audit failures by regulatory agencies or customers. In industries such as food and beverage, pharmaceuticals, and clean manufacturing, lubricant selection is driven by compliance with stringent safety standards due to heightened hygiene and contamination concerns. The selection process must carefully balance performance attributes with regulatory acceptance to ensure both operational efficiency and adherence to safety regulations.

Key selection criteria for lubricants in these environments include certifications such as NSF H1 for incidental contact and H2 for noncontact applications, along with ISO 21469, which ensures hygienic and safe lubricant manufacturing processes. Lubricants must exhibit resistance to washout and chemical cleaners to maintain efficacy in demanding conditions. The use of base oils and additives recognized as safe, such as white oils and synthetic esters, is critical to ensure compatibility with regulatory requirements.

Lubricants must be free from heavy metals, allergens, and prohibited substances to prevent contamination and ensure compliance with industry standards. OEMs catering to these sectors often provide lists of certified lubricants, which serve as a guide for compliance-driven selection. Any substitutions require thorough validation, accompanied by proper documentation and, where necessary, third-party certification to confirm compliance.

Using noncompliant lubricants can lead to severe consequences, including product recalls, contamination incidents, and failures in audits conducted by regulatory agencies or customers. Therefore, meticulous attention to compliance in lubricant selection is essential to safeguard product integrity and maintain regulatory and consumer trust.

OEM WARRANTY REQUIREMENTS: HIDDEN PITFALLS IN LUBRICATION PRACTICE

Warranty terms frequently contain clauses that invalidate coverage if nonapproved lubricants or procedures are used. These requirements are often buried in technical appendices or footnotes, creating risk for unwary maintenance teams.

Maintenance and legal teams must work together to interpret and enforce warranty conditions proactively. OEM warranty requirements often include specific clauses that can invalidate coverage if nonapproved lubricants or procedures are used. These critical stipulations are frequently embedded in technical appendices or footnotes, posing a risk to maintenance teams who may inadvertently overlook them.

Common pitfalls include using lubricants that are considered equivalent but are not listed as approved by the OEM. This extends to nonapproved drain intervals or product substitutions, which can lead to premature wear or failure. Mixing lubricant brands without verifying their compatibility is another risk, as it may affect performance and lead to equipment issues. Deviating from OEM-specified contamination control measures can also void warranties, as these controls are essential for maintaining equipment integrity.

To avoid these pitfalls and potential warranty voidance, it's crucial to maintain a central repository of warranty documents and lubricant specifications. This repository should be easily accessible to all relevant team members to ensure compliance with OEM requirements. Documenting all deviations with correspondence from the OEM is essential to provide a clear audit trail and justification for any changes. Engaging OEMs in the approval process for alternative lubricants or intervals is a proactive approach to ensure that any deviations are sanctioned and do not jeopardize warranty coverage.

Collaboration between the maintenance and legal teams is vital to interpreting and enforcing warranty conditions effectively. By working together, these teams can ensure that all warranty requirements are understood and adhered to, minimizing risks and maintaining coverage.

RECORDKEEPING STANDARDS TO SUPPORT AUDITS AND INSPECTIONS

Accurate and retrievable lubrication records are essential for internal audits, external inspections, and third-party certifications (e.g., ISO 9001, ISO 14001). These records demonstrate procedural adherence, product integrity, and traceability.

Best practices include:

- Keeping digital logs with technician sign-off and task verification
- Cross-referencing with inventory control systems
- Tracking lubricant batch numbers, suppliers, and lot traceability
- Retaining oil analysis results for defined retention periods

Systems should be tested regularly to ensure data integrity and completeness. Accurate and retrievable lubrication records are critical for supporting internal audits, external inspections, and third-party certifications such as ISO 9001 and ISO 14001. These records substantiate procedural adherence, product integrity, and traceability, serving as a foundation for demonstrating compliance and operational excellence.

Best practices for recordkeeping include maintaining digital logs with technician sign-off and task verification, ensuring accountability and accuracy in maintenance activities. These logs should be cross-referenced with inventory control systems to align lubricant usage with inventory levels, enhancing oversight and control.

Tracking lubricant batch numbers, suppliers, and lots is essential for pinpointing any issues related to specific batches and facilitating swift corrective actions if needed. Retaining oil analysis results for defined retention periods provides historical data that can be invaluable during audits, helping to demonstrate ongoing compliance and condition monitoring.

Audit-ready documentation simplifies responses to inquiries and mitigates the risk of nonconformance findings. To ensure reliability, systems should be tested regularly, verifying data integrity and completeness. This proactive approach to recordkeeping not only supports compliance but also enhances transparency and operational efficiency, building confidence with auditors and stakeholders.

USING LUBRICATION LOGS TO DEFEND AGAINST PRODUCT LIABILITY

In the event of a product failure that causes injury, property damage, or operational loss, lubrication logs can be pivotal in defending against liability claims. These records show that maintenance practices met industry standards and that failure was not due to negligence.

Defense strategies include:

➤ Presenting time-stamped PM logs showing task execution
➤ Submitting oil analysis data to prove that the lubricant was within spec
➤ Providing inspection and incident reports that confirm the cause of failure

Legal teams benefit from organized, real-time records that support expert witness testimony and forensic analysis. Proactive legal review of documentation practices reduces exposure and prepares organizations for future disputes. Lubrication logs play a crucial role in defending product liability claims in the event of a failure causing injury, property damage, or operational loss. These logs demonstrate that maintenance practices adhered to industry standards, providing evidence that the failure was not a result of negligence.

Key defense strategies involve presenting time-stamped preventive maintenance logs that verify task execution and adherence to scheduled maintenance. Submitting oil analysis data is essential to prove that lubricants were within specification, effectively countering claims of improper maintenance or product misuse. Providing detailed inspection and incident reports can help confirm the actual cause of failure, supporting the argument that it was not due to maintenance-related issues.

Legal teams benefit from having organized, real-time records that can support expert witness testimony and forensic analysis. These comprehensive records provide a robust foundation for defense, illustrating due diligence and attention to maintenance protocols. Proactively reviewing documentation practices with legal counsel can reduce exposure to liability and ensure that organizations are well prepared to address future disputes. This approach not only strengthens the defense in liability cases but also fosters a culture of meticulous recordkeeping and compliance.

COORDINATION BETWEEN LEGAL, MAINTENANCE, AND ENGINEERING FUNCTIONS

Effective management of warranties and compliance in lubrication practices necessitate structured collaboration among the legal, maintenance, and engineering departments. This alignment is crucial to ensure that operational decisions align with contractual, regulatory, and technical requirements.

Coordination strategies include:

> **Regular review of OEM specifications and updates.** Keeping all departments informed about the latest OEM specifications and updates ensures that everyone is on the same page regarding compliance and operational standards.
> **Legal participation in procurement and lubricant selection.** Involving the legal team in procurement and lubricant selection processes helps ensure that all products meet contractual and regulatory obligations, reducing the risk of noncompliance.
> **Engineering input into lubricant approvals and system compatibility.** Engineering expertise is vital in evaluating lubricant approvals and ensuring system compatibility, helping to optimize performance and prevent equipment issues.
> **Shared ownership of incident investigations and documentation standards.** When incidents occur, a collaborative approach to investigations and documentation ensures comprehensive analysis and accountability, promoting transparency and improvement.

Cross-functional training and shared dashboards can enhance communication, fostering a culture of collaboration and preventing siloed decision-making. Integrated document control platforms ensure that all departments have access to up-to-date information, reinforcing coordination and efficiency.

These measures support lubrication management within a broader risk management and quality assurance framework, ensuring compliance and enhancing operational resilience. By aligning the objectives and efforts of the legal, maintenance, and engineering teams, organizations can effectively manage warranties and compliance, safeguarding against risks and optimizing performance.

CHAPTER 18

MANPOWER PLANNING, ADMINISTRATION, STAFF TRAINING, AND CERTIFICATION

WORKFORCE MODELING FOR A PROACTIVE LUBRICATION PROGRAM

Building a proactive lubrication program requires strategic manpower planning and workforce modeling to ensure that personnel with the right capabilities are effectively assigned to lubrication tasks. This approach is crucial for maintaining reliability, safety, and efficiency.

Effective workforce modeling involves evaluating several key factors:

> **Equipment count and criticality.** Assessing the number and importance of equipment helps prioritize lubrication tasks based on their impact on operations.

> **Lubrication task frequency and complexity.** Understanding how often and how complex lubrication tasks are will guide staffing levels and skill requirements.

> **Plant layout and access logistics.** Considering the physical layout of the plant and access logistics can influence how tasks are assigned and scheduled.

➤ **Available technology.** Factoring in technology such as automation and condition monitoring can optimize manpower needs and improve task efficiency.

The workforce model should also consider nonlubrication responsibilities, such as inspections, documentation, and flushing, while providing a buffer for absenteeism or emergencies. Effective shift alignment, overtime management, and workload balancing are essential to maintain consistent lubrication practices and prevent bottlenecks or staff burnout.

Maintenance planners should define a clear lubrication full-time equivalent (FTE) requirement based on data-driven insights rather than relying on traditional roles. This ensures that staffing plans are aligned with current operational needs. It is advisable to review and adjust these plans annually, taking into account any asset expansion, process changes, or evolving reliability goals. By doing so, organizations can ensure that their manpower planning supports a robust and proactive lubrication program.

DEFINING CORE COMPETENCIES FOR LUBE TECHNICIANS AND ANALYSTS

Defining core competencies for lubrication technicians and analysts is essential for building a robust lubrication workforce. These competencies ensure that personnel have the necessary skills to maintain equipment reliability and operational efficiency.

For **lubrication technicians**, key competencies include:

➤ **Understanding of lubrication principles.** Technicians should have a solid grasp of lubrication principles, methods, and the specific equipment used in their work environment.

➤ **Identification and application.** They must be able to accurately identify lubrication points and apply the correct volumes and products to ensure optimal performance.

➤ **Tool familiarity.** Proficiency with tools such as grease guns, filter carts, and breathers is essential for effective task execution.

➤ **Compliance and safety.** Adhering to standard operating procedures and safety regulations is critical to ensure a safe and compliant work environment.

For **lubrication analysts,** essential competencies include:

➤ **Oil sampling and contamination control.** Analysts should practice best methods for oil sampling and maintain rigorous contamination control procedures.
➤ **Data interpretation.** They must be adept at interpreting oil analysis results and recognizing failure trends to provide actionable insights.
➤ **Root cause analysis.** Conducting thorough root cause analysis and generating detailed reports helps in addressing and preventing lubrication-related issues.
➤ **Effective communication.** Analysts must communicate effectively with engineering, operations, and suppliers to facilitate collaboration and problem-solving.

Across both roles, soft skills such as attention to detail, reliability, and the ability to document processes clearly are vital. These competencies should be clearly articulated in job descriptions, integrated into hiring practices, and evaluated during performance assessments. By embedding these competencies into organizational processes, companies can ensure that their lubrication workforce is well prepared to support maintenance goals and drive operational success.

TRAINING PATHWAYS: FROM ENTRY LEVEL TO SPECIALIST

Training programs should provide clear, progressive pathways for lubrication personnel to grow from basic competency to specialist roles. A tiered training model ensures that knowledge acquisition aligns with real-world experience and task complexity.

Typical training levels are:

➤ **Level 1.** Orientation and safety, basic lubrication theory, introduction to tools and procedures
➤ **Level 2.** Route-based lubrication, contamination control, task documentation, preventive maintenance
➤ **Level 3.** Advanced equipment servicing, automated systems, oil analysis sampling, troubleshooting
➤ **Level 4.** Specialist certification, program leadership, participation in root cause analysis, training of others

In a structured lubrication training program, the tiered approach is crucial to developing a workforce capable of handling increasingly complex tasks with confidence and precision.

Following is a more detailed look at each of the levels.

- ➤ **Level 1.** At Level 1, the foundational level, technicians are introduced to the core principles of lubrication, as well as to the safety protocols and to the basic tools and procedures essential for their everyday tasks. This initial phase lays the groundwork for more specialized knowledge and skills.
- ➤ **Level 2.** As technicians progress to the second level, the focus shifts to more practical applications, such as route-based lubrication and contamination control. Here, they learn to document tasks accurately and engage in preventive maintenance activities that are critical to minimizing equipment downtime and extending the life of the machinery. This stage helps technicians understand the importance of maintaining clean and efficient lubrication systems.
- ➤ **Level 3.** The third level of training delves into advanced equipment servicing and the integration of automated systems, reflecting the technological advancements in the field. Technicians gain expertise in oil analysis sampling, which is vital for monitoring the health of the machinery and preemptively identifying potential issues. Advanced troubleshooting skills are developed, enabling technicians to address and resolve complex lubrication problems efficiently.
- ➤ **Level 4.** At Level 4, the highest level, the focus is on specialization and leadership. Technicians who reach this stage are prepared for specialist certification, which signals their readiness to take on leadership roles within the lubrication program. They participate in root cause analysis to understand and rectify underlying issues that could lead to equipment failure. Additionally, these individuals are equipped to train others, passing on their expertise and ensuring the continuous development of the team.

Training delivery methods include classroom instruction, hands-on demonstrations, e-learning modules, and job shadowing. Certification goals (e.g., ICML MLT/ MLA, NLGI, STLE) should be built into the progression path.

The delivery of this training is diverse, combining classroom instruction with hands-on demonstrations to reinforce learning. E-learning modules offer flexibility and accessibility, while job shadowing provides real-world experience under the guidance of seasoned professionals. Certification goals are strategically integrated into the training path, ensuring that personnel not only acquire the necessary skills but also achieve industry-recognized qualifications that enhance their credibility and career prospects.

CROSS-TRAINING PROGRAMS TO MITIGATE SINGLE POINTS OF FAILURE

Overreliance on a single skilled technician creates vulnerability in lubrication programs. Cross-training ensures continuity, promotes knowledge sharing, and supports flexibility during outages, leave, or turnover.

Effective cross-training practices include:

- Maintaining a matrix of skills versus personnel and coverage gaps
- Rotating technicians through different lubrication routes or equipment types
- Assigning mentoring pairs or teams for peer-to-peer learning
- Building redundancy into scheduling and task allocation

In lubrication programs, cross-training plays a critical role in creating a resilient and flexible workforce, capable of adapting to changes and unexpected absences without compromising operational integrity. The process begins with a thorough assessment of current skills and potential coverage gaps within the team. By mapping these skills against personnel, managers can identify areas where additional training is needed to ensure that all critical lubrication tasks are covered, even in the absence of key technicians.

Rotating technicians through different lubrication routes or equipment types is an effective way to broaden their skill sets and prevent overreliance on specific individuals. This rotation not only enhances individual proficiency but also ensures that multiple team members are familiar with various aspects of the lubrication system, increasing the team's overall adaptability.

Mentoring pairs or teams are established to facilitate peer-to-peer learning, which is instrumental in transferring specialized knowledge and techniques. This collaborative approach not only enhances skill development but also fosters a culture of continuous improvement and teamwork.

Redundancy is built into scheduling and task allocation to ensure that all critical lubrication activities can be performed without interruption, even if a technician is unavailable. This redundancy is crucial for maintaining consistent equipment performance and preventing unexpected downtime.

Documenting cross-training efforts and validating them through competency assessments ensures that the training is effective and that technicians have genuinely acquired the necessary skills. This validation process is vital for maintaining high standards and ensuring that the team is well prepared to handle a variety of lubrication challenges.

Cross-training ultimately reduces operational silos, encouraging technicians to engage with different aspects of the lubrication process and collaborate more effectively. By developing broader capabilities and fostering a deeper understanding of the entire system, technicians become more engaged and motivated, contributing to a more dynamic and resilient lubrication program.

CERTIFICATION FRAMEWORKS AND THEIR STRATEGIC VALUE

Industry-recognized certifications validate technician expertise and elevate program credibility. They also support compliance, supplier relationships, and continuous improvement initiatives.

Relevant certification programs include:

- ➤ ICML (International Council for Machinery Lubrication): MLE, MLT I/II, MLA I/II/III
- ➤ STLE (Society of Tribologists and Lubrication Engineers): CLS, OMA
- ➤ NLGI (National Lubricating Grease Institute): CLGS

Certifications demonstrate:

- ➤ Commitment to professional standards
- ➤ Verified technical knowledge
- ➤ Capacity to interpret and apply best practices

Certifications in the lubrication field not only validate individual expertise but also significantly enhance the overall credibility and effectiveness of a lubrication program. By ensuring that technicians are certified, organizations demonstrate a commitment to maintaining high standards and adhering to industry best practices, which is crucial for both internal operations and external relationships.

Certifications in the lubrication field provide a structured framework for assessing and validating the skills and knowledge of technicians, ensuring that they meet industry standards. These credentials are not merely about individual achievement; they play a strategic role in enhancing the entire lubrication program's effectiveness and reputation.

Certifications like ICML's MLT and MLA series are designed to cover various aspects of lubrication, from basic principles to complex analysis and engineering. The MLT certifications focus on the practical skills needed for effective lubrication management, while the MLA certifications delve into the analytical skills required to interpret lubricant conditions and diagnose equipment health. The MLE certification extends this expertise into broader engineering applications, bridging the gap between technical execution and strategic planning.

STLE's CLS and OMA certifications are revered for their comprehensive coverage of lubrication science and oil monitoring techniques. These certifications ensure that technicians are well versed in the latest technologies and methodologies, enabling them to implement cutting-edge practices that improve equipment reliability and performance.

The NLGI's CLGS certification specifically targets expertise in grease lubrication, an area that requires specialized knowledge due to the unique properties and applications of grease compared with other lubricants.

For organizations, investing in these certifications ensures that the workforce is equipped with the latest skills and knowledge, fostering a culture of excellence and continuous improvement. This investment pays off in multiple ways, including enhanced compliance with industry regulations, improved supplier relationships due to a shared commitment to quality standards, and a stronger position in negotiations and partnerships.

By tracking certifications within HR systems, organizations can maintain a clear picture of their workforce's capabilities, identifying skill gaps and opportunities for development. This information is vital for succession planning, helping to ensure that the organization is prepared for future challenges and changes in personnel.

Ultimately, certifications represent a strategic investment in both people and processes, driving operational excellence and reinforcing the organization's reputation as a leader in the field. They are a testament to a commitment to quality, reliability, and professional growth, benefiting both the organization and its employees.

ASSESSING TECHNICIAN SKILL GAPS THROUGH TASK OBSERVATION

Direct observation of lubrication task execution reveals skill gaps that formal training may not detect. Observational assessments ensure that procedures are being followed correctly and that safety and quality standards are upheld.

Key elements of task observation include:

➤   Using standardized evaluation forms or digital checklists

➤ Focusing on technique, tool usage, PPE compliance, and procedural adherence
➤ Providing real-time coaching and feedback
➤ Documenting findings for individual development plans

These assessments confirm adherence to established procedures while ensuring compliance with safety and quality standards. The use of standardized evaluation forms or digital checklists streamlines the observation process, facilitating a structured approach to evaluating task execution. Key aspects of the observation include a concentrated focus on technique, tool usage, PPE compliance, and procedural adherence, which are critical for maintaining operational integrity.

Real-time coaching and feedback during the observation process enable immediate correction of identified issues, enhancing skill development on the spot. Documenting findings is essential for creating individual development plans tailored to each technician's needs, allowing for targeted training interventions. Periodic assessments foster a culture of accountability, assisting supervisors in prioritizing training investments based on observed performance gaps. These evaluations also highlight areas where standard operating procedures may lack clarity or practicality, presenting opportunities for procedural improvements.

An effective observation program operates on a nonpunitive basis, emphasizing continuous improvement and empowering technicians. This approach cultivates an environment where technicians feel supported in their development, leading to enhanced performance and operational reliability. Ultimately, the integration of direct observation into the lubrication process serves as a critical tool for identifying and addressing skill deficiencies, thereby improving overall task execution and adherence to established protocols.

Audit and Evaluation Form Checklist for Lubrication Tasks

The following checklist focuses on key performance indicators and compliance with safety and quality standards.

Technician Information

➤ Name:
➤ Date:
➤ Task Location:
➤ Equipment ID:
➤ Supervisor's Name:

Pre-Task Assessment

➤ Verification of work order and scope of work
➤ Review of Safety Data Sheets (SDS) for lubricants used
➤ Ensured availability of necessary tools and personal protective equipment (PPE)

Technique Evaluation

➤ Correct identification of lubrication points
➤ Appropriate selection of lubricant type
➤ Proper application technique used (e.g., pressure, volume)
➤ Adherence to manufacturer's specifications for lubrication

Tool Usage

➤ Correct tools used for the task (e.g., grease gun, oil can)
➤ Tools inspected for functionality prior to use
➤ Tools cleaned and maintained post-task

PPE Compliance

➤ Use of appropriate PPE (e.g., gloves, goggles, masks)
➤ Proper fit and condition of PPE
➤ PPE stored correctly after task completion

Procedural Adherence

➤ Followed standard operating procedures (SOPs) for lubrication
➤ No deviations from established protocols
➤ Completed all required documentation accurately

Real-Time Coaching and Feedback

➤ Immediate feedback provided by supervisor during the task
➤ Opportunities for improvement discussed
➤ Positive reinforcement of correct techniques offered

Post-Task Evaluation

➤ Task documented in maintenance management system
➤ Findings recorded for individual development plans
➤ Review of any incidents or near misses during the task

General Observations

➤ Overall technician performance rating (1–5 scale)
➤ Areas of strength identified
➤ Areas requiring further training noted

Follow-Up Actions

➤ Recommended training sessions
➤ Suggested updates to SOPs, if applicable
➤ Schedule for next evaluation or follow-up assessment

This checklist provides a structured approach to evaluating lubrication tasks, ensuring that observations are comprehensive and focused on continuous improvement.

BUILDING A CULTURE OF LUBRICATION EXCELLENCE

Culture building begins with leadership commitment and is sustained through recognition, engagement, and integration.

Culture-building strategies include:

> Publicly recognizing high-performing lubrication personnel
> Ensuring lubrication metrics are included in corporate KPIs and dashboards
> Sharing success stories and failure recoveries linked to lubrication efforts
> Embedding lubrication in operator care programs and cross-functional training

Sustained lubrication performance relies on fostering an organizational culture that prioritizes and supports excellence in lubrication practices. This cultural transformation begins with a strong commitment from leadership, which serves as the foundation for ongoing recognition, engagement, and integration of lubrication principles throughout the organization. Public recognition of high-performing lubrication personnel reinforces the importance of their contributions and motivates others to strive for excellence. Integrating lubrication metrics into corporate key performance indicators and dashboards provides visibility and accountability, ensuring that lubrication performance is monitored and valued alongside other critical operational metrics.

Sharing success stories and instances of failure recovery linked to lubrication efforts creates a narrative that highlights the impact of effective lubrication on overall operational performance. This practice not only celebrates achievements but also serves as a learning tool that can inform future strategies. Embedding lubrication practices into operator care programs and cross-functional training initiatives promotes a holistic understanding of the role of lubrication across various functions, ensuring that all employees recognize its significance.

Leadership must communicate the strategic value of lubrication in terms of cost control, reliability, safety, and regulatory compliance. This effort requires dispelling the misconception of lubrication as merely a low-skill or secondary task. By cultivating a sense of pride in lubrication practices and establishing professional pathways for skill development, organizations can transform lubrication from a tactical activity into a cornerstone of their reliability culture. This comprehensive approach ultimately enhances operational efficiency and supports long-term success in equipment performance and reliability.

Common Factors That Contribute to Organizational Culture Change

The following 10 factors contribute to changing the culture of the organization:

1. **Leadership commitment.** Strong and visible support from leadership is essential. Leaders must demonstrate their commitment to cultural change through their actions and decisions.
2. **Clear vision and goals.** Establishing a clear vision for the desired culture and specific goals helps align the organization's efforts and provides a road map for change.
3. **Communication.** Open and transparent communication about the reasons for the change, the expected outcomes, and the roles of employees fosters buy-in and reduces resistance.
4. **Employee involvement.** Actively involving employees in the change process encourages ownership and allows for diverse perspectives, which can enhance the effectiveness of the change initiatives.
5. **Training and development.** Providing training opportunities equips employees with the skills and knowledge necessary to adapt to the new culture. Continuous professional development reinforces the desired behaviors.
6. **Recognition and rewards.** Implementing recognition and reward systems that align with the new cultural values reinforces positive behaviors and encourages others to follow suit.
7. **Consistency in policies and practices.** Ensuring that organizational policies, procedures, and practices align with the desired culture helps to reinforce the change and promotes trust among employees.
8. **Feedback mechanisms.** Establishing channels for feedback allows employees to share their thoughts on the culture change process, helping to identify areas for improvement and making employees feel heard.
9. **Role models and champions.** Identifying and empowering champions within the organization can help to promote the desired culture through their influence and by modeling the expected behaviors.
10. **Monitoring and evaluation.** Regularly assessing the progress of culture change initiatives and making adjustments based on feedback and outcomes ensures that the organization stays on track and can adapt as needed.

These factors interact to create an environment conducive to cultural change, ultimately leading to a more engaged workforce and improved organizational performance.

Fostering teamwork in the workplace can be achieved through a variety of effective strategies. Here are several key approaches:

➤ **Clear objectives.** Establish and communicate clear, shared goals for teams. This creates a sense of purpose and direction, aligning individual efforts toward common outcomes.

➤ **Open communication.** Encourage open and honest communication among team members. This includes fostering an environment where feedback is welcomed and all voices are heard, enhancing collaboration.

➤ **Defined roles and responsibilities.** Clearly outline each team member's roles and responsibilities. This clarity helps prevent overlap and ensures that everyone understands their contributions to the team.

➤ **Team-building activities.** Organize team-building exercises that promote collaboration and strengthen relationships. These activities can range from workshops and retreats to casual social gatherings.

➤ **Cross-functional collaboration.** Facilitate opportunities for teams from different departments to work together on projects. This breaks down silos and encourages diverse perspectives and skill sets.

➤ **Empowerment and autonomy.** Empower teams by giving them the autonomy to make decisions related to their work. This fosters ownership and accountability, leading to increased motivation and engagement.

➤ **Recognition and rewards.** Acknowledge and celebrate team achievements, both big and small. Recognition reinforces positive behaviors and encourages continued collaboration.

➤ **Conflict resolution mechanisms.** Establish clear processes for addressing conflicts within teams. Providing tools and strategies for resolving disagreements can prevent issues from escalating and maintain a positive team dynamic.

➤ **Diversity and inclusion.** Promote diversity within teams to leverage a wide range of perspectives and ideas. An inclusive environment encourages participation and innovation.

➤ **Continuous learning and development.** Encourage ongoing professional development for team members. Providing training opportunities enhances skills and fosters a culture of continuous improvement.

By implementing these strategies, organizations can create a collaborative environment that enhances teamwork, improves productivity, and supports overall organizational success.

IMPACT OF ICML 55.1 ON ASSET RELIABILITY

ICML 55.1 is an international standard detailing best practices for lubricated asset management, aligned with ISO 55001 asset management principles. By addressing 12 key elements (from workforce skills and lubricant selection to contamination control and program metrics), ICML 55.1 provides a comprehensive framework to improve machinery lubrication management and, in turn, asset reliability. Implementing this standard can significantly boost equipment uptime and lifespan and program sustainability, but organizations must navigate certain challenges to realize its full benefits. Below, we analyze the standard's impact on reliability with a focus on implementation challenges, real-world case studies, cost-benefit considerations, and best practices for integration.

Implementation Challenges

Adopting ICML 55.1 often requires cultural change, investment, and new processes. Common obstacles include training gaps, upfront costs, and integration issues:

Workforce Training and Competency

A skilled team is crucial, yet many organizations lack formally trained lubrication personnel. In fact, only about 12% of industrial lubrication workers hold professional certification, leading to inconsistent practices. Building an ICML 55.1-compliant program demands significant training efforts to develop competencies in lubrication tasks, oil analysis, and reliability practices. For example, Weyerhaeuser's maintenance crew had to obtain Machine Lubrication Technician (MLT) certifications to understand and apply best practices. This training requirement can be daunting if employees are unaccustomed to advanced lubrication concepts.

Management Support and Culture

Securing management buy-in is often a make-or-break factor. Without leadership advocating for lubrication excellence, even the best technical plan may stall. Companies frequently face reluctance to change "the way we've always done it," making it hard to enforce new lubrication procedures or policies. As ICML notes, strong top-down support correlates with easier program implementation. In successful cases like Blue Buffalo's

Heartland plant, a management "champion" proactively prioritized lubrication from day one, funding training and resources before the first machine started up. Organizations lacking such support must overcome skepticism and illustrate the reliability improvements possible under ICML 55.1.

Cost of Compliance

Establishing a lubrication program to ICML 55.1 standards can entail significant initial costs. Expenses may include purchasing oil storage and handling equipment (e.g., proper containers, filters, breathers), retrofitting machines with sampling ports or filtration systems, hiring consultants, and paying for employee training/certification. There is also the cost of documentation and audits to achieve compliance. These upfront investments can be a barrier for budget-conscious operations. For instance, installing hardware for oil sampling and contamination control was a necessary step in one pharmaceutical plant's ICML 55.1-based program. Justifying these costs requires confidence that long-term savings (fewer failures, extended oil life) will outweigh the initial outlay.

Integration with Existing Systems

Implementing ICML 55.1 means embedding new practices into current maintenance workflows and systems. This integration can be challenging on both a technical and operational level. Legacy equipment may not be "lubrication-ready"—for example, OEM machines often lack convenient oil sample ports or proper drainage, hindering oil analysis efforts. Maintenance scheduling must be adjusted to include lubrication inspections, oil changes based on condition, and other tasks prescribed by the standard. Aligning these tasks with production is tricky; companies historically struggle to coordinate lubrication PMs without causing downtime. Additionally, existing Computerized Maintenance Management Systems (CMMS) need updates to track lubrication activities and metrics. Ensuring all this fits into the current maintenance regime (and possibly aligning with ISO 55001 management systems) requires careful planning.

STRATEGIES TO OVERCOME CHALLENGES

Successfully adopting ICML 55.1 calls for a structured approach and organizational commitment. Key strategies include:

Invest in Training and Certification

Close the skills gap by upskilling lubrication technicians and engineers. Formal training (e.g., ICML's MLT, MLA, or the advanced MLE certification) builds the expertise and confidence needed to execute the best practices. Weyerhaeuser's case showed that training staff on lubrication fundamentals enabled them to identify critical improvements and "focus on ... procedures whose modification would produce greatest ROI." Regular workshops and certification courses ensure the workforce can competently meet the standard's requirements.

Secure Management Buy-In

Educate leadership on the high cost of poor lubrication and the tangible benefits of ICML 55.1 compliance. Citing that improper lubrication causes up to 43% of mechanical failures and 70% of equipment failures can help make the case that lubrication is a strategic reliability issue, not just a maintenance detail. Presenting success stories (see next section) and quick wins builds executive support. Ideally, identify a management "champion" to advocate for the program's funding and to drive a culture that values proactive maintenance. Management support was the "necessary ingredient" that let Blue Buffalo implement a world-class lubrication program with "no hurdles" encountered.

Phase the Implementation

ICML 55.1 covers a broad scope, so it's wise not to tackle everything at once. Prioritize the most critical or high-impact areas first (the "crawl-walk-run" approach). As one expert noted, not all 12 elements merit equal weight initially, and they should not be implemented concurrently. Instead, use data and audits to identify pressing weaknesses; for example, tackle lubricant contamination control or staff training first if those are causing frequent failures. Achieving some quick wins (like reducing a known contamination issue or eliminating a redundant oil change) will build momentum and justify further efforts. Over time, expand the program to include all elements as the organization's maturity grows.

Leverage ICML 55.2 Guideline

To simplify integration, use the companion ICML 55.2 Guideline, which serves as a practical "how-to" blueprint for meeting 55.1 requirements. The guideline provides examples, step-by-step recommendations, and even punch lists of typical requirements auditors look for. This helps organizations translate the standard's requirements into actionable tasks and checkpoints. Essentially, ICML 55.2 breaks down abstract requirements into concrete practices, which can be gradually incorporated into existing maintenance processes.

Integrate with Maintenance Systems

Modify and augment your existing maintenance systems to embed lubrication management. Update the CMMS to include lubrication PM schedules, routes for inspections and oil sampling, and tracking for lubricant inventories. In the Eli Lilly IE43 facility, for instance, an oil analysis PM was added to the CMMS, complete with job plans for how to take samples and follow-up work orders for any issues. Standard operating procedures (SOPs) should be developed (or updated) to align with ICML 55.1, covering everything from how lubricants are received and stored to how lubrication tasks are executed in the field. Integrating these processes ensures lubrication activities become a routine, auditable part of maintenance, rather than ad-hoc tasks.

By anticipating these challenges and applying such strategies, organizations can smooth the transition to an ICML 55.1-compliant lubrication program and set the stage for significant reliability gains.

INDUSTRY CASE STUDIES

Real-world implementations of ICML 55.1 (or its best-practice principles) show clear improvements in asset reliability and maintenance efficiency. Below are examples of organizations that successfully upgraded their lubrication management, along with the benefits and lessons learned:

Eli Lilly (Pharmaceutical Manufacturing)—*Kinsale, Ireland*

An Eli Lilly facility (IE43) managing 40,000 assets revamped its lubrication program in line with ICML 55.1. A key challenge was enabling routine oil analysis on high-utilization equipment that lacked easy access for sampling. The team engineered custom technical solutions, installing sample ports, dry-break fittings, and desiccant breathers on critical machines to allow safe, repeatable, and representative oil samples even while equipment is running. This investment paid off quickly. By moving to condition-based oil maintenance, they cut oil sampling costs by 45% in the first year. Scheduled oil changes were eliminated on several assets—oil is now changed only when analysis indicates necessity, yielding an annual cost avoidance of €11,100 in oil purchase and disposal. Reliability and efficiency improved as well: 98% of targeted assets are now sampled on schedule, and the site avoided unnecessary downtime by consolidating what used to be 90 separate maintenance permits into just one coordinated procedure. Over four years, the facility's lubrication program went from baseline to "world class," as verified by an independent audit that showed a nine-fold increase in the audit score. This case demonstrates that applying ICML 55.1's practices (like engineered sampling, condition monitoring, and comprehensive procedures) can dramatically reduce maintenance effort and improve asset health. A crucial lesson was the importance of data-driven decision making, oil changes are now driven by actual condition data, not arbitrary intervals, thereby optimizing both reliability and cost.

Blue Buffalo Pet Food (Heartland Plant)—*Richmond, IN, USA*

Blue Buffalo's Heartland facility provides a model of proactive implementation. As a brand-new plant (opened in late 2018), management took the rare opportunity to build a lubrication program from scratch, following best practices aligned with ICML 55.1 even before production began. The plant manager, already familiar with world-class lubrication programs, acted as a champion to prioritize lubrication excellence. During construction and commissioning, the company invested in hiring a dedicated lubrication specialist, training staff, and obtaining ICML certifications. They set up lubrication procedures, a dedicated lube room, and color-coded lubricant storage and dispensing systems for mistake-proof handling. Thanks to this early planning, the site avoided many startup issues; in its first years, Blue Buffalo Heartland saw almost no lubrication-related downtime, with only a few "infant mortality" failures on new machines. Although exact reliability metrics weren't available (no "before" data for a new plant), the lubrication team attributes their high uptime to the robust system and engaged personnel. The plant even earned ICML's John R. Battle Award for lubrication excellence in 2020, just a couple of years into operation. The key lesson is the power of management support and early integration: by treating lubrication as a core part of asset management from day one, Blue Buffalo avoided problems that other facilities only discover after years of reactive maintenance. They also illustrated that even without historical data, you can justify a lubrication program by benchmarking against industry standards, management was convinced by what "other companies have incurred without quality lubrication programs" and strove to preempt those costs. This case underscores the value of a culture of reliability and continuous improvement; the team continues to seek training and stay abreast of new lubrication technologies to remain at a "world standard" level.

Weyerhaeuser (Forest Products)—*Longview, WA, USA*

An older facility experiencing frequent equipment failures used lubrication best practices to turn performance around. The plant had rising failures in hydraulic units (bearings, pumps, valves) that were causing costly downtime. Oil analysis revealed extremely high contamination levels (particle counts around ISO 21/19/16), pointing to poor lubricant quality as a root cause of failure. In response, Weyerhaeuser's maintenance crew sought formal training; several members became ICML-certified (MLT I and MLA I), which helped them identify critical gaps and optimal improvements. They proceeded to map out all lubrication points on 32 hydraulic systems and focused on contamination control, installing better filtration (including offline filters) and using pressure differential gauges to shift from time-based to condition-based filter changes. These changes had an immediate impact on both reliability and cost. With cleaner oil (one notorious machine's oil cleanliness improved from ISO 21/19/16 down to 16/14/11 after fixes), the team noted "significant improvement in hydraulic and bearing reliability." Breakdowns decreased, and the proactive maintenance approach paid financial dividends. They saved $40,000 per year by eliminating unnecessary scheduled filter replacements and a further $9,700 per year by reducing oil leaks and failure-related repairs. In total, roughly $50K/year of savings were realized, alongside improved uptime. Just as important, the success boosted morale; technicians took pride in their MLT certifications and were entrusted with larger projects by management as a result. Weyerhaeuser's case validates the cost-benefit of ICML 55.1 principles: relatively inexpensive measures (training, filtration units, and monitoring gauges) led to a substantial ROI and reliability boost. It also highlights the importance of contamination control and tailored maintenance intervals (changing filters when dirty, not just on a fixed schedule) as best practices for asset longevity.

These case studies spanning pharmaceuticals, food manufacturing, and heavy industry, all demonstrate measurable improvements in asset reliability and maintenance efficiency after implementing lubrication management aligned with ICML 55.1. Common threads include a shift to condition-based maintenance (avoiding arbitrary lube changes), improved contamination control, and a strong emphasis on training and procedure. The lessons learned point to the need for technical innovation (like custom sampling methods or better filters) and organizational support (training, culture change) to reap the full benefits of the standard.

COST-BENEFIT CONSIDERATIONS

Implementing ICML 55.1 requires investment, but the balance of costs and benefits often heavily favors the latter once the program is in place. Companies evaluating the standard should consider both the initial expenses and the long-term returns in reliability and efficiency:

Upfront Investment

The initial phase may involve costs such as training personnel, purchasing or upgrading lubrication equipment, and possibly hiring experts to assist with program development. There might also be expenses for establishing new oil storage facilities (e.g., climate-controlled lube rooms, fire-safe cabinets for oils/greases as seen at Blue Buffalo), buying tools like desiccant breathers, filtration systems, condition monitoring devices, and acquiring the ICML 55 standards and audit services. While these costs can be significant, they are usually a small fraction of what unreliable assets cost a business. Maintenance budgets typically allocate only 1–3% to lubrication, yet lubrication-related issues have an outsized impact on downtime and repair expenses. In other words, a modest increase in spending on excellent lubrication can prevent failures that would cost far more to fix. For example, retrofitting Eli Lilly's pumps with proper sampling hardware and breathers had an upfront cost, but it enabled the plant to avoid replacing hundreds of liters of oil and multiple unplanned shutdowns each year is a trade-off that quickly justifies itself.

Maintenance and Downtime Reduction

The primary financial benefit of ICML 55.1 comes from avoiding failures and extending the life of components. Numerous studies show that poor lubrication is a leading cause of equipment failure; for instance, manufacturers report it contributes to ~43% of mechanical failures and over half of all bearing failures. By systematically addressing lubrication (e.g., correct lubricant selection, proper intervals, contamination avoidance), companies can eliminate a major root cause of breakdowns. The result is fewer emergency repairs, lower spare parts consumption, and reduced maintenance labor. Weyerhaeuser's program, for instance, cut out chronic hydraulic failures, saving nearly $50,000 annually in parts and labor. Beyond direct maintenance savings, the avoidance of unplanned downtime leads to huge productivity gains, is production lines can keep running without interruption. Blue Buffalo's lubrication excellence, yielding virtually no lubrication-related downtime, means the facility can maximize output and meet orders without costly stops. Similarly, by switching to condition-based oil changes, Eli Lilly's team not only saved on oil costs but also prevented unnecessary downtime that would have been needed for routine oil drain/refill tasks. In many industries, a single hour of downtime on a critical asset can cost tens of thousands of dollars in lost production, so every failure prevented translates to significant monetary value.

Extended Asset Life and Energy Savings

A well-lubricated machine runs cooler, smoother, and with less wear. Over the long term, this extends the mean time between failures (MTBF) and delays capital expenditures for replacements. Proper lubrication can also improve energy efficiency (reducing friction losses), which lowers operating costs. While harder to quantify in the short term, these factors contribute to the return on investment (ROI). For instance, by maintaining cleaner oil, Weyerhaeuser improved their hydraulic systems' efficiency and longevity, likely deferring expensive overhauls. Another example is the elimination of frequent oil changes at Eli Lilly, which not only saved oil costs but also meant machines spent more time in optimal lubricated condition, reducing wear rates. Some studies even tie lubrication to energy consumption(e.g., well-lubricated bearings draw less power) providing utility cost savings and supporting sustainability goals (a point addressed by ICML 55.1's focus on energy conservation and environmental impact).

ROI and Payback Period

Companies often see a fast payback from reliability improvements. According to industry experts, a quality lubrication program provides a substantial financial opportunity, often yielding "remarkable ROI . . . eclipsing 1,000%" (10x return) and sometimes recouping investments within six months. While results vary, this underscores that lubrication initiatives can dramatically reduce costs. Real-world cases back this up: after implementing its oil analysis program, Eli Lilly's site reduced sampling costs by nearly half in the first year and projected further savings in year two, meaning the project likely paid for itself very quickly. Weyerhaeuser's $50,000 yearly savings similarly implies that the cost of training and new filters was recovered in short order. In essence, preventative lubrication care is far cheaper than reactive repairs, every bearing or gearbox saved from failure avoids not just part replacement cost but also secondary damage and downtime. Moreover, achieving ICML 55.1 compliance can have intangible benefits like improved safety (fewer catastrophic failures or oil spills) and environmental gains (less waste oil disposal), which reduce risk and potential regulatory costs.

Justifying the Cost

To build a business case for ICML 55.1 compliance, reliability teams should gather baseline data on lubrication-related downtime, maintenance expenses, and safety incidents. Quantifying the current cost of poor lubrication (e.g., failed components, labor hours on reactive fixes, lost production) creates a contrast for the expected improvements. Many organizations perform an initial lubrication audit to identify "low-hanging fruit" savings. For example, if analysis shows that a significant percentage of breakdowns are lube-related, one can project the savings from preventing those. It's also persuasive to reference industry benchmarks and success stories (e.g., "Company X saved $40,000 by switching to condition-based oil changes" or "Plant Y cut lube maintenance costs nearly in half by implementing a lubrication management system.") These examples, backed by the authoritative ICML 55 framework, help convince stakeholders that the program is not theoretical; it delivers real value. Additionally, compliance with ICML 55.1 can be pitched as a step toward broader asset management excellence (since it aligns with ISO 55001) and even as a marketplace differentiator. ICML representatives have noted that certified lubrication excellence gives companies a "demonstrable advantage in the marketplace" meaning fewer disruptions and a reputation for reliability that can win business. When all these factors are considered, the long-term ROI of implementing ICML 55.1 is typically very high, often covering initial costs many times over through sustained reductions in downtime and maintenance overhead.

While organizations must invest time and money to adopt ICML 55.1, the payoff comes in the form of significant maintenance cost savings, improved operational efficiency, extended equipment life, and risk reduction. The cost-benefit balance strongly favors implementation, and many companies should find that the program "pays for itself" quickly by shifting maintenance from a cost center into a source of profit through reliability gains.

BEST PRACTICES FOR EFFECTIVE INTEGRATION

To maximize asset reliability improvements, companies should integrate ICML 55.1's principles into their daily operations in a structured and sustainable way. Below are best practices and expert recommendations for aligning lubrication management with ICML 55.1 and ensuring continuous improvement:

Develop a Comprehensive Lubrication Management Plan

Begin by mapping out all assets and lubrication points and formulating a Lubrication Management Plan that addresses the 12 ICML 55.1 elements. This plan should cover areas such as personnel roles and skills, lubricant selection/specifications, preventive maintenance tasks (like relubrication intervals and inspections), contamination control measures, waste oil handling, and program governance. Essentially, treat lubrication as a formal sub-system of asset management. Document standard operating procedures (SOPs) for lubrication work, for example, how to properly grease bearings ("the rights of lubrication: right lubricant, right place, right amount, right frequency, right procedure"), how to take an oil sample, how to store and label lubricants, etc. At Eli Lilly's IE43 facility, a site-wide SOP covering all aspects of lubrication (from procurement of lubricants to their application in the field) was created based on ICML 55.1. This ensured consistency and clearly defined what needs to be done, forming the foundation of their program. A written plan and SOPs make training easier and allow for auditing compliance against the standard.

Aligning with Asset Management Frameworks (ISO 55001/PDCA)

ICML 55.1 is explicitly aligned with ISO 55000/55001, meaning it follows a plan-do-check-act cycle for continual improvement. Leverage this approach by integrating lubrication objectives into the broader asset management strategy of the organization. For instance, include lubrication performance in your asset management policy and objectives (Plan), execute the lubrication plan and training (Do), monitor results via audits and metrics (Check), and refine the program based on findings (Act). Ensure lubrication management is harmonized with other management systems, quality (ISO 9001), environmental (ISO 14001), and safety (ISO 45001) as these often intersect. For example, environmental policies will relate to handling waste oil (an element in ICML 55.1), and safety management will relate to ensuring lubrication tasks (like sampling or oil changes) are done without incident. By aligning lubrication initiatives with these frameworks, the program gains organization-wide visibility and avoids being siloed. It also helps in securing certification or recognition if that is a goal, since ICML 55.1 can enable or complement ISO 55001 certification efforts.

Use the ICML 55.2 Guideline and Audit Tools

To effectively implement ICML 55.1 requirements, utilize ICML 55.2 (Guideline) as a roadmap. This guide offers in-depth recommendations and even checklists for each of the twelve areas, essentially serving as a "blueprint" for building an audit-ready lubrication management system. Following the guideline helps translate high-level requirements into actionable steps. For example, it might detail what an auditor would expect to see for the "Lubricant Storage and Handling" element such as labeled containers, contamination controls, and tracking of lubricant shelf-life. Companies can perform an internal gap analysis against these checklists to see where current practices fall short. One best practice is to conduct an initial lubrication audit (either internally or via a third party like 5th Order Industry) to benchmark current performance. The audit results will highlight weaknesses (perhaps poor labeling, or missing inspection routines) that can then be prioritized. After implementing improvements, periodic audits should be repeated (e.g. annually) to measure progress. This was done in the Eli Lilly case after embedding ICML 55.1 practices, a follow-up audit scored the program nine times higher than the baseline from four years prior. Such audits create accountability and drive continuous improvement, ensuring the program doesn't become static. In addition, consider pursuing ICML's certification or awards for lubrication excellence as a framework: the criteria for these recognitions are aligned with best practices and can motivate the team to maintain high standards.

Leverage Technology and Data Analytics

Modern technology is a powerful enabler of lubrication management. Oil analysis and condition monitoring should be integral to the program as they provide the data needed to make informed decisions on oil change intervals, detect early signs of wear, and verify that practices are working (e.g. oil cleanliness levels). A mantra in reliability is "good data is king," and that holds true for lubrication. Invest in a robust oil analysis program: determine proper sampling frequencies for critical assets, use standardized procedures to get consistent samples (as the IE43 team emphasized, samples must be safe, repeatable, and representative), and trend the lab results. Many companies integrate oil analysis results into their maintenance software or use specialized data analytics tools to predict failures (for example, tracking rising particle counts or viscosity changes and linking them to predictive maintenance triggers). Additionally, consider online sensors where feasible, today there are IoT devices for continuous monitoring of oil condition (e.g., moisture, particle count) and automatic lubricators for bearings that dispense grease in controlled amounts. While continuous monitoring might not be practical for every asset, deploying it on the most critical machines can provide real-time alerts and further reduce reliance on routine time-based tasks. Technology also helps with integration: a CMMS can be set up to generate work orders when analysis indicates an out-of-spec condition, or to schedule lube routes and inspections at optimized intervals. Some organizations use mobile apps or RFID tagging to ensure every lube point is serviced and recorded properly, feeding data back for analysis. Analytics can identify trends like recurring contamination issues or differences in performance between lubricant brands, guiding continuous improvement decisions. In summary, use data to drive the lubrication program, from selecting the right lubricant (perhaps using software to match machine requirements to oil properties) to optimizing maintenance intervals and detecting issues before they cause downtime. This data-centric approach was key to Eli Lilly's success (shifting to condition-based oil changes and seeing clear trends of improvement in asset condition).

STANDARDIZED TOOLS AND PROCEDURES (LUBRICATION BEST PRACTICES)

Implement the best physical and procedural practices that underpin ICML 55.1's requirements. Some proven practices include:

Contamination Control

Keep lubricants clean and dry throughout their life cycle. Use desiccant breathers on oil reservoirs to prevent moisture ingress, and high-efficiency filtration (both on new oil before use and on machines through kidney-loop filters or filter carts). Establish target cleanliness codes for critical systems, and strive to meet them. Weyerhaeuser's turnaround illustrated how big an impact cleaner oil can have on reliability. Also, ensure all funnels, transfer containers, and grease guns are kept clean and dedicated to specific lubricant types to avoid cross-contamination. Simple practices like color-coding can be extremely effective and Blue Buffalo and Eli Lilly both employed color-coded tags and containers so that each lubricant is visually identified from storage to point-of-use. This prevents miscible lubricant mistakes (e.g., mixing incompatible oils) and ensures the right product goes into the right machine.

Proper Lubricant Selection and Management

Choose lubricants that meet or exceed OEM specifications and are suitable for the operating environment. Document these choices in the asset's lubrication plan. Manage inventory to avoid using degraded or expired lubricants for FIFO (first-in-first-out) stock rotation and periodic lab testing of stored oil can help. Under ICML 55.1, lubricant quality management is as important as the act of lubrication itself. Ensure any lubricant substitutions or consolidations are evaluated for compatibility.

Optimized Lubrication Intervals

Use reliability-centered maintenance (RCM) logic or vendor recommendations as a starting point but refine intervals with actual data. Over-greasing and under-greasing are both harmful on sensors or ultrasound tools can determine when a bearing needs grease, for example. For oil changes, the trend (as seen in case studies) is toward condition-based intervals: utilize oil analysis to safely extend oil life or conversely identify when oil is degrading faster due to harsh conditions. Eliminating arbitrary time-based tasks not only saves cost but also avoids disturbing systems unnecessarily. One success story had the plant "conducting condition-based rather than time-based filter changes" after implementing differential pressure gauges; a practice that saved money and ensured filters were only changed when truly needed.

Inspections and Preventive Tasks

ICML 55.1 emphasizes regular inspections (both machine condition and lubricant condition). Train operators or maintenance techs to perform routine lube checks: look for leaks, inspect lubricant levels and appearance, listen for sounds of poor lubrication, etc. Simple inspection routes can catch issues like a missing breather cap or an overheating gearbox before they escalate. Also, include lubricant-related tasks in PMs (e.g., checking automatic lubricators, cleaning breather filters, and retorquing oil sump bolts to stop leaks). These small tasks support reliability in the long run and are part of a disciplined lubrication regime. Blue Buffalo's team, for instance, incorporated thermal checks and routine greasing into their maintenance routine and scheduled bigger jobs (like yearly oil drain-and-fills) during planned downtime windows. Coordination of lubrication tasks with production schedules (like making oil changes during a monthly cleaning day) is a best practice to minimize any impact on operations.

Tools and Infrastructure

Provide the lubrication team with the right tools and workspace. A best-in-class program typically has a dedicated lubrication room or area, equipped with bulk oil storage tanks or drums on dispensing racks, filter carts, clean transfer containers (like "Oil Safe" containers, which can be color-coded), grease guns calibrated for the correct volume, and spill containment. The area should be organized for efficiency and cleanliness. At Blue Buffalo's site, for example, they use fire-safe cabinets for oil safe containers, labeled by specific oil type, illustrating attention to safety and organization. Moreover, having specialized tools such as oil analyzers (for quick on-site tests), infrared thermometers, ultrasonics, and vibration analyzers can greatly enhance the program's ability to detect lubrication issues early.

Continuous Training and Knowledge Sharing

Even after initial training and certification, make lubrication a subject of ongoing learning. Machinery lubrication evolves with new technologies (e.g., advances in synthetic lubricants or new sensor tech), so keep the team updated. Schedule periodic refresher courses or toolboxes on topics like "proper greasing techniques" or "interpreting oil analysis reports." Encourage lubrication technicians to attain higher ICML certifications (e.g., MLA II, MLE) and to participate in industry events or forums. Creating a culture where the lubrication team feels pride and ownership (as noted in Weyerhaeuser's case where certification "instilled players with pride") leads to better program sustainability. Also, share results with the broader maintenance and operations teams (e.g., show production supervisors the correlation between improved lubrication and fewer breakdowns, or celebrate reaching a cleanliness target in oil samples). This helps reinforce organizational buy-in.

Metrics and Continuous Improvement

Define key performance indicators (KPIs) for the lubrication program, and track them. Examples of useful metrics are: percent of PM compliance for lubrication tasks, oil analysis compliance rate (and number of alarms or out-of-spec conditions found), average oil cleanliness level achieved versus target, number of lubrication-related failures or work orders, and mean time between failure (MTBF) for critical assets before versus after program implementation. Also track cost metrics like lubricant consumption per month and any reductions in component replacement costs. By monitoring these, you can quantify improvements and identify any negative trends early. For continuous improvement, institute a routine (e.g., quarterly or biannual meetings) to review lubrication performance. If a failure occurred, perform a root cause analysis (RCA) with an eye on lubrication, so was it due to a missed lube, contamination, wrong oil, etc., and how can recurrence be prevented? ICML 55.1 includes program management and metrics as a core element, reinforcing that measuring results is integral to the standard. Blue Buffalo's team exemplified a forward-looking mindset: they committed to "continue to fine tune our machinery and processes" and acknowledged "there will always be room for improvement." This attitude, combined with hard metrics, keeps the program from stagnating. Additionally, consider external benchmarking and compare your program's metrics or practices with industry peers or award-winning programs (many publish case studies). This can spark ideas for further enhancements.

Implementing ICML 55.1 is not a one-time project but an ongoing journey. By following these best practices, organizations can effectively integrate the standard into their operations and create a self-sustaining lubrication program. The result is a virtuous cycle: better lubrication management leads to improved reliability and efficiency, which justifies continued support and refinement of the program. In the words of an industry expert, adherence to ICML 55.1 gives companies "optimized reliability" and a "foundational bedrock for programmatic sustainability" in maintenance. Through diligent application of the standard's principles supported by technology, training, and management commitment, companies can expect to see their asset reliability reach new heights and maintain those gains over the long term.

Sources:

International Council for Machinery Lubrication—ICML 55 Standards Overview

Ken Bannister (ICML)—ICML 55.1 Standard Book Intro

Paul Hiller (ICML)—ICML 55 Completion Announcement

Garrett FitzGerald (ICML)—Case Study: Eli Lilly IE43 Oil Analysis Program

Paul Hiller (ICML)—Case Study: Blue Buffalo Heartland Award-Winning Program

Paul Hiller (ICML)—Case Study: Weyerhaeuser Hydraulic Program

CHAPTER 19

LUBRICATION INFORMATION MANAGEMENT

STRUCTURING A LUBRICATION INFORMATION ARCHITECTURE

An effective lubrication program begins with a well-structured information architecture that defines how data is captured, stored, accessed, and used. This architecture must support both operational workflows and strategic decision-making.

Core components include:

➤ Master equipment lists with linked lubrication points
➤ Standardized lubrication procedures and task instructions
➤ Hierarchical storage of documents (e.g., SOPs, SDSs, spec sheets)
➤ Access controls by user role (technician, supervisor, analyst)

Centralizing this information into a unified platform ensures clarity, traceability, and alignment with reliability and quality programs. An effective lubrication program relies on a well-structured information architecture that delineates the processes for capturing, storing, accessing, and utilizing lubrication data. This architecture must effectively support both operational workflows and strategic decision-making. Essential components of this structure include master equipment lists that clearly identify all equipment along with linked lubrication points, ensuring that technicians can quickly locate the necessary information for each asset. Standardized lubrication procedures and task instructions are critical for maintaining consistency in execution and for training purposes, as they provide clear guidelines for technicians to follow.

The architecture should facilitate a hierarchical storage system for essential documentation, including standard operating procedures, Safety Data Sheets, and specification sheets. This organization allows for easy retrieval of documents while ensuring that the most current versions are readily accessible. Implementing access controls based on user roles, such as technician, supervisor, or analyst, enhances security and ensures that individuals can only access information relevant to their responsibilities.

Additionally, architecture must support consistent naming conventions and data validation rules to maintain data integrity and reliability. Integration with asset management systems is crucial for seamless data flow and for ensuring that lubrication practices align with broader reliability and quality programs. Centralizing lubrication information into a unified platform enhances clarity and traceability, promoting better decision-making and operational efficiency. This structured approach to lubrication information management lays the foundation for a robust lubrication program that contributes to overall equipment reliability and performance.

Structured Lubrication Information Architecture

To establish a structured lubrication information architecture, follow these systematic steps:

1. **Conduct a needs assessment.** Evaluate the current state of lubrication data management by identifying existing gaps and inefficiencies. Engage stakeholders, including technicians, supervisors, and analysts, to gather insights on their needs and challenges.
2. **Define objectives.** Establish clear goals for the lubrication information architecture. These objectives should align with organizational strategies and focus on enhancing data accuracy, accessibility, and integration with existing systems.
3. **Develop master equipment lists.** Create a comprehensive master equipment list that includes all assets requiring lubrication. For each piece of equipment, identify and document specific lubrication points, including details such as type of lubricant and frequency of application.
4. **Standardize procedures.** Develop standardized lubrication procedures and task instructions. Ensure that these documents outline best practices

for lubrication tasks, specifying the tools, techniques, and safety measures required.

5. **Create document hierarchies.** Organize all relevant documents, such as SOPs, SDSs, and spec sheets, into a hierarchical structure. This should facilitate easy navigation and retrieval of information while ensuring that the most current versions are accessible.

6. **Implement access controls.** Establish user role definitions and corresponding access controls within the information architecture. This ensures that individuals can only access data pertinent to their roles, enhancing security and data integrity.

7. **Adopt naming conventions.** Develop and enforce consistent naming conventions for all data entries, documents, and equipment identifiers. This practice improves clarity and reduces confusion when accessing and managing information.

8. **Integrate data validation rules.** Implement data validation rules within the information architecture to ensure the accuracy and reliability of the data being captured. This may involve setting parameters for acceptable values and formats for lubrication-related data.

9. **Leverage technology platforms.** Choose and configure a centralized platform for managing lubrication information. This platform should support integration with existing asset management systems and facilitate data sharing across departments.

10. **Train personnel.** Provide training for all users on the new lubrication information architecture, emphasizing the importance of data accuracy, adherence to procedures, and effective use of the centralized platform.

11. **Monitor and evaluate.** Establish metrics to monitor the effectiveness of the lubrication information architecture. Regularly evaluate its performance, and make adjustments based on user feedback and changing operational needs.

12. **Foster continuous improvement.** Foster a culture of continuous improvement by encouraging ongoing feedback from all stakeholders. Use this input to refine processes, update documentation, and enhance the overall lubrication information management system.

By following these steps, organizations can create a robust lubrication information architecture that enhances data management, supports operational workflows, and aligns with reliability and quality standards.

Audit Framework

An internal audit framework for evaluating the lubrication information architecture can be structured as follows:

- ➤ **Audit objectives.** Define the primary objectives of the audit, which may include assessing the effectiveness, accuracy, and compliance of the lubrication information management system with established standards and procedures.
- ➤ **Scope of the audit.** Determine the scope by identifying which aspects of the lubrication information architecture will be audited. This may include data capture processes, document management, access controls, and integration with asset management systems.
- ➤ **Audit criteria.** Establish criteria against which the audit will be evaluated. This may include compliance with internal policies, alignment with industry best practices, adherence to regulatory requirements, and effectiveness in supporting operational workflows.
- ➤ **Audit team.** Assemble a team with relevant expertise, including members with knowledge of lubrication practices, data management, and auditing processes. Clearly define roles and responsibilities within the audit team.
- ➤ **Data collection methods.** Identify the methods for data collection during the audit. This may include document reviews, interviews with personnel, direct observations of processes, and analysis of system outputs.
- ➤ **Audit schedule.** Develop a timeline for the audit, including key milestones and deadlines for each phase of the process. Ensure that the schedule allows ample time for data collection, analysis, and reporting.
- ➤ **Assessment of master equipment lists.** Evaluate the completeness and accuracy of the master equipment lists, ensuring that all equipment is documented and linked to the appropriate lubrication points.
- ➤ **Review of standardized procedures.** Assess the existence of and adherence to the standardized lubrication procedures and task instructions. Verify that they are up-to-date, accessible, and effectively communicated to all relevant personnel.
- ➤ **Document hierarchy evaluation.** Examine the hierarchical organization of documents related to lubrication. Ensure that SOPs, SDSs, and spec sheets are properly categorized and that users can easily locate the most current versions.

- ➤ **Access control assessment.** Review the access control measures in place. Verify that user roles are clearly defined and that access to sensitive information is appropriately restricted based on those roles.
- ➤ **Data validation and naming conventions.** Assess the implementation of data validation rules and consistent naming conventions. Evaluate whether these practices are effectively preventing errors and maintaining data integrity.
- ➤ **Technology platform review.** Evaluate the centralized platform used for managing lubrication information. Assess its functionality, integration capabilities, and user-friendliness.
- ➤ **Training and awareness.** Review the training programs provided to personnel regarding the lubrication information architecture. Assess whether employees are adequately informed and equipped to utilize the system effectively.
- ➤ **Monitoring and continuous improvement.** Examine the mechanisms in place for monitoring the effectiveness of the lubrication information architecture. Evaluate how feedback is collected and used for continuous improvement.
- ➤ **Reporting and action plan.** Compile findings into a comprehensive audit report, highlighting strengths, weaknesses, and areas for improvement. Include actionable recommendations, and establish a timeline for addressing identified issues.
- ➤ **Follow-up.** Plan for follow-up audits to assess the implementation of recommendations and the overall effectiveness of changes made to the lubrication information architecture.

This framework provides a systematic approach to auditing the lubrication information architecture, ensuring that it meets operational needs and aligns with reliability and quality objectives.

INTEGRATING LUBE DATA INTO CMMS, ERP, AND RELIABILITY PLATFORMS

Lubrication data such as PM completion, oil analysis results, consumption rates, and failure incidents must be integrated into enterprise systems to be actionable. This integration connects lubrication to broader asset management, procurement, and performance monitoring functions.

Key integration points include:

> **CMMS.** Maintenance schedules, work orders, asset history
> **ERP.** Inventory management, purchasing, vendor performance
> **Reliability platforms.** Condition monitoring, asset criticality, root cause analysis databases

Integrating lubrication data into computerized maintenance management systems, enterprise resource planning, and reliability platforms is crucial for making lubrication practices actionable within the broader context of asset management. This integration ensures a seamless flow of relevant data, facilitating informed decision-making across various organizational functions.

To achieve effective integration, key data points must be connected to specific enterprise systems. In the CMMS, lubrication data such as PM completion rates, work orders, and asset history should be embedded within maintenance schedules to provide technicians with timely insights into lubrication tasks. This integration enables automated alerts for overdue lubrication, allowing maintenance teams to prioritize actions and reduce equipment downtime.

In the ERP system, lubrication data plays a critical role in inventory management and procurement processes. By tracking lubricant consumption rates and oil analysis results, organizations can optimize inventory levels and streamline purchasing activities. Additionally, integrating vendor performance data with lubrication metrics allows for better evaluation of suppliers, ensuring that only high-quality lubricants are sourced.

Reliability platforms benefit from the integration of lubrication data through enhanced condition monitoring and failure analysis capabilities. By linking lubrication practices to asset criticality assessments and root cause analysis databases, organizations can identify trends and correlations between lubrication practices and equipment performance. This data-driven approach supports continuous improvement initiatives by enabling teams to analyze the impact of lubrication on reliability outcomes.

Centralized work order processing, facilitated by the integration of lubrication data, promotes cross-department visibility. This ensures that the procurement, engineering, and operations teams have access to shared data, fostering collaboration and informed decision-making regarding lubricant-related strategies. By breaking down silos, organizations can enhance their overall asset management strategies and improve operational efficiency. This integrated approach ultimately leads to a more proactive lubrication program that supports long-term reliability and performance objectives.

To implement an effective integration of lubrication data into CMMS, ERP, and reliability platforms, specific system requirements and a detailed implementation plan are essential.

System Requirements

System requirements include:

- ➤ **Compatibility.** The systems (CMMS, ERP, reliability platforms) must be compatible with each other to facilitate data exchange. This may require middleware or integration software that can harmonize data formats and protocols.
- ➤ **Data storage capacity.** Adequate storage capacity is necessary to manage the volume of lubrication data, including historical records, analysis results, and performance metrics.
- ➤ **User interface.** A user-friendly interface is essential to ensure that technicians and other personnel can easily access and input lubrication data across systems.
- ➤ **Data security.** Robust security measures must be in place to protect sensitive lubrication data. This includes user authentication, role-based access controls, and data encryption.
- ➤ **Real-time data processing.** The system should support real-time data processing capabilities to provide timely alerts and updates regarding lubrication tasks and statuses.
- ➤ **Reporting and analytics tools.** Integrated reporting and analytics functionality are necessary to analyze lubrication data, track performance metrics, and generate actionable insights.
- ➤ **Mobile access.** Mobile compatibility allows technicians to access lubrication data and submit updates from the field, enhancing flexibility and responsiveness.
- ➤ **Training and support.** The implementation plan should include provisions for training personnel on the integrated system and ongoing technical support.

Implementation Plan

Steps in the implementation plan include:

1. **Project kickoff.** Initiate the project with a kickoff meeting involving key stakeholders from maintenance, procurement, and engineering. Define project goals, timelines, and roles.

2. **Requirements gathering.** Conduct a thorough analysis of current lubrication practices, data management needs, and existing systems. Engage end users to gather functional requirements and identify integration challenges.

3. **System selection.** Evaluate and select the appropriate software solutions for CMMSs, ERP, and reliability platforms that meet the defined requirements. Consider factors such as scalability, user experience, and vendor support.

4. **Integration design.** Develop a detailed integration design that outlines data flow, mapping between systems, and the specific integration methods to be employed (e.g., APIs, middleware).

5. **Development and configuration.** Configure the chosen systems to support integration. This may involve customizing fields, setting up data validation rules, and establishing workflows for lubrication data entry and retrieval.

6. **Testing.** Conduct thorough testing of the integrated systems to ensure data accuracy, functionality, and performance. Perform unit testing, system testing, and user acceptance testing (UAT) with a focus on lubrication data processes.

7. **Training.** Develop and deliver comprehensive training programs for all users, ensuring they understand how to navigate the integrated systems and input or access lubrication data effectively.

8. **Deployment.** Roll out the integrated system across the organization, ensuring that all users have access and can utilize the system effectively. This may be done in phases to minimize disruption.

9. **Monitoring and feedback.** After deployment, monitor system performance and user feedback closely. Address any issues promptly, and gather insights for continuous improvement.

10. **Continuous improvement.** Establish a plan for ongoing evaluation and refinement of the integrated system. This includes regular reviews of lubrication data practices, user feedback sessions, and updates to the integration as needed.

By adhering to these system requirements and implementing the outlined plan, organizations can achieve a successful integration of lubrication data into their CMMS, ERP, and reliability platforms, ultimately enhancing asset management and operational efficiency.

TAGGING, INDEXING, AND VERSION CONTROL FOR LUBRICATION DOCUMENTS

Effective document control prevents confusion and errors due to outdated or misfiled lubrication procedures. Tagging and indexing support rapid retrieval, while version control ensures that technicians access the most current information.

For best practices:

- ➤ Use unique asset IDs, and tag all associated documents accordingly.
- ➤ Implement document management software with search functionality.
- ➤ Track revision history, author, approval, and effective date.
- ➤ Archive outdated versions with read-only access for audit trail purposes.

For multisite operations, centralized control with localized appendices ensures both standardization and site relevance. Indexing should enable sorting by equipment, lube type, task type, and date.

To implement a robust tagging system, each document should be associated with unique asset IDs that correspond to the equipment that the document pertains to. This practice facilitates precise categorization and ensures that technicians can quickly locate the necessary documents related to specific assets. Tagging should encompass all relevant documents, including standard operating procedures, Safety Data Sheets, and lubrication task instructions.

Management software with advanced search functionality is vital for efficient retrieval of lubrication documents. This software should allow users to search by various criteria, including asset ID, document type, and keywords, enabling rapid access to the most pertinent information. Additionally, the software should support comprehensive indexing, allowing users to sort documents by equipment, lubricant type, task type, and date. This indexing capability is particularly beneficial in multisite operations, where a centralized document control system can maintain standardization while allowing for localized appendices that address site-specific needs.

Version control is a crucial component of effective document management. It ensures that the most current versions of lubrication documents are readily available to technicians while maintaining a clear revision history. Each document should track details such as the author, approval status, and effective date to provide context for changes made. Outdated versions should be archived with read-only access, preserving an audit trail that allows for reference without compromising the integrity of current practices.

Centralized control of documentation, coupled with localized appendices, supports the need for both standardization across the organization and relevance to specific sites.

This approach ensures that while the overarching procedures remain consistent, individual sites can adapt documentation to meet their unique operational conditions. By implementing these best practices in tagging, indexing, and version control, organizations can enhance the reliability and accessibility of lubrication procedures, ultimately contributing to improved maintenance outcomes and equipment performance.

Categorizing

Common methods for categorizing digital files include the following approaches:

- **Folder structure.** Organizing files into a hierarchical folder system is one of the most traditional methods. Users create main folders for broad categories and subfolders for more specific topics, allowing for logical groupings based on project, department, or file type.
- **Naming conventions.** Establishing consistent naming conventions helps categorize files based on their content, purpose, or version. This may involve including dates, project names, or descriptive keywords in file names, making it easier to identify and locate files quickly.
- **Tags and metadata.** Using tags or metadata allows users to assign keywords or phrases to files, enabling more dynamic categorization. This method facilitates searching and sorting files based on specific attributes or themes, regardless of their physical location in the folder structure.
- **File types.** Categorizing files by their type (e.g., documents, spreadsheets, presentations, images) is a straightforward approach. This method helps users quickly identify the format of a file and locate the documents needed for specific tasks.
- **Date-based organization.** Files can be categorized based on dates, such as creation date or last modified date. This method is particularly useful for tracking project timelines or version histories or for archival purposes.
- **Project-based grouping.** For teams working on multiple projects, organizing files by project name or code can provide clear categorization. This method helps team members quickly access all relevant documents associated with a specific project.
- **Client- or customer-based organization.** In service-oriented environments, files can be categorized based on clients or customers. This approach helps maintain all documentation related to a specific client in one easily accessible location.

> **Access or permission levels.** Categorizing files based on access levels can enhance security and streamline workflows. Sensitive files may be stored in restricted folders, while general documents can be made readily available to all users.

> **Version control.** For documents that undergo frequent updates, implementing a version control system allows files to be categorized by version number or revision date. This method ensures that users can access the most current version while retaining older iterations for reference.

> **Collaboration tools.** Utilizing collaboration platforms often comes with built-in categorization methods that integrate tagging, folders, and metadata features, allowing teams to manage files collectively in a structured manner.

By employing these categorization methods, individuals and organizations can improve file management, enhance retrieval efficiency, and streamline collaboration efforts.

Implementation Plan

An implementation plan for categorizing digital files effectively involves several key steps to ensure a structured and efficient approach. The following plan outlines the stages of implementation:

1. **Define objectives.** Clearly outline the goals for file categorization. Objectives may include improving retrieval times, enhancing collaboration, or ensuring compliance with data management policies.

2. **Conduct an inventory.** Assess the current state of digital files. This includes identifying existing files, their types, and how they are currently organized. Engage stakeholders to understand their needs and challenges related to file management.

3. **Establish a categorization framework.** Develop a framework that outlines the methods for categorizing files. This framework should incorporate folder structures, naming conventions, tagging systems, and any specific criteria relevant to your organization.

4. **Design a folder structure.** Create a logical folder hierarchy based on categories such as department, project, client, or file type. Ensure that this structure aligns with the goals defined in the first step and is intuitive for users.

5. **Develop naming conventions.** Establish consistent naming conventions for files that incorporate key information such as dates, project names, or descriptive keywords. Provide guidelines for users to follow when naming new files.

6. **Implement tags and metadata.** Determine the relevant tags and metadata fields that will be used for categorization. Develop a standardized list of tags to ensure consistency across all files. Consider using software that supports tagging capabilities.

7. **Initiate a migration strategy.** Plan for the migration of existing files into the new categorization system. This may involve reorganizing files into the new folder structure, renaming files according to the established conventions, and applying tags and metadata.

8. **Provide user training.** Develop and conduct training sessions for all users on the new categorization system. Issue clear instructions on how to organize, tag, and search for files effectively, and be sure to stress the importance of adhering to the established framework.

9. **Implement technology solutions.** If using document management software, set it up to support the new categorization methods. Ensure that it includes features such as advanced search functionality, tagging, and user access controls.

10. **Do a pilot test and address feedback.** Before full implementation, conduct a pilot test with a small group of users to identify any issues or areas for improvement. Gather feedback, and make necessary adjustments to the categorization system.

11. **Do a full rollout.** Once testing is complete and any adjustments have been made, roll out the new categorization system organization-wide. Provide ongoing support and resources to ensure a smooth transition.

12. **Monitor and evaluate.** After implementation, monitor the effectiveness of the new categorization system. Gather feedback from users on their experience, and assess retrieval times and overall file management efficiency.

13. **Foster continuous improvement.** Establish a process for ongoing evaluation and refinement of the categorization system. Regularly review the framework, naming conventions, and tagging practices to ensure they remain relevant and effective.

By following this implementation plan, organizations can create a structured and efficient system for categorizing digital files, leading to improved accessibility, collaboration, and data management practices.

VISUAL DASHBOARDS FOR MAINTENANCE DECISION SUPPORT

Dashboards transform raw lubrication data into actionable insights. Designed for technicians, supervisors, and reliability engineers, these tools display trends, alerts, and KPIs at a glance.

Common dashboard elements include:

- ➤ Compliance with PM lubrication tasks (OTIF [on time in full])
- ➤ Lubricant consumption rates by department or machine
- ➤ Oil analysis status: overdue, flagged, or normal
- ➤ Asset-level lubrication health ratings

Dashboards should be tailored by role and refresh in real time or on a regular schedule. Integration with mobile devices enables field technicians to receive alerts, report completions, or log issues directly from the route.

Visual representation enhances engagement and supports proactive maintenance decision-making. Creating visual dashboards for maintenance decision support is an effective way to transform raw lubrication data into actionable insights. These dashboards can significantly enhance the decision-making process for technicians, supervisors, and reliability engineers by providing a clear and concise overview of KPIs and trends.

Implementing Visual Dashboards

Here's a breakdown of the important elements and considerations for implementing visual dashboards:

- ➤ **Compliance with PM Lubrication Tasks (OTIF)**
  - • Visualize adherence to planned maintenance schedules, highlighting tasks completed on time and in full.
  - • Use color-coded indicators (e.g., green for compliant, red for overdue) to quickly communicate status.
- ➤ **Lubricant Consumption Rates**
  - • Display consumption rates by department or machine, enabling users to identify trends and anomalies.
  - • Include charts or graphs that illustrate usage patterns over time, helping to optimize inventory levels.

➤ **Oil Analysis Status**
- Present the status of oil analysis results, categorizing them as overdue, flagged for attention, or normal.
- Use visual indicators (e.g., traffic lights or warning icons) to alert users to potential issues that require action.

➤ **Asset-Level Lubrication Health Ratings**
- Provide a health rating for each asset based on lubrication practices, condition monitoring results, and historical performance.
- Use gauges or scorecards to represent health ratings, allowing for quick assessments of equipment status.

➤ **Dashboard Customization**
- **Role-based customization.** Tailor dashboard views according to user roles to ensure that technicians, supervisors, and reliability engineers see the most relevant information for their responsibilities. This enhances usability and focuses attention where it is needed most.
- **Real-time or scheduled refresh.** Use dashboards that are designed to refresh in real time or on a regular schedule, thereby ensuring that users have access to the latest data and insights. This timeliness is crucial for proactive decision-making.

➤ **Integration with Mobile Devices**
- **Mobile accessibility.** Ensure that dashboards are accessible on mobile devices, allowing field technicians to view critical information while on the go. This capability supports immediate decision-making and enhances responsiveness.
- **Alerts and reporting.** Integrate features that enable technicians to receive alerts regarding overdue tasks or issues directly on their mobile devices. They should also be able to report task completions or log issues from the field, facilitating seamless communication and documentation.

➤ **Visual Engagement**
- **Data visualization.** Utilize charts, graphs, and visual indicators to represent data clearly and engagingly. Effective visual representation enhances user engagement and comprehension, making it easier to identify trends and take action.
- **Interactive elements.** Consider incorporating interactive features that allow users to drill down into specific data points for more detailed analysis. This functionality empowers users to explore the underlying data behind the metrics displayed.

> **Implementation Considerations**
>   • **User feedback.** Involve end users in the design process to gather feedback on their needs and preferences. This collaborative approach ensures that the dashboard meets their requirements.
>   • **Training and support.** Provide training to users on how to navigate and utilize the dashboards effectively. Ongoing support should also be available to address any questions or challenges that arise.
>   • **Continuous improvement.** Regularly review and update dashboard elements based on user feedback and evolving business needs. This ensures that the dashboards remain relevant and effective over time.

By implementing visual dashboards with these elements, organizations can enhance maintenance decision support, improve operational efficiency, and foster a proactive maintenance culture. The actionable insights derived from these dashboards will ultimately contribute to better lubrication practices and overall asset reliability.

Visual dashboards for maintenance decision support are powerful tools that enhance the ability of technicians, supervisors, and reliability engineers to manage lubrication and maintenance tasks effectively. Here's a deeper dive into various aspects of creating and utilizing these dashboards:

> **Benefits of Visual Dashboards**
>   • **Improved decision-making.** By presenting data visually, dashboards help users quickly identify trends, anomalies, and areas needing attention. This enables informed decision-making and prioritization of maintenance tasks.
>   • **Enhanced communication.** Dashboards foster better communication among teams by providing a common platform where everyone can view the same data and metrics. This shared understanding helps align efforts across departments.
>   • **Proactive maintenance.** With real-time data and alerts, maintenance teams can shift from reactive to proactive maintenance strategies. Identifying potential issues before they escalate reduces downtime and maintenance costs.
>   • **Performance monitoring.** Dashboards allow organizations to monitor KPIs continuously. This ongoing evaluation helps identify opportunities for improvement and ensures that maintenance practices align with organizational goals.

➤ **Key Considerations for Dashboard Design**
- **User-centric design.** The dashboard should be intuitive and user-friendly. Consider the technical expertise of the end users and design the interface accordingly. Use clear labels, legends, and tooltips to aid understanding.
- **Data sources integration.** Ensure that the dashboard integrates seamlessly with existing systems such as CMMSs, ERP, or reliability platforms. This integration allows for a comprehensive view of lubrication data alongside other relevant maintenance information.
- **Customization and flexibility.** Dashboards should allow users to customize their views according to specific needs or preferences. This flexibility enables individuals to focus on the metrics that matter most to them.
- **Historical data analysis.** Incorporating historical data into the dashboard can provide valuable context for current performance. Users can identify trends over time, helping to forecast future maintenance needs and budget accordingly.
- **Alerts and notifications.** Implement a robust alert system that notifies users of critical issues such as overdue lubrication tasks or abnormal oil analysis results. Notifications can be sent via email, SMS, or push notifications to ensure timely responses.

➤ **Types of Visualizations**
- **Graphs and charts.** Use line graphs to display trends over time, bar charts to compare data across categories, and pie charts to show proportions of total usage or compliance rates.
- **Heat maps.** Heat maps can visually represent areas of concern, such as machines with high lubricant consumption or those flagged for maintenance, allowing for quick identification of hotspots.
- **Gauges and scorecards.** These visual elements can provide a quick snapshot of key metrics, such as asset health ratings or compliance levels, helping users assess overall performance at a glance.
- **Data tables.** While visualizations are important, data tables can provide detailed information that users may want to explore further. Including sortable tables allows for deeper dives into specific data points.

➤ **Ongoing Maintenance of Dashboards**
- **Regular updates.** Ensure that the dashboard is updated regularly to reflect the latest data, trends, and insights. This might involve

scheduled data pulls or real-time updates depending on system capabilities.

- **User training and support.** Continuous training and resources should be provided to ensure that users are comfortable navigating the dashboards and interpreting the data. This can include training sessions, user manuals, or online resources.
- **Feedback loop.** Create a mechanism for users to provide ongoing feedback on the dashboard's functionality and design. Regularly review this feedback to make necessary adjustments and improvements.
- **Performance review**. Periodically assess the effectiveness of the dashboards in meeting organizational objectives. Evaluate whether the insights provided are actionable and beneficial for decision-making.

By implementing visual dashboards for maintenance decision support, organizations can create a culture of data-driven decision-making. These dashboards not only enhance the visibility of lubrication and maintenance data but also empower teams to take proactive actions that improve equipment reliability and operational efficiency. Ultimately, investing in well-designed dashboards can lead to significant benefits in maintenance management and organizational performance.

LEVERAGING LUBRICATION DATA FOR ROOT CAUSE FAILURE ANALYSIS

Lubrication-related data is essential in root cause failure analysis. Trends in oil degradation, contamination, application errors, and maintenance intervals often precede equipment failures.

Data sources to leverage include:

- ➤ Oil analysis reports (viscosity, wear metals, particle counts)
- ➤ Lubrication PM logs (frequency, task notes, technician ID)
- ➤ Asset operating conditions (temperature, load, uptime)
- ➤ Failure logs annotated with lube-related causes (seal failure, varnish, dry run)

RCFA teams should include lubrication specialists who can interpret fluid trends and connect symptoms to procedural or environmental factors. Integrating this data supports corrective actions that prevent recurrence, such as adjusting relube intervals, changing products, or retraining technicians.

Identifying trends in oil degradation, contamination, application errors, and maintenance intervals frequently precedes equipment failures. Analyzing oil analysis reports provides insights into viscosity changes, wear metals, and particle counts, which are essential indicators of lubricant condition and machine health. Regular lubrication PM logs offer data on task frequency, technician notes, and the specific personnel responsible for maintenance, enabling the identification of patterns or inconsistencies in lubrication practices.

Asset operating conditions, such as temperature, load, and uptime, further contextualize lubrication data. These parameters can significantly impact lubricant performance and longevity, making it crucial for RCFA teams to correlate them with failure events. Failure logs annotated with lubrication-related causes, including seal failures, varnish formation, and dry runs, provide direct evidence linking lubrication practices to equipment malfunctions.

Incorporating lubrication specialists into RCFA teams is vital, as they possess the expertise to interpret fluid trends and connect symptoms to procedural or environmental factors. Their ability to analyze the data collected from various sources allows for a comprehensive understanding of the lubrication system and its impact on equipment reliability. By integrating this data effectively, organizations can implement corrective actions that prevent the recurrence of failures. Adjusting relube intervals, changing lubricant products, or retraining technicians based on identified issues ensures that lubrication practices align with the operational needs of the equipment.

Establishing a systematic approach to leveraging lubrication data in RCFA allows for a proactive maintenance strategy. It not only addresses existing failures but also enhances the overall reliability of equipment by preventing future incidents. This data-driven methodology fosters a culture of continuous improvement in lubrication practices, ultimately leading to greater operational efficiency and reduced downtime.

Implementation

Organizations can implement several strategies to enhance their lubrication maintenance practices, thereby improving equipment reliability and reducing downtime:

➤ Establishing a comprehensive lubrication management program
   is essential. This program should include clear guidelines on
   lubrication standards, procedures, and responsibilities. Developing

detailed lubrication schedules based on equipment manufacturer recommendations and operational conditions ensures that relubrication tasks are performed at optimal intervals.

➤ Organizations should invest in training and certification programs for maintenance personnel. Providing technicians with training on lubrication best practices, equipment-specific requirements, and the importance of proper lubrication can significantly reduce errors and improve the overall effectiveness of lubrication tasks.

➤ Implementing a robust oil analysis program can provide valuable insights into lubricant condition and equipment health. Regularly analyzing oil samples for viscosity, wear metals, and contamination levels allows for timely interventions before failures occur. Utilizing the results from oil analysis to adjust lubrication intervals or change lubricant types based on performance data enhances proactive maintenance efforts.

➤ Adopting advanced technology solutions, such as condition monitoring systems, can further improve lubrication practices. These systems can provide real-time data on equipment conditions, alerting personnel to potential lubrication issues before they escalate into failures. Integrating these systems with existing maintenance management software enables a more holistic view of asset performance and maintenance needs.

➤ Standardizing lubrication procedures across the organization ensures consistency and accountability. Creating detailed lubrication work instructions, including step-by-step procedures and checklists, helps technicians perform tasks uniformly, reducing the risk of oversight or errors.

➤ Regular audits and reviews of lubrication practices can identify areas for improvement. Conducting periodic assessments of lubrication tasks, equipment performance, and compliance with established procedures helps organizations recognize gaps and implement corrective actions.

➤ Fostering a culture of continuous improvement is crucial. Encouraging feedback from maintenance personnel on lubrication practices and actively involving them in the decision-making processes can lead to innovative solutions and enhancements in lubrication strategies. By implementing these strategies, organizations can significantly improve their lubrication maintenance practices, resulting in greater equipment reliability and reduced operational costs.

Audit

An internal audit of lubrication maintenance practices is a systematic evaluation aimed at identifying deficiencies and opportunities for improvement. The audit process typically follows several key steps to ensure comprehensive coverage of all relevant aspects of lubrication management.

The first step involves defining the scope of the audit. This includes identifying the specific assets, lubrication procedures, and operational areas to be evaluated. The audit should focus on critical machinery and systems where lubrication plays a significant role in reliability and performance.

Next, auditors should gather relevant documentation, including lubrication schedules, preventive maintenance logs, oil analysis reports, and training records. This documentation serves as the foundation for assessing compliance with established lubrication standards and procedures.

Auditors perform a thorough review of lubrication procedures to evaluate their completeness and effectiveness. This includes examining the clarity of work instructions, consistency in naming conventions, and adherence to manufacturer specifications. The review should also assess whether lubrication tasks are performed at the recommended intervals and whether any deviations from established practices are documented and justified.

On-site inspections form a crucial part of the audit process. Auditors should observe lubrication tasks being performed to evaluate technician compliance with established procedures. This includes verifying that the correct lubricants are used, checking for proper application techniques, and ensuring that equipment is cleaned and prepared appropriately before lubrication. Observations should also focus on the condition of lubrication equipment, such as pumps and grease guns, to identify any maintenance needs or deficiencies.

Interviews with maintenance personnel provide valuable insights into the effectiveness of lubrication practices. Auditors should engage with technicians, supervisors, and lubrication specialists to gather feedback on existing challenges, knowledge gaps, and suggestions for improvement. This qualitative data complements the quantitative findings from documentation reviews and on-site inspections.

The audit should also include an evaluation of the oil analysis program. Auditors need to assess whether oil samples are being collected and analyzed regularly, whether the results are reviewed and acted upon, and whether the findings are integrated into maintenance decision-making processes. Analyzing historical data for trends in lubricant condition and equipment performance can further highlight areas of concern.

After completing the assessment, auditors compile their findings into a comprehensive report detailing identified deficiencies and areas for improvement. This report should categorize issues based on severity and provide recommendations for corrective actions. Common deficiencies may include noncompliance with lubrication schedules, inadequate technician training, lack of documentation, and insufficient monitoring of lubricant quality.

Finally, the organization should establish a follow-up process to ensure that audit findings are addressed. This may include setting timelines for corrective actions, assigning responsibilities, and conducting subsequent evaluations to verify improvements. By implementing a structured internal audit process for lubrication maintenance practices, organizations can identify deficiencies, enhance operational efficiency, and ultimately improve equipment reliability.

The following are the steps for conducting an internal audit of lubrication maintenance practices:

1. **Define the scope.** Identify the specific assets, lubrication procedures, and operational areas to be evaluated.

2. **Gather documentation.** Collect relevant documents such as lubrication schedules, preventive maintenance logs, oil analysis reports, and training records.

3. **Review lubrication procedures.** Evaluate the completeness and effectiveness of lubrication procedures, including clarity of work instructions and adherence to manufacturer specifications.

4. **Conduct on-site inspections.** Observe lubrication tasks being performed to verify compliance with established procedures, proper lubricant usage, and equipment condition.

5. **Interview maintenance personnel.** Engage with technicians, supervisors, and lubrication specialists to gather feedback on challenges and suggestions for improvement.

6. **Evaluate the oil analysis program.** Assess the regularity of the oil sampling and analysis, review the results, and determine whether the findings are integrated into maintenance decision-making.

7. **Analyze historical data.** Review historical data for trends in lubricant condition and equipment performance to identify areas of concern.

8. **Compile findings.** Create a comprehensive report detailing identified deficiencies and areas for improvement, categorizing issues by severity.

9. **Provide recommendations.** Offer actionable recommendations for corrective actions based on the audit findings.

10. **Establish a follow-up process.** Set timelines for corrective actions, assign responsibilities, and plan for subsequent evaluations to verify improvements.

DIGITAL TRANSFORMATION
OF LUBE ROUTES AND SCHEDULES

Digitizing lubrication routes replaces manual paperwork with mobile-friendly, data-rich task management tools. This transformation improves traceability, reduces human error, and supports dynamic scheduling.

Key features of digital route systems include:

➤ GPS-tagged equipment maps with color-coded lube points
➤ Route optimization based on task proximity and frequency
➤ Interactive task prompts with embedded instructions and videos
➤ Digital sign-off and real-time exception reporting

Data from digital routes feeds back into the CMMS, updating task compliance metrics and supporting analytics. Systems can also auto-generate tasks based on sensor data or oil condition reports. The digital transformation of lubrication routes and schedules is a significant advancement that replaces traditional manual processes with mobile-friendly, data-rich task management tools. This shift enhances traceability, minimizes human error, and allows for the dynamic scheduling of lubrication tasks, leading to improved operational efficiency.

Digital route systems feature GPS-tagged equipment maps that provide visual representations of lubrication points, color-coded for easy identification. This visual aspect allows technicians to quickly locate and prioritize the points that require attention, streamlining the lubrication process. Route optimization algorithms analyze task proximity and frequency, enabling the efficient planning of routes that reduce travel time and maximize productivity.

Interactive task prompts within these systems are designed to guide technicians through lubrication activities. These prompts can include embedded instructions and instructional videos that ensure that tasks are performed correctly and consistently. This level of detail supports technicians in adhering to best practices and reduces the potential for errors during execution.

The digital platforms also incorporate features such as digital sign-off and real-time exception reporting. Digital sign-off provides a verifiable record of task completion, while real-time exception reporting alerts supervisors to any deviations from established protocols or unexpected issues encountered during lubrication tasks. This immediate feedback loop allows for prompt corrective actions and enhances overall compliance with lubrication schedules.

Data generated from digital lubrication routes feeds back into the CMMS, automatically updating task compliance metrics and supporting advanced analytics. This integration allows organizations to track performance trends, identify areas for improvement, and make data-driven decisions regarding lubrication practices. Furthermore, systems can auto-generate tasks based on sensor data or oil condition reports, ensuring that timely maintenance actions are taken without manual intervention.

The digitization of lubrication routes not only reduces the reliance on paperwork but also enhances compliance with established maintenance standards. By modernizing the workforce through these digital tools, organizations can improve efficiency, increase accountability, and foster a culture of continuous improvement in lubrication practices. Overall, the transition to digital lubrication management represents a critical step in optimizing maintenance operations and ensuring the longevity and reliability of equipment.

Implementation

Implementing a digital lubrication management system with features such as GPS-tagged equipment maps, route optimization, interactive task prompts, and digital sign-off involves several structured steps:

1. **Do a needs assessment.** Conduct a detailed analysis of current lubrication processes, and identify specific requirements for digital transformation. Engage stakeholders, including maintenance personnel, supervisors, and IT, to gather input on desired features and functionality.
2. **Select a digital solution.** Research and evaluate digital lubrication management systems that offer the necessary features. Consider aspects such as compatibility with existing systems (e.g., CMMSs), ease of use, scalability, and vendor support. Choose a solution that aligns with organizational goals and user needs.
3. **Map equipment locations.** Create GPS-tagged maps of all equipment requiring lubrication. This involves physically locating each asset and inputting its coordinates into the digital system. Color-code lubrication points based on criteria such as task frequency, type of lubricant, or criticality to enhance visual identification.
4. **Design route optimization algorithms.** Work with the software provider to develop algorithms that optimize lubrication routes based on task proximity and frequency. This may involve configuring settings within the system to prioritize tasks effectively and reduce travel time for technicians.

5. **Develop interactive task prompts.** Collaborate with subject-matter experts to create detailed task prompts for each lubrication point. Embed step-by-step instructions and instructional videos within the digital system to guide technicians through the lubrication process. Ensure that the content is clear and concise and is tailored to the specific equipment.

6. **Implement a digital sign-off mechanism.** Configure the digital system to enable digital sign-off for completed tasks. This feature should allow technicians to confirm task completion electronically, creating a verifiable record. Determine the process for supervisors to review and approve sign-offs as needed.

7. **Set up real-time exception reporting.** Integrate real-time exception reporting capabilities into the digital system. This feature should allow technicians to report any issues or deviations encountered during lubrication tasks immediately. Establish protocols for handling these exceptions, including notifying supervisors or triggering corrective actions.

8. **Do pilot testing.** Conduct a pilot implementation of the system within a controlled environment or on a limited set of equipment. Gather feedback from users to identify any issues or areas for improvement. Adjust configurations or content based on this feedback to enhance usability.

9. **Provide training and onboarding.** Develop and deliver training programs for all users of the digital lubrication management system. Ensure that technicians are familiar with GPS-tagged maps, route optimization features, interactive prompts, and digital sign-off processes. Provide ongoing support and resources to facilitate smooth adoption.

10. **Carry out full deployment.** Roll out the digital lubrication management system across the organization, ensuring that all relevant equipment and personnel are included. Monitor the implementation closely to address any challenges that arise during the transition.

11. **Foster continuous improvement.** After deployment, establish a feedback loop to gather ongoing input from users regarding the system's effectiveness. Regularly review performance metrics to identify areas for further optimization and enhancement of the lubrication practices.

By following these steps, organizations can effectively implement a digital lubrication management system that enhances operational efficiency, reduces errors, and supports a proactive maintenance culture.

DATA VALIDATION PROTOCOLS
TO PREVENT MISLEADING METRICS

Accurate decision-making depends on valid, trustworthy data. Lubrication programs must establish validation protocols to detect and correct errors in data entry, interpretation, or automation.

Validation strategies include:

➤ Cross-checking oil analysis anomalies against asset operating conditions
➤ Reviewing technician entries for outliers or duplicate records
➤ Implementing alerts for inconsistent task intervals or fluid usage
➤ Establishing approval workflows for critical data uploads (e.g., spec changes)

Automated validation rules in CMMSs or analytics platforms can flag suspect data. Regular audits of lubrication data ensure that metrics reflect real operating conditions and not procedural or recording errors.

Training personnel in data integrity and root cause analytics reinforces a culture of precision and reliability. Accurate decision-making in lubrication programs hinges on the establishment of robust data validation protocols to ensure that metrics are valid and trustworthy. These protocols are essential for detecting and correcting errors in data entry, interpretation, or automation, thereby enhancing the reliability of insights derived from lubrication data.

One effective validation strategy involves cross-checking oil analysis anomalies against asset operating conditions. By comparing unexpected results from oil analysis with relevant operational parameters, such as temperature and load, discrepancies can be identified and investigated. This process helps ensure that any reported issues are legitimate and not the result of data entry errors or misinterpretations.

Another important strategy is to review technician entries for outliers or duplicate records. Regularly examining data entered by the technicians can reveal inconsistencies that may indicate problems with data collection practices. Identifying outliers—data points that significantly deviate from established norms—can prompt further investigation to determine their validity. Additionally, addressing duplicate records is crucial to maintain the integrity of the dataset, as duplicates can skew analysis and lead to misleading conclusions.

Implementing alerts for inconsistent task intervals or fluid usage serves as another critical validation measure. Setting thresholds within the lubrication management system can trigger notifications when tasks are performed outside expected intervals or when fluid consumption deviates significantly from historical averages. These alerts facilitate timely reviews of lubrication practices and enable corrective actions before issues escalate.

Establishing approval workflows for critical data uploads, such as changes to lubricant specifications, enhances data integrity. By requiring designated personnel to review and approve significant data changes, organizations can minimize the risk of errors that could impact lubrication performance and asset reliability.

Automated validation rules within computerized maintenance management systems or analytics platforms can further bolster data integrity by flagging suspect data for review. These rules can be configured to identify inconsistencies based on predefined criteria, allowing for proactive management of potential data issues.

Regular audits of lubrication data are essential to ensure that reported metrics accurately reflect real operating conditions rather than procedural or recording errors. Systematic audits allow organizations to assess the effectiveness of data validation protocols and make necessary adjustments to improve accuracy.

Training personnel in data integrity and root cause analytics is vital for fostering a culture of precision and reliability within lubrication programs. By equipping staff with the knowledge and tools to recognize and address data quality issues, organizations can enhance their overall decision-making capabilities and ensure that lubrication practices are based on trustworthy metrics.

Implementation

An implementation plan for establishing data validation protocols in lubrication programs, focusing on cross-checking oil analysis anomalies, reviewing technician entries, implementing alerts, and establishing approval workflows, involves several structured steps:

1. **Define objectives.** Establish clear objectives for the implementation plan, outlining the desired outcomes for each validation protocol, including improved data accuracy, enhanced reliability of metrics, and streamlined data management processes.
2. **Assemble a project team.** Form a cross-functional project team that includes representatives from maintenance, data management, IT, and quality assurance. This team will be responsible for overseeing the implementation of data validation protocols from inception to completion.

3. **Develop validation protocols.** Create detailed procedures for each validation strategy:

    - **Cross-checking oil analysis anomalies.** Define specific operating conditions to be monitored alongside oil analysis results. Establish criteria for what constitutes an anomaly, and outline the process for conducting cross-checks, including responsible personnel and frequency.

    - **Reviewing technician entries.** Create guidelines for reviewing technician data entries, specifying what constitutes an outlier or duplicate record. Designate personnel responsible for conducting these reviews, and establish a regular review schedule.

    - **Implementing alerts.** Identify key metrics for task intervals and fluid usage that will trigger alerts. Work with IT to configure alert thresholds within the CMMS or analytics platform, ensuring that alerts are actionable and directed to the appropriate personnel.

    - **Establishing approval workflows.** Define the approval process for critical data uploads, such as changes to lubricant specifications. Identify the stakeholders who must approve changes, and outline the steps required for submitting and reviewing data changes.

4. **Select tools and technology.** Choose the appropriate software tools necessary for implementing the validation protocols. This may include enhancements to the existing CMMS or analytics platform to support automated alerts and approval workflows.

5. **Do pilot testing.** Conduct a pilot test of the validation protocols within a controlled environment or on a limited set of equipment. Monitor the effectiveness of each protocol, gather feedback from users, and make necessary adjustments based on the findings.

6. **Train personnel and emphasize data integrity.** Develop and deliver training sessions for all relevant personnel on the new data validation protocols. Ensure that the technicians and data management staff understand their roles and responsibilities in maintaining data integrity. Provide documentation and resources for ongoing reference.

7. **Carry out full deployment.** After successful pilot testing and training, roll out the data validation protocols across the organization. Ensure that all relevant personnel are informed of the changes and understand the importance of adhering to the new protocols.

8. **Monitor and evaluate.** Continuously monitor the effectiveness of the implemented validation protocols. Establish KPIs to assess data accuracy and integrity over time. Regularly review the performance metrics to identify areas for improvement.

9. **Do regular audits and reviews.** Schedule regular audits of data to ensure compliance with the validation protocols. Use audit findings to refine and improve the protocols as necessary, addressing any challenges or deficiencies identified during the review process.

10. **Foster continuous improvement.** Foster a culture of continuous improvement by encouraging feedback from personnel on the data validation processes. Regularly update protocols and training materials based on user experiences and evolving best practices to ensure ongoing effectiveness.

By following this implementation plan, organizations can establish robust data validation protocols that enhance the accuracy and reliability of lubrication program metrics, ultimately leading to better decision-making and improved operational efficiency.

CHAPTER 20

OIL ANALYSIS COORDINATION

COORDINATING SAMPLING INTERVALS WITH MACHINE CRITICALITY

Effective oil analysis programs begin with strategic sampling intervals that align with equipment criticality. High-speed, high-value, or safety-critical assets require more frequent monitoring than redundant or nonessential machinery. Key considerations for interval planning include:

➤ Machine criticality (based on FMEA, downtime cost, safety risk)
➤ Operating conditions (temperature, load, duty cycle)
➤ Lubricant type and service life
➤ Historical failure and wear trends

High-critical machines may require biweekly or monthly sampling, whereas low-risk assets can follow quarterly or semiannual schedules. Interval determination should be dynamic, adjusted based on trend deviations, lubricant age, or change in service conditions.

A criticality matrix and sampling calendar should be maintained in the CMMS or lubrication management software, ensuring adherence and visibility across departments.

Prioritizing oil sampling intervals based on machine criticality optimizes resource allocation and enhances reliability. High-criticality machines, often integral to operations, require more frequent analysis due to their potential impact on production and safety if they fail. These machines are typically assessed using failure mode and effects analysis, which helps identify potential failure modes and their consequences.

Operating conditions play a significant role in determining sampling frequency. Machines operating under high temperatures, heavy loads, or variable duty cycles may experience accelerated lubricant degradation, necessitating more frequent checks. The type of lubricant and its service life are also crucial; synthetic oils might offer longer intervals than mineral oils under similar conditions.

Historical data is invaluable for interval planning. Analyzing past failure and wear trends provides insights into optimal sampling frequencies. Machines with a history of rapid wear or failures may need closer monitoring, even if not classified as high criticality.

Implementing a dynamic approach to interval determination allows for adjustments in response to observed trend deviations, aging lubricants, or changes in service conditions. This flexibility ensures that the program remains responsive and effective.

Centralizing this information within a criticality matrix and sampling calendar in CMMS or lubrication management software ensures that all departments have access to the necessary data, fostering compliance and coordination across the organization. This structured approach minimizes the risk of oversight and enhances overall machine reliability.

Implementation Plan for Launching an Oil Analysis Program

Steps in the implementation plan include:

1. **Define the Objectives and Scope**
   - Establish clear objectives for the oil analysis program. Align with organizational goals, such as reducing downtime, extending equipment life, and improving safety.
   - Determine the scope, including types of machinery, criticality levels, and specific performance metrics to monitor.
2. **Assess the Current State**
   - Conduct a comprehensive assessment of current lubrication practices and existing data.
   - Identify gaps in current processes, and gather historical data on machinery performance, lubricant usage, and failure rates.
3. **Develop a Criticality Matrix**
   - Create a criticality matrix to categorize equipment based on FMEA, downtime cost, and safety risks.

- Assign criticality levels to each machine, prioritizing those requiring more frequent monitoring.

4. **Establish Sampling Intervals**
   - Define sampling intervals based on machine criticality, operating conditions, lubricant type, and historical trends.
   - Create a dynamic schedule that allows for adjustments based on real-time data and trend deviations.

5. **Select and Train Personnel**
   - Identify personnel responsible for sampling, testing, and data analysis.
   - Provide training on sampling techniques, equipment handling, and data interpretation to ensure consistency and accuracy.

6. **Integrate with CMMS or Lubrication Management Software**
   - Implement or update existing CMMS or lubrication management software to include the criticality matrix and sampling schedule.
   - Ensure system functionality to track adherence, generate alerts for upcoming samples, and store historical data.

7. **Select Analytical Methods and Equipment**
   - Determine appropriate analytical methods (e.g., viscosity measurement, particle count, spectrometric analysis) based on objectives and machinery types.
   - Acquire necessary testing equipment, or partner with a qualified laboratory for sample analysis.

8. **Do a Pilot Test**
   - Conduct a pilot test of the program with a select group of machines representing different criticality levels.
   - Gather the data, refine the processes, and make the necessary adjustments based on pilot results.

9. **Do a Full-Scale Implementation**
   - Roll out the program across all applicable machinery.
   - Ensure that all departments are informed and aligned with program goals and procedures.

10. **Ensure Continuous Monitoring and Improvement**
    - Regularly review program performance using KPIs such as equipment uptime, failure reduction, and cost savings.
    - Adjust sampling intervals, techniques, and processes based on continuous data analysis and feedback.

11. **Prepare Documentation and Reports**
    - Maintain comprehensive documentation of sampling results, trend analysis, and corrective actions.
    - Provide regular reports to stakeholders highlighting program achievements, challenges, and future plans.
12. **Review and Adapt**
    - Schedule periodic reviews to assess the program's effectiveness, and make strategic adaptations in response to technological advancements or changing operational needs.

This implementation plan ensures a structured, data-driven approach to launching an effective oil analysis program, enhancing machinery reliability and operational efficiency.

Audit Checklist for Oil Analysis Program

An audit checklist for an oil analysis program serves as a comprehensive tool to evaluate the program's effectiveness and identify areas for improvement. It encompasses pre-implementation and post-implementation phases to ensure thorough preparedness and ongoing performance assessment.

Before implementation, the checklist examines the clarity of objectives, scope alignment, current lubrication practices, criticality matrix development, sampling interval planning, personnel readiness, software integration, and selection of analytical methods.

Post-implementation, it assesses the alignment of program objectives, adherence to the criticality matrix and sampling schedule, personnel performance, data management efficacy, software functionality, equipment reliability, achievement of key performance indicators, thoroughness of documentation, and effectiveness of reporting.

Ultimately, the checklist aims to foster continuous improvement by guiding strategic adaptations in response to operational needs and technological advancements, ensuring that the program enhances machinery reliability and operational efficiency.

Following are examples of a pre-implementation checklist for establishing gaps and a post-implementation checklist to assess the effectiveness of the program.

Pre-Implementation Audit: Establishing Gaps

1. Objectives and Scope

- ➤ Are the objectives of the oil analysis program clearly defined?
- ➤ Is the scope of the program comprehensive and aligned with organizational goals?

2. Current State Assessment

- ➤ Has a thorough review of current lubrication practices been conducted?
- ➤ Is historical data on machine performance, lubricant usage, and failure rates available?

3. Criticality Matrix Development

- ➤ Is there a clear methodology for categorizing machinery based on criticality?
- ➤ Are all machines assigned appropriate criticality levels?

4. Sampling Interval Planning

- ➤ Are sampling intervals defined based on criticality, conditions, and lubricant type?
- ➤ Is there flexibility to adjust intervals based on real-time data?

5. Personnel and Training

- ➤ Have personnel responsible for sampling and analysis been identified?
- ➤ Is there a training plan in place for sampling techniques and data interpretation?

6. Software Integration

➤ Is there a CMMS or lubrication management software to track the program?
➤ Are criticality matrices and sampling schedules integrated into the software?

7. Analytical Methods and Equipment

➤ Are appropriate analytical methods selected for the program's objectives?
➤ Is necessary testing equipment available, or is there a plan to partner with a laboratory?

Post-Implementation Audit: Assessing Effectiveness

1. Program Objectives and Alignment

➤ Are the program objectives being met?
➤ Is the program aligned with current organizational goals and priorities?

2. Criticality Matrix and Sampling Schedule

➤ Are the criticality matrix and sampling schedule up-to-date and adhered to?
➤ Are adjustments to intervals being made based on data trends?

3. Training and Personnel Performance

➤ Are personnel effectively performing their roles in sampling and analysis?
➤ Is ongoing training provided to adapt to new techniques or tools?

4. Data Management and Analysis

- Is data from oil analysis being accurately recorded and analyzed?
- Are trends and deviations being identified and acted upon promptly?

5. Software and System Integration

- Is the CMMS or lubrication management system fully functional and utilized?
- Are alerts and schedules effectively managed through the software?

6. Equipment and Methodology

- Is the analysis equipment functioning correctly and producing reliable results?
- Are analytical methods still suitable for current operational needs?

7. Program Outcomes and KPIs

- Are key performance indicators such as equipment uptime and failure reduction being met?
- Has the program led to measurable improvements in reliability and cost savings?

8. Documentation and Reporting

- Is documentation thorough and up-to-date?
- Are reports being generated and communicated effectively to stakeholders?

9. Continuous Improvement

- Is there a process for reviewing and improving the program based on audit findings?
- Are strategic adaptations being made in response to technological advancements or changing needs?

These checklists ensure that both the planning and implementation phases of the oil analysis program are comprehensive and effective, identifying gaps and assessing ongoing performance to drive continuous improvement.

Role of Oil Analysis in Predictive Maintenance

Oil analysis plays a critical role in predictive maintenance strategies by providing insights into the condition of both the lubricant and the machinery it services. It involves regularly monitoring and analyzing lubricant properties to detect signs of wear, contamination, and degradation, enabling early identification of potential issues. This proactive approach helps in predicting equipment failures before they occur, allowing for timely maintenance interventions.

Key roles of oil analysis in predictive maintenance include:

➤ **Wear monitoring.** Detects and quantifies wear particles in the oil, helping to identify abnormal wear patterns or component failures, such as bearings or gears, before they lead to more significant damage

➤ **Contaminant detection.** Identifies contaminants such as dirt, water, fuel, or coolant, which can indicate seal failures, leaks, or operational issues, allowing for corrective actions to prevent further damage

➤ **Lubricant condition assessment.** Evaluates the oil's physical and chemical properties, such as viscosity, acidity, and additive depletion, to determine whether the oil is still effective or needs replacement

➤ **Trend analysis.** Provides historical data that enables trend analysis, helping to predict future failures and optimize maintenance schedules based on actual equipment condition rather than time-based intervals

➤ **Cost reduction.** Minimizes unplanned downtime and repair costs by addressing issues before they escalate, extending equipment life, and optimizing lubricant change intervals

➤ **Safety and reliability enhancement.** Improves equipment reliability and safety by ensuring that the machinery operates within specified parameters and by reducing the risk of catastrophic failures

By integrating oil analysis into predictive maintenance, organizations can achieve more efficient and effective maintenance practices, leading to better utilization of resources, reduced operational costs, and enhanced overall equipment performance.

Common Test Methods

Common methods used for oil analysis in industrial applications include:

- **Spectrometric analysis.** Identifies and quantifies wear metals, contaminants, and additive elements by measuring the concentration of various elements in the oil using techniques such as inductively coupled plasma or atomic emission spectroscopy (AES).
- **Viscosity measurement.** Determines the oil's resistance to flow, which can indicate contamination, degradation, or changes due to temperature and pressure. This is often measured using viscometers or rheometers.
- **Particle count analysis.** Measures the number and size of particles in the oil to assess contamination levels. This is typically performed using automatic particle counters or microscopy.
- **Fourier transform infrared spectroscopy.** Analyzes the chemical composition of the oil, identifying contaminants, oxidation products, and additive depletion by measuring infrared light absorption at different wavelengths.
- **Total Acid Number and Total Base Number.** TAN measures the acidity of the oil, indicating oxidation and degradation. TBN assesses the oil's alkalinity reserve, reflecting its ability to neutralize acids.
- **Water content analysis.** Determines the amount of water in the oil, which can cause corrosion and affect lubrication. Techniques include Karl Fischer titration and crackle tests.
- **Ferrous density.** Measures the concentration of ferrous (iron-containing) particles to assess wear and contamination levels using techniques like magnetometry.
- **Flash point testing.** Evaluates the temperature at which oil vapors ignite, indicating contamination with volatile substances or degradation.
- **Oxidation and nitration testing.** Assesses the extent of oxidation and nitration, which can degrade oil performance and indicate engine or system issues.

These methods provide valuable insights into the condition of the oil and the machinery it lubricates, helping to predict failures, optimize maintenance schedules, and extend equipment life.

Oil Condition Monitoring Techniques

Oil condition monitoring techniques vary widely in terms of accuracy and reliability, each offering distinct advantages and limitations. Understanding these differences is key to selecting the right approach for a particular application.

- ➤ **Ferrography.** This technique analyzes wear particles in the oil to assess the condition of the machinery. It is highly accurate in identifying the size, shape, and composition of particles, which helps in diagnosing wear types and potential failure modes. Its reliability is well regarded for detecting abnormal wear, but it requires skilled interpretation and may not provide real-time insights.
- ➤ **Fourier-transform infrared spectroscopy.** FTIR is used to identify chemical changes in the oil, such as oxidation, nitration, and contamination with water or fuel. It is accurate in providing a snapshot of the oil's chemical condition and is reliable for routine monitoring. However, it may not be as effective in detecting mechanical wear particles.
- ➤ **Viscometry.** Measuring the oil's viscosity offers insights into its lubricating properties and degradation. Viscosity measurements are accurate and reliable for assessing oil health, but they may not directly indicate mechanical wear or contamination unless correlated with other tests.
- ➤ **Particle counting.** This method quantifies the number and size of particles in the oil, providing a general indication of cleanliness and contamination. While highly reliable for monitoring contamination levels, it does not differentiate among particle types, so it may not accurately reflect wear conditions without supplementary analysis.
- ➤ **Atomic emission spectroscopy.** AES (ICP) detects and quantifies metal particles in oil, indicating wear of specific components. It is accurate for elemental analysis and reliable for trend monitoring, but like particle counting, it may require additional tests to provide a complete picture of wear mechanisms.
- ➤ **Remaining Useful Life Evaluation Routine.** RULER assesses the remaining life of oil additives by measuring their depletion rate. It is accurate for predicting when oil will no longer perform effectively, making it reliable for maintenance scheduling. However, it does not directly measure wear or contamination.

Each technique offers unique insights, and combining multiple methods often yields the most comprehensive understanding of oil condition and machinery health. The choice of techniques should be guided by specific monitoring objectives, the type of machinery, and the criticality of the operation.

CHAIN OF CUSTODY: PRESERVING THE INTEGRITY OF OIL SAMPLES

Sample integrity is essential for reliable oil analysis. Errors in sampling technique, labeling, handling, or storage can result in misleading data, misdiagnosis, and costly interventions.

Ensuring sample integrity is fundamental for accurate oil analysis. Missteps in sampling techniques, labeling, handling, or storage can lead to erroneous data, incorrect diagnoses, and unnecessary costs.

To maintain integrity:

- ➤ **Use proper containers.** Utilize precleaned, sealed bottles approved by the laboratory to avoid contamination.
- ➤ **Take care in labeling.** Clearly label each sample with essential details such as date, equipment ID, lubricant type, technician name, and running condition at the time of sampling.
- ➤ **Document the sampling location.** Record the exact sampling location, whether upstream of a filter, at a drain port, or within a live zone.
- ➤ **Prevent contamination.** Purge sample points before collection to remove any contaminants that could skew the results.
- ➤ **Properly store samples.** Store samples upright in insulated containers to maintain stability until shipment.

Accompany each sample batch with a documented chain of custody form, detailing the collection process, handoffs, and delivery. Implementing digital logging through mobile apps or barcodes can enhance traceability, ensuring that every step is recorded and easily retrievable, ultimately streamlining the reporting process.

INTERPRETING OIL ANALYSIS BEYOND THE LAB REPORT

While laboratory reports provide useful flagging indicators (green/yellow/red) and comments, the true value of oil analysis is realized when site reliability engineers or lubrication specialists interpret the data within the broader context of equipment operation and history.

Key interpretation tactics include:

➤ **Historical trend analysis.** Rather than focusing solely on whether current values exceed a single threshold, compare them against historical data to identify emerging trends and patterns that may indicate developing issues.

➤ **Machine-specific context.** Correlate wear-metal concentrations with the specific metallurgy and load patterns of the machinery to understand wear mechanisms and potential sources of contamination.

➤ **Viscosity and contamination correlation.** Examine shifts in viscosity in conjunction with temperature records and contamination data to diagnose issues such as dilution, oxidation, or additive breakdown.

➤ **Additive and chemical analysis.** Investigate additive depletion in relation to oxidation levels and TBN or TAN metrics to assess oil life and potential degradation.

Advanced interpretation may involve trend modeling and statistical outlier detection to predict future failures. By integrating oil analysis with other diagnostic techniques such as vibration analysis, thermography, or ultrasonic testing, the goal is to shift from descriptive diagnostics to predictive maintenance, enabling timely interventions and minimizing unplanned downtime.

SELECTING THE RIGHT LABORATORY: COST VERSUS CAPABILITY

Laboratories vary significantly in technical capabilities, turnaround time, reporting formats, and customer support. Selection should weigh cost against service quality and program needs.

Criteria for lab selection include:

➤ ISO 17025 or other relevant certifications

➤ Range and depth of testing methods offered (e.g., ferrography, FTIR, RULER)

➤ Sample processing time and availability of expedited options

➤ Data reporting format (digital, API compatibility, dashboard access)

➤ Technical support availability (phone consultation, recommendations)

Labs that act as analytical partners offering trend insights, training, and customized alerts provide more value than low-cost labs offering basic pass/fail metrics. Site visits and blind sample testing can validate laboratory consistency and accuracy.

Selecting the right oil condition analysis laboratory is a critical component of maintaining the reliability and efficiency of the machinery. The decision should be guided by a comprehensive assessment of the laboratory's technical capabilities, the breadth of services offered, and how these align with the specific needs of the program, balanced against cost considerations.

A laboratory's technical credentials are a primary consideration. ISO 17025 certification is a fundamental requirement, as it signifies adherence to internationally recognized standards of testing and calibration. This certification provides assurance of the laboratory's competence in delivering accurate and reliable test results. Beyond certification, the variety of testing methods available is crucial. Advanced techniques such as ferrography, FTIR, and RULER are essential for a detailed analysis of oil condition. These methods provide insights into wear particles, chemical properties, and the remaining additive life of lubricants, respectively.

Turnaround time for sample processing is another critical factor. Efficient laboratories offer expedited processing options, which can significantly reduce equipment downtime and allow for quicker decision-making. The ability to receive rapid results is particularly important in situations where operational continuity is at risk.

The format in which data is reported is a vital aspect of laboratory services. Modern operations often require digital data formats that are compatible with existing systems. API compatibility and dashboard access can facilitate seamless integration with maintenance management systems, enabling real-time monitoring and analysis of oil condition data. This integration allows for proactive maintenance strategies and can significantly enhance operational efficiency.

Technical support provided by the laboratory should extend beyond simple data reporting. Access to expert consultations by phone and the provision of actionable recommendations can greatly enhance the value of the service. Laboratories that offer insights into trends, provide training for interpreting results, and set up customized alerts for specific conditions act as true partners in maintaining equipment reliability.

It's also vital to evaluate the laboratory's consistency and accuracy, which can be verified through site visits and blind sample testing. Site visits allow for a firsthand assessment of the laboratory's facilities, processes, and staff expertise. Blind sample testing can further validate the laboratory's ability to consistently produce accurate results under real-world conditions.

In conclusion, while cost is an important factor in selecting an oil condition analysis laboratory, it should not be the sole consideration. The laboratory's ability to act as an analytical partner, providing comprehensive insights and support, offers significant advantages over lower-cost options that may only deliver basic pass/fail metrics. A strategic approach to laboratory selection, focused on technical capability and service alignment with operational needs, will enhance overall equipment reliability and performance.

Laboratory Audit

To assess a laboratory effectively, follow these audit steps to ensure comprehensive evaluation of its capabilities and reliability:

- **Pre-Audit Preparation**
    - Review the laboratory's accreditation and certifications, such as ISO 17025, to confirm compliance with international standards.
    - Collect background information on the laboratory's reputation, client feedback, and past performance records.
- **Documentation Review**
    - Examine the laboratory's quality management system documentation, including standard operating procedures for testing methods.
    - Review records of proficiency testing and interlaboratory comparison results to assess accuracy and consistency.
- **Facility Inspection**
    - Conduct an on-site visit to inspect the laboratory's facilities, equipment, and environmental controls.
    - Verify the calibration status of the analytical instruments, and observe maintenance procedures.
- **Personnel Evaluation**
    - Assess the qualifications, training, and experience of laboratory personnel.
    - Observe the staff as they perform sample analyses to evaluate adherence to SOPs and technical competence.

➤ **Test Method Assessment**
  - Review the range of testing methods offered, ensuring that they align with your specific needs (e.g., ferrography, FTIR).
  - Evaluate the laboratory's capability to handle complex analyses and its flexibility in adopting new techniques.

➤ **Sample Handling and Processing**
  - Observe sample handling procedures to ensure a proper chain of custody and sample integrity.
  - Evaluate sample processing times, including the availability of expedited services.

➤ **Data Reporting and Interpretation**
  - Review the laboratory's data reporting formats, and ensure compatibility with your systems (digital formats, API).
  - Assess the clarity and comprehensiveness of reports, including any insights and recommendations provided.

➤ **Technical Support and Customer Service**
  - Evaluate the availability and quality of technical support, including consultation services and responsiveness to inquiries.
  - Consider the laboratory's willingness to collaborate and act as a partner in your oil condition monitoring efforts.

➤ **Consistency and Accuracy Validation**
  - Conduct blind sample testing to verify the laboratory's accuracy and consistency in delivering results.
  - Compare the laboratory's test results with those from other laboratories for benchmarking purposes.

➤ **Post-Audit Evaluation**
  - Compile the findings, and identify areas of strength and opportunities for improvement.
  - Provide feedback to the laboratory, and discuss potential actions for addressing any identified gaps.

By following these steps, you can thoroughly assess a laboratory's capability to meet your oil condition monitoring needs and ensure reliable and accurate testing services.

QMS Audit

Evaluating the effectiveness of a laboratory's quality management system involves several methods to ensure that it meets the required standards and consistently produces reliable results. Here are some common methods used in this evaluation:

- ➤ **Internal audits.** Conduct regular internal audits to assess compliance with the laboratory's QMS procedures and policies. These audits help identify any deviations from established protocols and provide opportunities for corrective actions.
- ➤ **Proficiency testing.** Participate in proficiency testing programs where the laboratory's results are compared with those from other laboratories. This external validation helps assess the accuracy and consistency of the test results.
- ➤ **Management reviews.** Perform regular management reviews to evaluate the overall performance of the QMS. These reviews involve analyzing audit findings, customer feedback, and corrective actions to ensure continuous improvement.
- ➤ **Corrective and preventive action (CAPA).** Implement a robust CAPA process to address nonconformities and prevent their recurrence. Monitoring the effectiveness of CAPA is crucial in maintaining the integrity of the QMS.
- ➤ **Customer feedback.** Collect and analyze feedback from clients regarding the quality and reliability of the laboratory's services. Customer satisfaction surveys can provide insights into areas needing improvement.
- ➤ **Document control.** Review the laboratory's document control system to ensure that all procedures, manuals, and records are up-to-date, properly maintained, and accessible to authorized personnel.
- ➤ **Training and competency assessment.** Evaluate the training programs and competency assessments for laboratory staff to ensure that they have the necessary skills and knowledge to perform their duties effectively.
- ➤ **Risk management.** Assess the laboratory's risk management processes to identify potential risks to quality and implement strategies to mitigate them.
- ➤ **Traceability and calibration.** Verify the traceability of measurements to national or international standards, and ensure that all equipment is regularly calibrated and maintained.

➤ **External audits and accreditation.** Engage external auditors to review the laboratory's QMS as part of the accreditation assessments. Achieving and maintaining accreditation (e.g., ISO 17025) is a strong indicator of an effective QMS.

By employing these methods, a laboratory can evaluate and enhance the effectiveness of its quality management system, ensuring consistent delivery of high-quality and reliable test results.

If the lab of choice is deficient in certain areas, an external assessment can be very valuable. Implementing a quality management system in a laboratory setting can present several challenges. Here are some of the most common to consider:

➤ **Resource allocation.** Implementing a QMS requires significant resources, including time, money, and personnel. Laboratories often struggle with allocating sufficient resources without disrupting regular operations.

➤ **Staff resistance.** Employees may resist changes brought by the QMS due to fear of increased workload or a lack of understanding of the benefits. Overcoming this resistance requires effective communication and training.

➤ **Training and competency.** Ensuring that all staff members are adequately trained and competent in new procedures can be challenging. Continuous training programs are essential but can be resource-intensive.

➤ **Document control.** Maintaining up-to-date documentation and ensuring accessibility can be cumbersome. Laboratories must establish robust systems for document control to manage revisions and approvals effectively.

➤ **Process standardization.** Achieving consistency across various processes and procedures is challenging, especially in laboratories with diverse testing methods. Standardizing procedures while allowing flexibility for different tests requires meticulous planning.

➤ **Risk management.** Identifying and managing risks associated with laboratory operations is complex. Developing a proactive risk management strategy that addresses potential quality issues is essential but can be difficult to implement effectively.

➤ **Cultural change.** Instilling a quality-focused culture across the laboratory requires a shift in mindset. This cultural change needs strong leadership and commitment from all levels of the organization.

➤ **Continuous improvement.** Establishing mechanisms for continuous improvement within the QMS can be challenging. Laboratories must create systems for regularly reviewing and enhancing processes based on performance data and feedback.

➤ **Compliance and accreditation.** Meeting the requirements for accreditation (such as ISO 17025) can be demanding. Laboratories must continuously monitor compliance with standards, which can be resource-intensive and complex.

➤ **Integration with existing systems.** Integrating the QMS with existing laboratory information management systems and workflows can present technical challenges. Ensuring compatibility and seamless operation requires careful planning and execution.

By addressing these challenges with strategic planning and ongoing commitment, laboratories can successfully implement a QMS that enhances quality, reliability, and operational efficiency.

ALIGNING OIL ANALYSIS FINDINGS WITH WORK ORDER GENERATION

Oil analysis should be seamlessly tied to maintenance workflows. When exceptions are detected, such as high particle count, water content, or wear metals, a corresponding work order should be auto-generated or recommended for review.

This linkage includes:

➤ Automatic flag-to-CMMS integration for high-priority findings
➤ Recommended actions embedded in analysis dashboards (e.g., resample, flush, inspect bearing)
➤ Maintenance planner review protocols for flagged results
➤ Feedback loops to confirm task execution and resolution

Integrating oil analysis findings with work order generation is a critical practice for optimizing maintenance workflows. This integration ensures that when oil analysis detects exceptions such as high particle count, elevated water content, or excessive wear metals, immediate and appropriate actions are triggered within the maintenance management system.

The process begins with the automatic integration of flagged high-priority findings into the CMMS. This integration facilitates the immediate generation of work orders or alerts maintenance personnel to review the findings. Such seamless communication between oil analysis results and the CMMS is vital for prioritizing and addressing critical issues promptly.

Incorporating recommended actions directly into analysis dashboards further enhances this process. When an exception is detected, the dashboard should provide clear, actionable recommendations, such as resampling the oil, flushing the system, or inspecting specific components like bearings. These recommendations guide the maintenance teams in making informed decisions quickly.

It is essential for maintenance planners to have established review protocols for flagged results. This involves evaluating the severity of the findings, considering the operational context, and determining the appropriate course of action. By having a structured review process, planners can ensure that high-priority issues are addressed consistently and efficiently.

Feedback loops are a crucial component of this alignment, providing a mechanism to confirm task execution and resolution. Once maintenance tasks are completed, feedback should be documented in the CMMS to validate that the work has been carried out and the issue resolved. This documentation not only ensures accountability but also serves as a valuable resource for assessing the accuracy of the oil analysis and the effectiveness of maintenance actions.

By aligning oil analysis findings with work order generation, organizations can achieve rapid, documented responses to potential issues, enhancing the reliability and efficiency of their maintenance operations. This integration also provides critical insights into the performance of both the oil analysis program and the maintenance strategies, enabling continuous improvement and optimization of asset management practices.

Implementation

Aligning oil analysis findings with work order generation involves a systematic process that ensures that detected anomalies in oil analysis are effectively translated into actionable maintenance tasks. Here's a step-by-step outline of this process:

1. **Data Collection and Analysis**
   - Conduct regular oil analysis using appropriate techniques to monitor the condition of the machinery.
   - Utilize advanced diagnostic tools to identify anomalies such as high particle count, water content, and wear metals.

2. **Integration with the CMMS**
   - Implement a seamless integration between the oil analysis system and the CMMS.
   - Set up automatic flagging for high-priority findings, ensuring immediate alerts or work order generation within the CMMS.

3. **Development of Actionable Recommendations**
   - Embed clear, actionable recommendations in the analysis dashboards. These should include suggested actions such as resampling, flushing systems, or inspecting specific components.
   - Ensure that recommendations are easily accessible for maintenance personnel.

4. **Maintenance Planner Review**
   - Establish structured review protocols for maintenance planners to evaluate flagged results.
   - Assess the severity and implications of the findings, determine corrective actions, and prioritize tasks based on operational criticality.

5. **Work Order Generation**
   - Automatically generate work orders in the CMMS for flagged findings, including detailed information on the issue and recommended actions.
   - Ensure that work orders are prioritized appropriately and assigned to relevant personnel.

6. **Execution of Maintenance Tasks**
   - Carry out the maintenance tasks as specified in the work orders.
   - Ensure that maintenance personnel follow the recommended actions, and document any additional observations or deviations.

7. **Feedback Loop Implementation**
   - Implement feedback loops to confirm the execution and resolution of maintenance tasks.
   - Record completion and outcomes in the CMMS to provide accountability and a historical record.

8. **Continuous Monitoring and Improvement**
   - Regularly review feedback and completed work orders to assess the accuracy of oil analysis and the effectiveness of maintenance actions.
   - Use insights from the feedback loops to refine analysis techniques and improve maintenance strategies.

9. **Training and Communication**
    - Provide ongoing training for the staff on interpreting oil analysis results and integrating findings with maintenance workflows.
    - Foster open communication between the analysis teams and maintenance personnel to ensure a cohesive approach to equipment management.

By following these steps, organizations can align oil analysis findings with work order generation, ensuring prompt and efficient responses to potential issues, thereby enhancing the reliability and performance of their machinery.

Training

Training maintenance personnel on oil analysis interpretation is crucial for ensuring that they can effectively understand and act on the findings. Here are some best practices to consider when developing a training program:

- ➤ **Comprehensive curriculum.** Develop a curriculum that covers the basics of oil analysis, including the types of tests performed, the significance of different parameters (e.g., particle count, water content, wear metals), and how these relate to equipment health.
- ➤ **Hands-on training.** Incorporate practical, hands-on training sessions where personnel can engage with real-world oil samples and analysis tools. This experiential learning approach helps reinforce theoretical knowledge.
- ➤ **Use of real data.** Utilize actual oil analysis reports from the organization to teach interpretation skills. Real data provides context and relevance, allowing trainees to relate learning to their everyday tasks.
- ➤ **Simulation and case studies.** Implement simulation exercises and case studies that present various scenarios and require trainees to interpret data, identify potential issues, and determine appropriate actions.
- ➤ **Cross-disciplinary training.** Encourage collaboration among maintenance personnel, reliability engineers, and oil analysis experts. Cross-disciplinary training sessions can enhance understanding and improve communication among teams.
- ➤ **Focus on critical indicators.** Emphasize the importance of key indicators that are critical to equipment performance and reliability. Teach how to prioritize findings based on severity and potential impact.

➤ **Regular updates and refresher courses.** Offer periodic refresher courses and updates on new techniques, technologies, and best practices to ensure that personnel remain informed and skilled in the latest developments.

➤ **Certification and recognition.** Provide certification for completing training programs to acknowledge competence and motivate personnel. This recognition can encourage continued learning and professional development.

➤ **Feedback and evaluation.** Solicit feedback from the trainees to continuously improve the training program. Conduct evaluations to assess the effectiveness of the training, and identify areas for enhancement.

➤ **Integration with maintenance workflows.** Train personnel on how to integrate oil analysis findings into maintenance workflows, including using the CMMS, generating work orders, and documenting actions taken.

➤ **Mentorship and support.** Establish a mentorship program where experienced personnel can guide and support less-experienced team members. Having a resource to turn to for questions and guidance can be invaluable.

By following these best practices, organizations can ensure that their maintenance personnel are well equipped to interpret oil analysis results effectively, leading to better decision-making and improved equipment reliability.

LUBRICANT FLAGGING LIMITS:
HOW TO SET AND WHEN TO REASSESS

Flagging limits are thresholds that trigger alerts for anomalies in wear, contamination, or chemical degradation. These limits must be tailored to lubricant type, equipment class, and operating context.

Common parameters are:

➤ Particle count (ISO 4406)
➤ Water content (ppm or %)
➤ Wear metals (Fe, Cu, Pb)
➤ Oxidation, nitration, and sulfation indices
➤ Viscosity at 40°C and 100°C

Flagging levels (normal, marginal, critical) should be reviewed semiannually or following major operating changes (e.g., new lubricant, increased load, different climate). Rigid adherence to outdated thresholds may obscure the true risk or generate unnecessary alarms.

Collaboration with labs and OEMs supports appropriate limit setting. Some organizations use machine learning models to refine flags based on evolving data trends.

Setting and periodically reassessing lubricant flagging limits is vital for effective condition monitoring and maintenance planning. These limits help identify potential issues related to wear, contamination, or chemical degradation before they lead to equipment failure. Here's a guide to how to set these limits and when to reassess them:

Setting Flagging Limits

- ➤ **Understand equipment and lubricant specifics.** Tailor flagging limits to the specific equipment type, operational conditions, and lubricant being used. Factors such as machine design, load, speed, and environment should be considered.
- ➤ **Identify key parameters.** Focus on critical parameters such as particle count (ISO 4406), water content (ppm or %), wear metals (e.g., Fe, Cu, Pb), oxidation, nitration, sulfation indices, and viscosity at 40°C and 100°C.
- ➤ **Collaborate with experts.** Work with lubricant manufacturers, laboratories, and original equipment manufacturers to set appropriate thresholds. Their expertise can provide insights into industry standards and best practices.
- ➤ **Establish flagging levels.** Define clear flagging levels: normal, marginal, and critical. This graded approach helps prioritize responses based on the severity of the condition.
- ➤ **Incorporate machine learning models.** Consider leveraging machine learning models to refine flagging limits. These models can analyze historical data and evolving trends to set more dynamic and accurate thresholds.

When to Reassess Flagging Limits

> **Regular reviews.** Conduct semiannual reviews of the flagging limits to ensure they remain relevant and effective. This routine evaluation helps in adapting to gradual changes in operating conditions or lubricant performance.
> **Major operational changes.** Reassess the limits following significant changes, such as the introduction of a new lubricant, changes in operational loads, or shifts in climate, or after major equipment overhauls.
> **Feedback from analysis and maintenance.** Use feedback from oil analysis results and maintenance activities to adjust limits. If frequent false alarms or missed detections occur, it may indicate the need for adjustment.
> **Continuous monitoring and data analysis.** Continuously monitor trends and analyze data to identify shifts in equipment behavior. This proactive approach can signal when limits need adjustment to reflect the current state of operations.
> **Technological advancements.** Stay informed about advancements in oil analysis techniques and technologies. New methods or tools might offer insights that warrant reassessment of flagging limits.

By setting and regularly reassessing lubricant flagging limits, organizations can maintain optimal machine health, avoid unnecessary maintenance actions, and reduce the risk of equipment failure. This dynamic approach ensures that flagging limits evolve in line with changing operational conditions and technological advancements.

COMMUNICATING ANALYSIS RESULTS ACROSS THE ORGANIZATION

Oil analysis findings must be effectively communicated to drive corrective action and long-term planning. Different audiences require tailored summaries, visualizations, and access levels.

Communication strategies include:

> Summary dashboards for leadership (asset health status, high-risk trends)
> Detailed reports with technician comments for maintenance teams

➤ Email alerts for flagged conditions with recommended actions
➤ Regular reviews with engineering for program insights

Training should be provided on how to read and respond to analysis reports. Engaging operators in basic interpretation fosters ownership and early issue reporting.

Effective communication of oil analysis findings is essential for driving corrective actions and supporting long-term planning. Tailoring the communication strategy to different audiences within the organization ensures that each group receives the information it needs in a format that can be readily understood and acted upon. Here are some strategies for communicating analysis results across the organization.

Summary Dashboards for Leadership

➤ **Purpose.** Provide high-level insights into asset health, highlight high-risk trends, and support strategic decision-making.
➤ **Content.** Summarize key metrics and trends using visualizations such as charts and graphs. Highlight critical issues that require immediate attention.
➤ **Format.** Interactive dashboards that allow leaders to drill down into specific areas of interest.

Detailed Reports for Maintenance Teams

➤ **Purpose.** Offer comprehensive insights into oil analysis findings, including specific details and technician comments.
➤ **Content.** Include data on parameters like wear metals, contamination levels, and chemical degradation. Provide context and recommendations for corrective actions.
➤ **Format.** Detailed digital reports that maintenance teams can access and update, ensuring a complete record of actions taken.

Email Alerts for Flagged Conditions

➤ **Purpose.** Ensure rapid response to urgent issues by notifying relevant personnel of flagged conditions.

- ➤ **Content.** Briefly describe the issue, its potential impact, and recommended actions. Include links to detailed reports for further information.
- ➤ **Format.** Automated email alerts that are sent immediately when critical thresholds are exceeded.

Regular Reviews with Engineering

- ➤ **Purpose.** Facilitate collaboration between the maintenance and engineering teams, leveraging oil analysis insights for program improvement.
- ➤ **Content.** Review trends, discuss findings, and explore ways to enhance equipment reliability and performance.
- ➤ **Format.** Scheduled meetings or workshops with shared access to analysis data and insights.

Training on Analysis Interpretation

- ➤ **Purpose.** Equip the staff with the skills to understand and act on analysis reports, fostering a sense of ownership and proactive issue identification.
- ➤ **Content.** Teach basic interpretation skills, explain common findings, and demonstrate how to use dashboards and reports.
- ➤ **Format.** In-person or virtual training sessions, supplemented with guides and reference materials.

Engaging Operators

- ➤ **Purpose.** Involve operators in the oil analysis process to promote early detection and reporting of potential issues.
- ➤ **Content.** Provide basic training on identifying warning signs and reporting concerns to the maintenance teams.
- ➤ **Format.** Informal sessions, quick-reference guides, and integrating feedback mechanisms into daily routines.

Digital Platforms for Broad Visibility

- ➤ **Purpose.** Ensure secure, widespread access to oil analysis results and related insights across the organization.
- ➤ **Content.** Host dashboards, reports, and alerts on a centralized platform with role-based access controls.
- ➤ **Format.** Cloud-based systems that offer real-time access to data and facilitate collaboration.

By implementing these communication strategies, organizations can ensure that oil analysis findings are effectively integrated into their maintenance and reliability culture. This shared understanding of analysis outcomes supports informed decision-making, enhances equipment performance, and contributes to long-term organizational success.

Visualizing data effectively in reports is crucial for conveying information clearly and enabling informed decision-making. Here are some of the most common methods for visualizing data in reports:

- ➤ **Bar Charts**
  - Ideal for comparing quantities across different categories
  - Useful for displaying trends over time or differences between groups
- ➤ **Line Graphs**
  - Effective for showing trends and changes over time
  - Useful for illustrating the progression of variables and identifying patterns
- ➤ **Pie Charts**
  - Best used for depicting proportions and percentages within a whole
  - Useful for showing the distribution of categories in a dataset
- ➤ **Histograms**
  - Used to represent the distribution of numerical data
  - Helpful for displaying the frequency of data points within specified ranges or bins
- ➤ **Scatter Plots**
  - Useful for examining the relationship between two numerical variables
  - Ideal for identifying correlations or patterns in data
- ➤ **Heat Maps**
  - Used in visualizing data using color coding to represent different values
  - Useful for displaying complex data, such as correlations in large datasets

➤ **Box Plots**
  • Used to show the distribution of data based on a five-number summary (minimum, first quartile, median, third quartile, maximum)
  • Useful for identifying outliers and understanding data spread

➤ **Area Charts**
  • Similar to line graphs but with the area below the line filled in
  • Useful for illustrating cumulative totals over time

➤ **Gantt Charts**
  • Used for project scheduling and tracking progress
  • Helpful for visualizing timelines and dependencies between tasks

➤ **Bullet Graphs**
  • Used to display progress toward a target or goal
  • Useful for performance measurements, such as KPIs

➤ **Tree Maps**
  • Used to show hierarchical data using nested rectangles
  • Useful for visualizing the structure and relative sizes within a dataset

➤ **Network Diagrams**
  • Effective in representing relationships between interconnected entities
  • Useful for visualizing social networks or communication flows

Selecting the appropriate visualization method depends on the nature of the data, the message you wish to convey, and the audience's familiarity with different visualization types. Combining multiple methods within a report can often provide a more comprehensive view of the data.

CHAPTER 21

FAILURE MODES EFFECTS ANALYSIS, ROOT CAUSE FAILURE ANALYSIS, AND TROUBLESHOOTING

COMMON FAILURE MODES IN LUBRICATION SYSTEMS AND HOW TO MODEL THEM

Lubrication system failures are among the most frequent and preventable causes of equipment downtime. Modeling these failure modes through FMEA provides early insight into risk factors and guides preventive action.

Typical lubrication-related failure modes include:

➤ Insufficient lubricant volume (underlubrication)
➤ Contamination ingress (particles, water, chemicals)
➤ Improper lubricant type or viscosity
➤ Overlubrication causing seal blowout or heat generation
➤ Loss of lubricant due to leaks or system bypass
➤ Degradation due to oxidation, thermal breakdown, or additive depletion

Modeling includes identifying each failure mode and its effects on the system—rating the effects for severity (S), likelihood of occurrence (O), and likelihood of detection

(D) and calculating a risk priority number (RPN). High RPNs indicate priority areas for redesigning, procedure changes, or monitoring enhancements.

By embedding lubrication-specific FMEAs within broader equipment assessments, maintenance strategies become more targeted and proactive. Failure modes and effects analysis is a structured approach used to identify potential failure modes within a system, assess their impact, and prioritize actions to mitigate risks.

Machinery Failure

Machinery failure in industrial settings can result from a variety of factors, often leading to costly downtime and repairs. Understanding these common causes can help in developing preventive maintenance strategies and improving equipment reliability. Here are some of the most frequent causes of machinery failure:

➤ **Poor Lubrication**
  - Insufficient or improper lubrication can lead to increased friction, wear, and overheating.
  - Contaminated lubricants can introduce particles that contribute to accelerated wear and tear.

➤ **Contamination**
  - Ingress of contaminants such as dust, dirt, water, and chemicals can compromise machinery components.
  - Contamination often results in abrasion, corrosion, and degradation of lubricants.

➤ **Overloading**
  - Operating machinery beyond its design limits can cause excessive stress and lead to component failure.
  - Overloading often results from poor planning or unexpected demand spikes.

➤ **Fatigue and Wear**
  - Repeated stress cycles can lead to material fatigue and eventual failure of components.
  - Normal wear over time can degrade the performance of parts, leading to failure if not addressed.

➤ **Improper Installation or Alignment**
  - Misalignment or incorrect installation of components can cause uneven wear and increased stress on the machinery.

- • This often leads to vibration issues and premature failure of bearings and other critical parts.
- ➤ **Neglected Maintenance**
  - • Failure to perform regular maintenance can lead to the accumulation of minor issues that escalate over time.
  - • Lack of proactive maintenance can result in undetected deterioration and impending failure.
- ➤ **Electrical Issues**
  - • Electrical faults such as short circuits, power surges, and insulation breakdowns can damage machinery.
  - • These issues may lead to motor failures, overheating, or fires.
- ➤ **Corrosion**
  - • Exposure to corrosive environments or substances can degrade metal components, leading to failure.
  - • Corrosion is often exacerbated by moisture, chemicals, and extreme temperatures.
- ➤ **Human Error**
  - • Mistakes during operation, maintenance, or repair can introduce new issues or exacerbate existing ones.
  - • Inadequate training or supervision often contributes to human error.
- ➤ **Vibration and Imbalance**
  - • Excessive vibration can result from imbalances, misalignment, or worn components.
  - • Vibration accelerates wear and can lead to catastrophic failures if not addressed.
- ➤ **Thermal Stress**
  - • Excessive heat can cause thermal expansion and stress, leading to deformation or failure of components.
  - • Inadequate cooling or overheating can exacerbate thermal stress.
- ➤ **Aging and Obsolescence**
  - • As machinery ages, parts can become obsolete, making repairs difficult and costly.
  - • Aging equipment may not meet current operational demands or standards.

Addressing these common causes through regular monitoring, preventive maintenance, and timely interventions can significantly reduce the risk of machinery failure in industrial settings, ensuring smoother operations and enhanced productivity.

Common Failure Modes in Lubrication Systems

When applied to lubrication systems, FMEA helps in diagnosing common lubrication-related failures and guiding preventive maintenance strategies. Here's a closer look at how to model these failure modes:

- ➤ **Insufficient Lubricant Volume (Underlubrication)**
  - **Effects.** Increased friction, wear, and potential seizure of moving parts.
  - **Modeling.** Identify potential causes like inadequate maintenance schedules or faulty delivery systems. Assess the severity, likelihood of occurrence, and likelihood of detection.
- ➤ **Contamination Ingress (Particles, Water, Chemicals)**
  - **Effects.** Accelerated wear, corrosion, and degradation of lubricant properties.
  - **Modeling.** Examine sources of contamination such as poor seals or environmental exposure. Rate the severity, likelihood of occurrence, and likelihood of detection.
- ➤ **Improper Lubricant Type or Viscosity**
  - **Effects.** Reduced lubrication efficiency, increased friction, and potential component damage.
  - **Modeling.** Evaluate the selection process for lubricants and adherence to specifications. Assign S, O, and D ratings.
- ➤ **Overpressurization**
  - **Effects.** Seal blowout, excessive heat generation, and energy loss.
  - **Modeling.** Identify root causes such as human error or improper equipment settings. Assess the severity, likelihood of occurrence, and likelihood of detection.
- ➤ **Loss of Lubricant (Leaks or System Bypass)**
  - **Effects.** Insufficient lubrication, increased wear, and potential equipment failure.
  - **Modeling.** Investigate potential leak points and system vulnerabilities. Determine S, O, and D ratings.
- ➤ **Degradation (Oxidation, Thermal Breakdown, Additive Depletion)**
  - **Effects.** Reduced lubricant effectiveness, increased wear, and potential for system failure.
  - **Modeling.** Analyze environmental conditions and operational stresses contributing to degradation. Rate severity, occurrence, and detection.

Modeling Lubrication-Related Failure Modes

Here's how to model lubrication-related failure modes:

➤ **Identify Failure Modes**
- Document each potential failure mode within the lubrication system.
- Consider both internal (e.g., equipment design) and external (e.g., environmental factors) causes.

➤ **Assess the Effects on the System**
- Determine the impact of each failure mode on equipment performance and reliability.
- Consider direct and indirect consequences.

➤ **Assign Ratings**
- **Severity.** Rate the seriousness of the failure mode's effects on a scale (e.g., 1 to 10).
- **Occurrence.** Estimate the likelihood of the failure mode occurring.
- **Detection.** Evaluate how easily the failure mode can be detected before it causes significant harm.

➤ **Calculate the Risk Priority Number**
- Use the formula RPN = S × O × D.
- Higher RPNs indicate priority areas that require immediate attention.

➤ **Develop Mitigation Strategies**
- Prioritize actions based on RPNs, focusing on high-risk areas.
- Consider redesigning components, revising maintenance procedures, or enhancing monitoring and diagnostics.

➤ **Embed FMEAs Within Broader Assessments**
- Incorporate lubrication-specific FMEAs into overall equipment assessments to create a comprehensive maintenance strategy.
- Ensure continuous review and updating of FMEAs as operational conditions and technologies evolve.

By modeling lubrication-related failure modes through FMEA, organizations can proactively address potential issues, improve equipment reliability, and reduce downtime. This structured approach ensures that maintenance strategies are targeted, data-driven, and aligned with overall operational goals.

Preventing Machinery Failure

Preventing machinery failure in industrial environments requires a proactive approach that combines regular maintenance, monitoring, and strategic planning. Here are some of the most effective methods for preventing machinery failure:

- ➤ **Preventive Maintenance**
  - Schedule regular maintenance activities to inspect, clean, and replace parts before they fail.
  - Follow the manufacturer's guidelines for maintenance intervals and procedures.
- ➤ **Condition Monitoring**
  - Use technologies like vibration analysis, thermography, and oil analysis to monitor the condition of machinery in real time.
  - Detect early signs of wear or failure, and address them before they escalate.
- ➤ **Predictive Maintenance**
  - Implement predictive maintenance using data analytics and machine learning to predict when equipment is likely to fail.
  - Schedule maintenance activities based on the actual equipment condition rather than fixed intervals.
- ➤ **Proper Lubrication**
  - Ensure that the machinery is properly lubricated with the correct type and amount of lubricant.
  - Regularly check for contamination, and replace lubricants as needed.
- ➤ **Training and Education**
  - Provide comprehensive training for operators and maintenance personnel to ensure that they are skilled in operating and maintaining equipment.
  - Educate the staff on recognizing early signs of potential failures.
- ➤ **Root Cause Failure Analysis**
  - Conduct root cause failure analysis to understand the underlying causes of failures.
  - Implement corrective actions to prevent recurrence.
- ➤ **Standard Operating Procedures**
  - Develop and enforce SOPs for operating and maintaining machinery.
  - Ensure consistency in how equipment is used and cared for.

➤ **Regular Inspections**
  - Perform regular inspections to identify wear, misalignment, leaks, and other issues.
  - Use checklists to ensure that thorough inspections are conducted.

➤ **Equipment Upgrades and Modernization**
  - Upgrade older equipment with modern technologies that offer improved performance and reliability.
  - Consider retrofitting the existing machinery with advanced monitoring systems.

➤ **Inventory Management**
  - Keep a well-stocked inventory of critical spare parts to minimize downtime during repairs.
  - Use inventory management systems to track parts usage and reorder as needed.

➤ **Environmental Controls**
  - Maintain proper environmental conditions, such as temperature, humidity, and cleanliness, to prevent external factors from affecting machinery performance.
  - Use protective enclosures or barriers to shield equipment from harsh conditions.

➤ **Collaboration with OEMs**
  - Work closely with original equipment manufacturers to stay informed about updates, recalls, and best practices.
  - Utilize OEM resources for training and technical support.

By implementing these methods, industrial environments can significantly reduce the risk of machinery failure, enhance equipment reliability, and maintain smooth and efficient operations.

INCORPORATING LUBRICATION VARIABLES INTO SYSTEM-LEVEL FMEAS

Incorporating lubrication variables into system-level failure modes and effects analyses is essential for creating a comprehensive framework for failure prevention. By addressing lubrication as a critical function, organizations can gain deeper insights into how lubrication practices impact overall system reliability. Here's how to effectively integrate lubrication variables into system-level FMEAs:

Key Lubrication Variables

- ➤ **Lubricant Delivery Method**
  - **Manual, automated, or centralized systems.** Identify the method used and the potential failure modes associated with each, such as missed lubrication points in manual systems or pump failures in automated systems.
- ➤ **Source of Lubricant**
  - **Type and compatibility.** Ensure that the lubricant type is appropriate for the application and compatible with the materials in the system.
  - **Contamination risk.** Assess the potential contamination sources during storage, handling, and application.
- ➤ **Human Factors**
  - **Accuracy of task execution.** Evaluate the risk of human error in lubrication tasks, including incorrect application or volumes.
  - **Clarity of standard operating procedures.** Ensure SOPs are clear and comprehensive, reducing the likelihood of errors.
- ➤ **Environmental Factors**
  - **Dust, humidity, vibration.** Consider how environmental conditions affect lubrication effectiveness, such as increased contamination risk in dusty environments or lubricant degradation due to heat and vibration.

Scenarios to Model

- ➤ **Missed Preventive Maintenance**
  - Model the impact of missed lubrication schedules on equipment performance and reliability.
  - Consider potential consequences like increased wear or overheating.
- ➤ **Lubricant Cross-Contamination**
  - Examine scenarios where different lubricant types are mixed, leading to compatibility issues and reduced effectiveness.
  - Analyze the impact on component performance, such as seal degradation or increased friction.
- ➤ **Expired Product Use**
  - Assess the risks associated with using expired or degraded lubricants, including reduced protective properties and increased wear.

➤ **Degraded Fluid Properties**
  • Model the effects of lubricant degradation due to oxidation, contamination, or thermal breakdown.
  • Connect these scenarios to potential component failures like bearing seizure, gear pitting, or pump cavitation.

Benefits of Integration

➤ **Holistic view.** Incorporating lubrication variables provides a complete picture of the potential failure modes, considering not just the mechanical aspects but also how lubrication practices contribute to or prevent failures.

➤ **Targeted mitigation strategies.** With a comprehensive understanding of lubrication-related risks, teams can develop targeted mitigation strategies that address both mechanical and lubrication-specific issues.

➤ **Enhanced communication.** By integrating lubrication variables, cross-functional teams can communicate more effectively about the role of lubrication in system reliability and performance.

➤ **Proactive maintenance.** Identifying lubrication-related failure modes facilitates proactive maintenance actions, reducing the likelihood of unplanned downtime and extending equipment life.

Overall, integrating lubrication variables into system-level FMEAs enhances the ability to predict, prevent, and mitigate failures, leading to more reliable and efficient operations.

FORENSIC LUBRICATION: WHAT USED OIL TELLS US ABOUT ROOT CAUSE

Forensic lubrication is a powerful method for diagnosing machinery health by analyzing used oil. By examining the molecular "fingerprint" of the oil, experts can gain insights into the machine's condition, operational stresses, and potential root causes of failure. Here's how used oil analysis contributes to understanding these factors:

Key Forensic Indicators

➤ **Wear Metals** (Iron, Copper, Lead)
- **Iron.** Indicates wear from ferrous components such as gears and bearings.
- **Copper and lead.** Suggest wear from nonferrous components like bushings and bearings.

Analysis of these metals helps identify which components are wearing and at what rate.

➤ **Silicon and Aluminum**
- **Silicon.** Often points to dirt or sand ingress, indicating a breach in filtration systems or seals.
- **Aluminum.** May suggest wear from aluminum components or contamination from external sources.

➤ **Elevated Water or Fuel Levels**
- **Water.** Can indicate issues with seals, condensation, or coolant leaks.
- **Fuel.** Points to problems with injectors or incomplete combustion, leading to dilution.

➤ **Viscosity Deviations**
- Changes in viscosity suggest thermal stress, contamination, or dilution.
- High viscosity might indicate oxidation, while low viscosity could imply fuel dilution or shearing.

➤ **Oxidation and Nitration Indices**
- **Oxidation.** Reflects chemical degradation, often due to high temperatures or prolonged use.
- **Nitration.** Indicates combustion by-products or interactions between nitrogen compounds and oil.

Techniques for Forensic Analysis

➤ **Analytical Ferrography**
- Utilizes microscopic examination to study wear particles' size, shape, and composition.
- Helps differentiate among abrasive, corrosive, and adhesive wear mechanisms.

➤ **Fourier-Transform Infrared Spectroscopy**
  • Analyzes chemical compounds in the oil, providing insights into oxidation, nitration, and contamination levels.

Mapping Forensic Findings

➤ **Failure timelines.** Match forensic evidence with historical failure events to identify patterns or recurring issues.
➤ **Operating conditions.** Consider factors like load, speed, temperature, and environment when interpreting forensic findings.
➤ **Historical interventions.** Compare oil analysis data before and after maintenance interventions to assess their effectiveness.

Applications of Forensic Lubrication

➤ **Root cause failure analysis.** Supports RCFA by providing concrete evidence of wear mechanisms and contamination sources.
➤ **Procedural changes.** Informs adjustments to maintenance procedures, such as improving filtration or modifying lubrication schedules.
➤ **Warranty or liability cases.** Provides objective evidence in disputes over warranty claims or liability issues.

Forensic lubrication is a critical tool in predictive maintenance and reliability-centered maintenance strategies. By understanding what used oil reveals about machinery condition, organizations can make informed decisions to enhance equipment reliability, optimize maintenance practices, and prevent future failures.

DIAGNOSTIC CHECKLISTS FOR LUBE-RELATED EQUIPMENT FAILURES

Creating diagnostic checklists for lube-related equipment failures is an effective way to standardize the investigation process and ensure thoroughness in identifying potential causes. These checklists help maintain consistency, improve communication, and provide valuable insights for future prevention efforts. Here's how to structure a lube-focused diagnostic checklist:

Elements of a Lube-Focused Diagnostic Checklist

1. **Correct Lubricant Usage**
   - Verify that the lubricant used matches the specifications required for the equipment.
   - Check for compatibility with the materials and operating conditions.
2. **Volume Specifications**
   - Confirm that the correct amount of lubricant was applied according to the manufacturer's guidelines.
   - Assess whether overlubrication or underlubrication occurred.
3. **Lubricant Condition Testing**
   - Determine whether the lubricant condition was regularly tested prior to the failure.
   - Evaluate the results of tests for contaminants, wear metals, and chemical degradation.
4. **PM Schedule**
   - Check if all preventive maintenance tasks, including lubrication, were completed on schedule.
   - Review maintenance records for any missed or delayed tasks.
5. **Breathers, Filters, and Seals Integrity**
   - Inspect the condition of breathers, filters, and seals for signs of wear or damage.
   - Ensure that these components were functioning properly to prevent contamination ingress.
6. **Operating Conditions**
   - Assess whether the equipment was operating within its design limits (e.g., load, speed, temperature).
   - Consider how deviations from these limits may have contributed to the failure.

Implementation and Use

➤ **Asset-Specific Customization**
   - Tailor the checklist to address the specific characteristics and requirements of each asset type.
   - Include any unique factors relevant to the particular equipment or operational environment.

➤ **Application During Inspections and Analysis**
  - Use the checklist during post-failure inspections to systematically evaluate potential lubrication-related causes.
  - Incorporate the checklist as a tool in RCFA meetings and field audits.

➤ **Documentation and Archiving**
  - Document findings from each investigation to create a comprehensive record of lubrication-related issues.
  - Archive the documentation for trend analysis and to inform continuous improvement in lubrication practices.

➤ **Trend Analysis and Program Refinement**
  - Analyze documented findings to identify recurring issues or trends in lubrication-related failures.
  - Use insights from the analysis to refine lubrication programs, adjust maintenance schedules, and implement corrective actions.

By using lube-focused diagnostic checklists, organizations can effectively identify and address lubrication-related issues, enhancing equipment reliability and reducing the likelihood of future failures. This structured approach also supports learning and improvement across maintenance and reliability programs.

RCFA FACILITATION: LEADING CROSS-FUNCTIONAL TEAMS TO TECHNICAL TRUTH

Facilitating root cause failure analysis in a cross-functional team setting requires a careful balance of leadership, technical expertise, and interpersonal skills. The goal is to lead the team to uncover the technical truth about failures and develop effective corrective actions. Here are best practices for conducting lubrication-related RCFA, along with a list detailing the facilitator's responsibilities:

Best Practices in Lubrication-Related RCFA

➤ **Structured Analytical Tools**
  - **Five whys.** Use this iterative questioning technique to drill down from the proximal to the root cause.

- **Ishikawa (fishbone) diagram.** Take advantage of this structured tool to help visualize and categorize potential causes of failure, including lubrication factors.
- **Fault tree analysis.** Employ this logical, diagrammatic method to identify and analyze the conditions leading to failure.

➤ **Cross-Functional Participation**
  - Involve the lubrication technicians, reliability engineers, maintenance staff, and representatives from the original equipment manufacturers.
  - Encourage diverse perspectives to ensure a comprehensive analysis of the failure.

➤ **Evidence-Based Focus**
  - Gather and analyze relevant data, including oil samples, inspection reports, and PM records.
  - Prioritize evidence over assumptions to ensure the analysis is grounded in factual data.

➤ **Blame-Free Environment**
  - Foster a culture that focuses on system, process, or communication failures rather than individual fault.
  - Encourage open discussion and collaboration to identify systemic issues.

➤ **Distinguishing Causes**
  - Differentiate between proximate causes (immediate events leading to failure) and root causes (underlying systemic issues).
  - Ensure the analysis addresses both kinds of causes to develop effective long-term solutions.

Facilitator Responsibilities

➤ **Guide the Process**
  - Lead the team through data gathering, cause mapping, hypothesis testing, and corrective action planning.
  - Ensure that discussions remain focused and productive.

➤ **Document Assumptions and Opinions**
  - Record all assumptions made during the analysis and any dissenting opinions.
  - Note any data gaps or uncertainties that may affect the conclusions.

➤ **Develop Preventive Systems**
  - Collaborate with the team to design corrective actions that not only address the immediate issue but also prevent future occurrences.
  - Use insights gained from the analysis to improve organizational learning and enhance preventive maintenance systems.
➤ **Build Organizational Learning**
  - Share findings and lessons learned across the organization to promote continuous improvement.
  - Encourage a culture of learning and adaptation by integrating RCFA insights into training and development programs.

By following these best practices, facilitators can lead cross-functional teams to uncover technical truths, develop effective solutions, and enhance overall organizational resilience and reliability. The focus should always be on building systems and processes that prevent future failures and promote continuous improvement.

LUBRICATION-TRIGGERED CHAIN REACTIONS IN SYSTEM FAILURES

Recognizing these chain reactions helps in both prevention and root cause tracing. Examples:

➤ Dry bearing leads to increased friction → motor overload → electrical trip
➤ Varnish buildup restricts valve movement → pressure spike → seal rupture
➤ Grease overfill causes excessive heat → premature bearing failure → misalignment

Chain reactions can be modeled in fault trees or event sequences to highlight interdependencies. This approach supports hazard analysis and prioritization of monitoring for early detection. Understanding the amplification effect of lubrication failures reinforces the importance of precision and consistency in lube practices.

A small lubrication error can trigger a cascade of failures that affect mechanical, electrical, and process systems. Modeling these chain reactions in fault trees or event sequences highlights interdependencies, supporting hazard analysis and prioritization of monitoring for early detection. Understanding the amplification effect of lubrication failures underscores the critical importance of precision and consistency in lubrication practices.

In industrial systems, lubrication errors can have far-reaching effects that extend beyond the immediate component, impacting the entire operational ecosystem. A dry bearing, for instance, isn't just a simple oversight; it creates increased friction, which can overload a motor. This overload may lead to an electrical trip, causing a sudden halt in operations and potential damage to the electrical systems. Similarly, varnish buildup within valves acts as a silent threat, gradually restricting movement until a critical moment when it causes a pressure spike. This spike can be severe enough to rupture seals, leading to fluid leaks and potential contamination of other system components.

Grease overfill is another subtle but significant error. Excessive grease can cause heat buildup in bearings, which not only shortens the bearing's lifespan but also leads to misalignment of connected components, potentially affecting the entire drive system. These scenarios illustrate how a simple lubrication error can set off a domino effect of failures.

By modeling these interactions using fault trees or event sequences, engineers can visualize and analyze the cascading effects of lubrication failures. This approach aids in understanding the interdependencies between components and systems, enabling more effective hazard analysis. Early detection becomes a priority, allowing for timely interventions before minor issues escalate into major failures.

Precision in lubrication practices is therefore not just recommended—it is essential. Consistent application of the right lubricant, in the correct amount, and at the proper intervals can prevent these chain reactions, safeguarding equipment health and ensuring operational reliability. This highlights the need for rigorous training and adherence to lubrication protocols, as well as the integration of advanced monitoring technologies to catch deviations early.

RCFA CASEBOOK: RARE FAILURES
AND WHAT THEY TAUGHT US

Case studies of rare lubrication failures serve as critical learning tools for understanding the complexities of failure mechanisms and improving reliability programs. One such example involves a paper machine press roll that experienced bearing seizure despite adherence to regular greasing schedules. The root cause analysis revealed that the failure had been due to the mixing of incompatible polyurea and lithium complex greases. This incompatibility led to oil separation and soap degradation within the grease, compromising its lubricating properties and ultimately causing the bearing to seize.

The key lesson learned from this case is the importance of strict adherence to grease compatibility guidelines. Incompatible greases can react chemically, losing their effectiveness and leading to equipment failure. This case underscores the need for clear and accurate grease compatibility charts that are easily accessible to maintenance personnel. Additionally, proper labeling of transfer equipment is crucial to prevent accidental mixing of incompatible lubricants. Ensuring that all lubrication tools and containers are clearly marked can significantly reduce the risk of such failures.

This case study highlights the importance of analytical rigor in root cause failure analysis, as well as the need for continuous education and awareness among the maintenance teams about the properties and compatibility of different lubricants. By incorporating these lessons into lubrication protocols, organizations can enhance their preventive maintenance strategies and reduce the likelihood of similar failures in the future.

Following are three examples of failures and lessons learned, starting with the first case study mentioned above.

Example 1: Polyurea Grease Incompatibility

- ➤ **Equipment.** Paper machine press roll
- ➤ **Failure.** Bearing seizure despite regular greasing
- ➤ **Root cause.** Mixing of polyurea and lithium complex greases, resulting in oil separation and soap degradation
- ➤ **Lesson.** Strict grease compatibility charts and transfer equipment labeling are essential.

In the paper machine press roll case, the mixture of polyurea and lithium complex greases led to a series of chemical reactions that degraded the grease's performance. Greases are formulated to provide specific properties, such as temperature resistance, water resistance, and mechanical stability. When two incompatible greases are combined, their thickening agents, polyurea and lithium complex, in this case, can interact adversely. This interaction causes oil to separate from the thickener, leading to a breakdown of the grease structure and a loss of lubricating capability.

Such degradation may not be immediately evident, as the visual appearance of the grease might seem unchanged. However, the chemical integrity is compromised, resulting in increased friction and wear. Over time, this can lead to bearing failure, as seen in the press roll scenario, where the bearing seized due to inadequate lubrication.

Preventing such failures requires a multifaceted approach:

1. **Compatibility charts.** The maintenance teams must have access to accurate and detailed grease compatibility charts. These charts provide essential information on which greases can be safely mixed and which cannot. Regular training sessions should be held to ensure that all team members understand how to interpret these charts.

2. **Labeling and segregation.** All lubrication equipment, including grease guns, containers, and storage areas, should be clearly labeled with the type of lubricant they contain. Implementing a robust labeling system helps prevent accidental mixing. Additionally, segregating storage areas for different types of lubricants can further reduce the risk of cross-contamination.

3. **Standard operating procedures.** Develop and enforce SOPs that outline the correct procedures for handling and applying lubricants. These procedures should include steps for verifying compatibility and ensuring that the correct lubricant is used for each application.

4. **Monitoring and testing.** Regular monitoring and testing of lubricant condition can help detect issues before they lead to equipment failure. Techniques such as oil analysis and grease sampling can identify early signs of incompatibility or degradation, allowing for timely corrective action.

The lessons learned from this case emphasize the importance of attention to detail and adherence to best practices in lubrication management. By understanding the risks associated with lubricant incompatibility and implementing preventive measures, organizations can enhance equipment reliability and extend the life of their assets.

Example 2: Electrostatic Spark Discharge in Filters

➤ **Equipment.** Turbine oil system
➤ **Failure.** Rapid oil darkening and varnish formation
➤ **Root cause.** Nonconductive filter media generating static discharges in dry oil conditions
➤ **Lesson.** Select filters based on dielectric behavior; monitor spark-induced degradation.

In a turbine oil system, an unusual and rapid onset of oil darkening and varnish formation posed a significant challenge. The root cause was traced back to electrostatic spark discharge occurring within the system's filters. This phenomenon was particularly prevalent in conditions where nonconductive filter media were used and the oil was dry. Static electricity accumulated due to the friction between the oil and the filter media, eventually discharging as sparks. These sparks were sufficient to cause oxidative degradation of the oil, leading to its darkening and the formation of varnish deposits on critical components.

The primary lesson from this case is the critical importance of selecting filter media with appropriate dielectric properties. Filters must be chosen not only for their mechanical filtration capabilities but also for their ability to dissipate static charges. Conductive or anti-static filter media can help prevent the buildup of static electricity, thereby reducing the risk of spark-induced oil degradation.

Monitoring for electrostatic discharge is also essential in preventing such failures. Implementing regular checks for signs of static buildup and discharge can alert the maintenance teams to potential issues before they result in significant oil degradation. Additionally, maintaining optimal oil moisture levels can mitigate the risk of static discharge, as moisture can enhance the conductivity of the oil, reducing static accumulation.

This case underscores the need for a comprehensive understanding of the interactions between lubricants and filtration systems. By considering both the chemical and electrical properties of filter materials, the maintenance teams can prevent unexpected failures and extend the operational life of turbine systems.

The phenomenon of electrostatic spark discharge within turbine oil systems is a complex interplay of materials science and fluid dynamics. When nonconductive filter media are used in these systems, especially under dry oil conditions, static electricity can accumulate due to the movement of the oil through the filter. This is because as the oil flows, it rubs against the filter media, creating friction. In nonconductive materials, this friction doesn't dissipate easily, leading to a buildup of static charge.

Once the static charge reaches a critical level, it discharges in the form of sparks. These sparks, though tiny, generate enough localized heat to initiate oxidation reactions in the oil. This process rapidly degrades the oil, causing it to darken and form varnish. Varnish is particularly problematic, as it adheres to turbine components, leading to reduced efficiency, increased wear, and potential failure of moving parts.

To mitigate these risks, it's crucial to select filter media that can either conduct electricity or are specifically designed to minimize static buildup. Conductive filters work by allowing the static charge to dissipate naturally through the filter material, preventing the accumulation of charges that lead to sparking.

In addition to using the right filter media, monitoring systems can be implemented to detect static discharge events. Sensors capable of detecting electrical discharges can provide early warnings of potential issues, allowing the maintenance teams to take corrective action before significant oil degradation occurs.

Furthermore, maintaining a slight level of moisture in the oil can also help, as moisture increases the conductivity of the oil, facilitating the dissipation of static charges. However, this must be balanced carefully, as too much moisture can lead to other forms of oil degradation and component corrosion.

Overall, this case highlights the importance of understanding the full range of physical and chemical interactions within lubrication systems. By carefully selecting materials and monitoring for unexpected behaviors like electrostatic discharges, operators can prevent costly failures and maintain optimal performance of turbine systems.

Example 3: Seasonal Viscosity Mismatch

- ➤ **Equipment.** Mobile hydraulic system
- ➤ **Failure.** Cavitation during cold-weather start-up
- ➤ **Root cause.** Use of summer-grade hydraulic oil year-round
- ➤ **Lesson.** Align lubricant selection with regional climate profiles and start-up conditions.

In a mobile hydraulic system, cavitation during cold-weather start-up was traced back to a seasonal viscosity mismatch. The root cause was the use of summer-grade hydraulic oil throughout the year. This oil, suited to higher temperatures, becomes too viscous in cold conditions, impeding proper flow through the system. During start-up, the thickened oil couldn't circulate effectively, leading to cavitation, a condition where vapor bubbles form and collapse, causing significant damage to pumps and other components.

This failure underscores the necessity of aligning lubricant selection with regional climate profiles and specific start-up conditions. In colder climates, using winter-grade hydraulic oils with lower viscosity ensures that the oil remains fluid enough to circulate properly, even at lower temperatures. This proactive approach prevents cavitation and the associated mechanical damage.

Documenting such cases contributes to institutional memory, allowing organizations to refine diagnostic checklists and enhance training programs. Sharing these insights helps prepare engineering teams to anticipate and mitigate similar issues, ensuring reliability across various operating conditions. By incorporating climate considerations into lubricant selection, organizations can optimize equipment performance and longevity, particularly in systems exposed to significant temperature fluctuations.

The issue of seasonal viscosity mismatch in mobile hydraulic systems highlights the critical importance of selecting the right lubricant for varying environmental conditions. Hydraulic systems rely heavily on the fluid's ability to circulate efficiently, providing the necessary pressure and lubrication to operate components smoothly. In colder temperatures, the viscosity of summer-grade hydraulic oils increases significantly, transforming what is normally a free-flowing fluid into a thicker, more resistant substance.

This increase in viscosity during cold weather results in a slow start-up, as the hydraulic pump struggles to draw the thickened oil from the reservoir, leading to reduced flow rates and insufficient lubrication. The situation is exacerbated during start-up, when the system is at its most vulnerable due to the lack of heat generated by operation. Cavitation occurs when the pump attempts to move oil more quickly than it can flow, creating low-pressure zones where vapor bubbles form. The collapse of these bubbles generates shock waves that can erode metal surfaces and damage internal components, leading to premature failure.

To prevent such issues, it is essential to use hydraulic oils that are specifically formulated for the expected operating temperature range. Winter-grade oils have lower viscosity indices, meaning they remain fluid even at lower temperatures, ensuring that the hydraulic system can start smoothly and operate efficiently. This requires careful planning and possibly changing the oil type as seasons shift, particularly in regions with significant temperature variations.

Moreover, organizations should incorporate seasonal lubricant changes into their maintenance schedules. This proactive strategy helps avoid potential downtime and reduces the risk of damage caused by inappropriate lubricant viscosities. Training programs should emphasize the importance of understanding lubricant properties and the effects of temperature on viscosity, equipping the technicians with the knowledge to make informed decisions.

By documenting such cases and integrating them into training and maintenance protocols, organizations can create a robust framework for preventing similar failures. This approach not only strengthens institutional memory but also enhances the diagnostic capabilities of the maintenance teams, ensuring that they are well prepared to handle seasonal challenges effectively.

CHAPTER 22

MANAGEMENT REPORTING AND PERFORMANCE METRICS

DEFINING KPIs THAT REFLECT LUBRICATION PROGRAM MATURITY

Key performance indicators provide measurable insight into the health, execution, and impact of a lubrication program. Mature programs track KPIs that go beyond basic compliance, reflecting integration with reliability, safety, and operational goals.

Effective lubrication KPIs include:

➤ PM task completion rate (on time in full)
➤ Relubrication compliance versus schedule
➤ Percentage of assets within target cleanliness levels (ISO 4406)
➤ Frequency of lubricant-related equipment failures
➤ Mean time between lubrication-related failures
➤ Oil analysis response time and resolution rate

Advanced programs also track lubricant consumption efficiency, training/certification coverage, and work order backlog related to lubrication. KPIs must be tailored to the asset base, organizational priorities, and system complexity.

Management reporting and performance metrics in lubrication programs require structured measurement through quantifiable data points. Performance indicators directly correlate to program effectiveness when properly implemented. Primary metrics track completion rates of preventive maintenance tasks with specific focus on schedule adherence and technical execution quality. Current industry standards mandate monitoring relubrication compliance against established schedules while maintaining detailed records of asset cleanliness levels according to ISO 4406 specifications.

Technical tracking of lubricant-related equipment failures provides critical trending data for reliability improvements. Mean time between lubrication-related failures serves as a core reliability metric, enabling data-driven decision-making for maintenance intervals and lubricant selection. Oil analysis response metrics require monitoring of both initial sample processing time and technical resolution implementation speed.

Advanced program metrics incorporate consumption efficiency ratios, technical certification coverage across maintenance personnel, and backlog analysis of lubrication-specific work orders. Metric selection requires calibration to specific asset configurations, operational parameters, and system complexity levels. Program maturity correlates directly with the depth and breadth of implemented performance tracking.

Operational KPIs must demonstrate clear links between lubrication activities and overall equipment reliability targets. Data collection systems require configuration to automatically capture relevant metrics while minimizing manual input requirements. Regular analysis of trend data enables the identification of systemic issues and optimization opportunities. Integration with existing maintenance management systems ensures consistent tracking and reporting capabilities. Performance metrics drive continuous improvement through quantifiable feedback on program effectiveness.

In addition, performance metrics drive operational decision-making through data correlation and failure pattern analysis. Asset reliability directly links to precision lubrication execution, requiring the systematic measurement of program effectiveness.

Contamination control metrics track particulate levels, moisture content, and lubricant degradation rates across critical equipment. Integration of oil analysis results with vibration data provides early detection of developing failure modes. Storage and handling procedures require verification through regular audits of lubricant receiving, dispensing, and disposal processes.

Technical competency assessments measure maintenance personnel capability in executing precision lubrication tasks. Documentation systems track root cause analysis findings to identify recurring issues in lubrication practices. Inventory management metrics monitor stock levels, consumption rates, and product obsolescence. Cross-functional communication effectiveness measures ensure proper information flow among operations, maintenance, and reliability teams.

Cost metrics evaluate program expenditures against reliability improvements and downtime reduction. Equipment-specific lubrication routes require optimization based on condition monitoring data and failure history. Contamination exclusion effectiveness measures the success of sealing systems and breather installations. Oil sampling frequency and locations require adjustment based on trending analysis and criticality assessments. Program maturity indicators track the implementation of precision lubrication tools and technologies.

Standardization of procedures across multiple facilities ensures consistent execution and comparable metrics. Integration with computerized maintenance management systems enables automated tracking of key performance indicators. Technical training programs require regular updates based on equipment modifications and new technology adoption.

Reliability improvement projects track the impact of enhanced lubrication practices on mean time between failures. Environmental compliance metrics monitor waste oil disposal and spill prevention measures. Cost-benefit analysis quantifies the return on investment for lubrication program improvements.

VISUALIZING LUBRICATION DATA FOR EXECUTIVE STAKEHOLDERS

Presenting lubrication data to executive stakeholders requires translation from technical metrics to strategic insight. Visualization tools such as dashboards, heat maps, and performance trends convert raw data into meaningful narratives.

Elements of effective visualization include:

➤ High-level summary KPIs with drill-down capability
➤ Color-coded health indicators for assets or departments
➤ Trend lines for lubricant consumption, failure rates, and compliance
➤ Comparative performance by site, shift, or business unit

Executives respond to clarity, brevity, and impact. So it's important to focus on how lubrication practices support uptime, cost control, risk reduction, and regulatory compliance. Storytelling with data tying events to outcomes adds relevance and drives engagement.

Visualizing lubrication data for executive stakeholders necessitates the transformation of technical metrics into strategic insights. Employ visualization tools like dashboards, heat maps, and performance trends to convert raw data into coherent narratives. Effective visualization begins with high-level summary KPIs that offer drill-down capabilities for detailed analysis. Utilize color-coded health indicators to provide quick status assessments for assets or departments, enabling the immediate identification of areas requiring attention. Implement trend lines to track lubricant consumption, equipment failure rates, and compliance levels over time, highlighting patterns and anomalies.

Comparative performance metrics across sites, shifts, or business units offer a broader perspective on operational effectiveness and facilitate benchmarking. Executives require clarity, brevity, and impact in presentations. Prioritize illustrating how lubrication practices enhance equipment uptime, manage costs, mitigate risks, and ensure regulatory compliance. Use data storytelling to connect events with outcomes, making the information relevant and engaging, which fosters informed decision-making.

Visualizing lubrication data for executive stakeholders involves leveraging advanced analytical tools to ensure that the data is not only accurate but also accessible and actionable. Dashboards should be designed to display key performance indicators at a glance, with options to delve deeper into specific areas as needed. This flexibility allows stakeholders to focus on both the big picture and the intricate details that could impact operational decisions. Color-coded health indicators can be employed to represent asset conditions or departmental performance, providing immediate visual cues that facilitate quick decision-making. For instance, green might indicate optimal conditions, yellow could suggest caution, and red would signal the need for immediate action.

Trend lines are essential for illustrating changes in lubricant consumption, failure rates, and compliance over time, which can reveal underlying issues or improvements in the lubrication program. By presenting comparative performance data across different sites, shifts, or business units, stakeholders can identify best practices and areas for improvement, driving a culture of continuous enhancement.

The ultimate goal is to demonstrate how effective lubrication practices contribute to broader business objectives such as maximizing equipment availability, controlling costs, reducing risks, and maintaining compliance with regulations. By tying data to specific outcomes through compelling narratives, stakeholders can see the direct impact of lubrication strategies on business performance. This approach not only enhances engagement but also supports strategic planning and investment decisions.

LINKING LUBRICATION METRICS
TO ASSET PERFORMANCE INDICES

Lubrication metrics should be aligned with broader asset performance indices to demonstrate value and identify performance levers. These linkages justify investment and support predictive maintenance initiatives.

Relevant correlations include:

➤ Lubrication PM compliance versus asset uptime or MTBF
➤ Oil cleanliness versus bearing or gearbox life extension

> ➤ Timely oil analysis interventions versus avoided unplanned downtime
> ➤ Lubricant cost per operational hour versus maintenance spend

Analytical tools such as regression modeling and root cause databases help demonstrate causal relationships. Cross-functional collaboration with the reliability engineers enhances interpretation and supports continuous improvement. Linking lubrication metrics to asset performance indices is essential for illustrating the value of lubrication efforts and identifying key performance drivers. These linkages provide justification for further investment in lubrication practices and bolster support for predictive maintenance initiatives. Important correlations include lubrication preventive maintenance compliance versus asset uptime or mean time between failures. This relationship highlights how timely and effective lubrication maintenance directly impacts equipment reliability and availability. Monitoring oil cleanliness and its impact on bearing or gearbox life extension demonstrates the role of lubrication in prolonging component life and reducing replacement costs. Timely oil analysis interventions are crucial for avoiding unplanned downtime, showcasing the preventive power of regular oil monitoring and prompt corrective actions. Evaluating lubricant cost per operational hour in relation to overall maintenance expenditure provides insights into cost efficiency and optimization opportunities.

To uncover these correlations, employ analytical tools such as regression modeling and root cause analysis databases. These tools help establish and validate causal relationships between lubrication metrics and asset performance. Collaboration with reliability engineers is vital to interpreting data accurately and developing actionable insights that drive continuous improvement. Working cross-functionally ensures that lubrication metrics are not only linked to asset performance but also integrated into broader maintenance strategies, maximizing the impact of lubrication on overall operational efficiency and reliability.

Process for Linking Lubrication Metrics to Asset Performance Indices

Linking lubrication metrics to asset performance indices involves a systematic approach to data collection, analysis, and interpretation. Here are the detailed steps:

1. **Identify relevant lubrication metrics.** Begin by selecting key lubrication metrics such as lubrication PM compliance, oil cleanliness, oil analysis intervention timeliness, and lubricant cost per operational hour. Ensure these metrics are consistently tracked and recorded.

2. **Define asset performance indices.** Establish performance indices like asset uptime, mean time between failures, bearing or gearbox life expectancy, and maintenance costs. These indices should reflect the broader operational goals and priorities.

3. **Collect and integrate data.** Collect data for both lubrication metrics and asset performance indices. Ensure data accuracy by utilizing automated systems where possible to minimize manual entry errors. Integrate this data into a centralized database or system for ease of access and analysis.

4. **Do a correlation analysis.** Use statistical methods to analyze the relationships between lubrication metrics and asset performance indices. Employ tools like regression modeling to identify correlations and potential causations. This step helps in understanding how changes in lubrication practices impact asset performance.

5. **Do a root cause analysis.** Conduct a root cause analysis on any identified issues to understand the underlying reasons for performance deviations. This will help in pinpointing specific lubrication-related factors contributing to asset performance.

6. **Collaborate with cross-functional teams.** Work with reliability engineers and other relevant stakeholders to interpret the data. Their expertise can provide valuable insights into the technical implications of the findings and suggest practical improvements.

7. **Develop actionable insights.** Translate the analytical findings into actionable insights. Identify specific areas for improvement in lubrication practices that could lead to enhanced asset performance.

8. **Implement changes and monitor impact.** Act on the insights by implementing changes in lubrication processes, such as updating lubrication schedules, improving oil cleanliness standards, or enhancing training programs. Continuously monitor the impact of these changes on asset performance indices.

9. **Foster continuous improvement.** Regularly review and refine the linkage process. As new data emerges or operational priorities shift, update the metrics and indices as necessary to ensure ongoing alignment with asset performance goals.

10. **Report and communicate.** Present the findings and improvements to stakeholders. Use clear visualization techniques to communicate how lubrication metrics are influencing asset performance. This step reinforces the value of lubrication efforts and supports strategic decision-making.

By following these steps, organizations can effectively link lubrication metrics to asset performance indices, driving improvements in reliability and operational efficiency.

REPORTING STRUCTURES THAT SUPPORT DATA-DRIVEN ACTION

Well-designed reporting structures ensure that lubrication data is not just collected but acted upon. Reports must be timely, role-specific, and integrated into existing operational routines. Recommended reporting formats include:

➤ Daily dashboards for maintenance leads (task status, alerts)
➤ Weekly summaries for the reliability teams (trends, flagging analysis)
➤ Monthly reports for the operations managers (uptime, cost metrics)
➤ Quarterly reviews for the executives (strategic alignment, improvement plans)

Automated report generation through CMMSs or BI (business intelligence) platforms reduces administrative burden and ensures consistency. Include call-to-action sections with recommendations, responsible parties, and deadlines. Effective reporting structures are essential for transforming lubrication data into actionable insights. Reports must be designed to fit seamlessly into the workflow of different roles within the organization, ensuring that the right information reaches the right people at the right time. Daily dashboards tailored for maintenance leads provide real-time updates on task status and alerts, enabling immediate responses to emerging issues. These dashboards facilitate quick decision-making and help in prioritizing maintenance activities.

Weekly summaries for the reliability teams focus on identifying trends and conducting flagging analysis. These reports offer a deeper dive into data, allowing the teams to spot patterns and potential problem areas before they escalate. By focusing on trends, the reliability teams can proactively address issues and optimize lubrication practices.

Monthly reports for the operations managers integrate uptime statistics and cost metrics, offering a comprehensive view of how lubrication impacts operational efficiency and financial performance. These reports help the managers make informed decisions regarding resource allocation and operational adjustments.

Quarterly reviews for the executives emphasize strategic alignment and outline improvement plans. These high-level reports provide a snapshot of how lubrication strategies contribute to overall business objectives, supporting long-term planning and investment decisions.

To reduce administrative burden and maintain consistency, automated report generation through CMMSs or BI platforms is recommended. These systems can streamline the reporting process, ensuring timely and accurate data dissemination. Each report should include a call-to-action section that outlines specific recommendations, assigns responsibilities to relevant parties, and sets deadlines for action. This structure ensures that reports not only inform but also drive meaningful improvements and accountability across the organization.

Report Structure

Outlining the structures of these reviews involves specifying the key components and layout for each type of report to ensure that they are effective and actionable. Here's how each report can be structured:

- ➤ **Daily Dashboards for Maintenance Leads**
  - **Overview section.** Quick summary of the day's key metrics, including task completion rates and any outstanding tasks
  - **Alerts and notifications.** Critical alerts that require immediate attention, such as missed lubrication tasks or equipment showing signs of potential failure
  - **Task status.** Detailed list of scheduled versus completed tasks, noting any deviations from the schedule
  - **Prioritized action items.** List of tasks prioritized by urgency and impact on operations
  - **Real-time data visualization.** Graphs or charts showing real-time data on lubrication activities and equipment status
- ➤ **Weekly Summaries for Reliability Teams**
  - **Trend analysis section.** Insights into lubrication trends over the past week, including lubricant consumption and compliance rates
  - **Flagging analysis.** Identification of any potential issues or anomalies detected, with suggested areas for further investigation
  - **Performance metrics.** Detailed breakdown of KPIs such as MTBF and oil cleanliness levels
  - **Actionable insights.** Recommendations based on the week's data, including suggested changes or interventions
  - **Collaboration notes.** Input from the cross-functional teams, highlighting any collaborative efforts or feedback

➤ **Monthly Reports for Operations Managers**
- **Executive summary.** High-level overview of the month's lubrication performance and its impact on operations
- **Uptime and downtime analysis.** Detailed analysis of equipment uptime and any instances of downtime related to lubrication issues
- **Cost metrics.** Breakdown of lubrication-related costs and their impact on overall operational expenses
- **Resource allocation.** Insights into resource utilization and efficiency, including labor and materials
- **Recommendations and next steps.** Strategic recommendations for operational improvements and resource allocation for the upcoming month

➤ **Quarterly Reviews for Executives**
- **Strategic alignment overview.** Description of how lubrication strategies align with broader business objectives and goals
- **Performance highlights.** Key achievements and milestones reached in the past quarter
- **Improvement plans.** Outline of proposed improvements and strategic initiatives for the next quarter
- **Financial impact.** Analysis of the financial implications of lubrication activities, including ROI and cost savings
- **Future outlook.** Projections and expectations for the future, including potential challenges and opportunities
- **Call to action.** Specific actions required from the executives, with assigned responsibilities and deadlines for implementation

These structured reviews ensure not only that data is reported, but that it also drives actionable and strategic decisions across different organizational levels.

COMMON PITFALLS IN LUBRICATION KPI INTERPRETATION

Misinterpreting or misapplying lubrication KPIs can undermine decision-making and obscure underlying issues. Here are the common pitfalls:

➤ Overemphasis on compliance rate without verifying task quality
➤ Treating oil analysis flagging as binary (pass/fail) rather than trend-based

➤ Misjudging lubricant consumption increases as a sign of better care, when it may indicate overlubrication or leakage

➤ Using generic KPI thresholds that ignore asset context

Interpreting lubrication KPIs requires careful consideration to avoid common pitfalls that can lead to flawed decision-making and obscure real issues. An example of a major pitfall is placing too much emphasis on compliance rates without assessing the quality of completed tasks. Merely achieving a high compliance rate does not guarantee that lubrication tasks are performed correctly or effectively. It's essential to verify the quality of work through regular audits and inspections.

Another common mistake is treating oil analysis flagging as a simple pass/fail indicator. This binary approach overlooks the importance of monitoring trends over time. Analyzing trend data can reveal gradual changes in oil condition that might indicate potential problems before they lead to equipment failure.

Misjudging increases in lubricant consumption as indicative of better maintenance care is another pitfall. While it might seem that increased lubrication indicates attentiveness, it could actually signal issues like overlubrication or leaks. It's crucial to investigate the root cause of any changes in consumption patterns.

Additionally, the use of generic KPI thresholds that do not account for the specific context and requirements of each asset can lead to misleading conclusions. KPIs should be tailored to reflect the unique operational conditions and performance expectations of the equipment being monitored.

To avoid these pitfalls, triangulating data from multiple sources is essential. This involves cross-referencing lubrication data with insights from other maintenance activities, operational data, and historical performance records. Conduct regular audits to ensure data accuracy and task quality, and involve subject-matter experts in the analysis process to provide nuanced insights. Periodically review and adjust KPIs to ensure they remain relevant and aligned with evolving operational realities and business objectives.

Avoiding pitfalls in lubrication KPI interpretation requires a systematic approach that integrates data triangulation, regular audits, expert involvement, and continuous review. Here's a step-by-step process to effectively manage this:

1. **Data Collection and Integration**
   - Gather lubrication data, including compliance rates, oil analysis results, and lubricant consumption figures.
   - Collect data from other maintenance activities, such as maintenance schedules, repair logs, and equipment performance reports.
   - Compile operational data, such as production schedules, environmental conditions, and equipment load factors.

- Assemble historical performance records, including past failures, maintenance interventions, and lubrication adjustments.

2. **Data Triangulation**
   - Cross-reference lubrication data with maintenance and operational data to identify correlations and discrepancies.
   - Use historical performance records to provide context and identify trends over time.
   - Analyze data from multiple angles to ensure a comprehensive understanding of lubrication performance and its impact on asset reliability.

3. **Regular Audits**
   - Conduct audits of lubrication processes and tasks to ensure compliance with established standards and procedures.
   - Verify the quality of lubrication tasks by inspecting equipment and reviewing task documentation.
   - Audit data accuracy regularly to identify and rectify any discrepancies or errors in data collection and reporting.

4. **Involve Subject-Matter Experts**
   - Engage lubrication specialists and reliability engineers in the analysis process to provide expert insights.
   - Use their expertise to interpret complex data patterns and identify underlying issues that might not be immediately apparent.
   - Collaborate with cross-functional teams to gain a holistic view of lubrication performance and its operational impact.

5. **Periodic KPI Review and Adjustment**
   - Schedule regular reviews of lubrication KPIs to assess their relevance and effectiveness.
   - Adjust KPI thresholds and targets based on changes in operational conditions, asset requirements, and business objectives.
   - Ensure that KPIs reflect the specific context of each asset and align with broader organizational goals.

6. **Feedback and Continuous Improvement**
   - Establish feedback loops to gather input from the maintenance teams, operators, and other stakeholders.
   - Use feedback to refine lubrication processes, adjust KPIs, and drive continuous improvement.
   - Document lessons learned and best practices to enhance future lubrication management efforts.

By following this structured process, organizations can effectively avoid common pitfalls in lubrication KPI interpretation, ensuring that data-driven insights lead to improved maintenance practices and asset reliability.

BUILDING A LUBRICATION SCORECARD FOR CONTINUOUS IMPROVEMENT

A lubrication scorecard provides a structured, repeatable format for evaluating program maturity and driving improvement initiatives.

Scorecard categories include:

➤ **Execution.** PM compliance, task accuracy, technician certification
➤ **Condition.** Cleanliness targets, oil condition trends, contamination incidents
➤ **Response.** Issue resolution time, root cause analysis completion, work order closeout
➤ **Sustainability.** Consumption efficiency, waste disposal metrics, re-refining use

Each category includes weighted KPIs, performance bands (e.g., red/yellow/green), and improvement targets. Scorecards support internal benchmarking, coaching, and recognition programs. Use the scorecard in quarterly reviews to guide strategic planning and prioritize budget allocations for tools, training, or system upgrades.

Building a lubrication scorecard is a powerful way to evaluate program maturity, identify improvement opportunities, and ensure continuous progress. Here's how to develop and utilize such a scorecard effectively:

Step-by-Step Process for Building a Lubrication Scorecard

1. **Define Scorecard Categories**
   - **Execution.** Focus on how well lubrication tasks are carried out. Include metrics like preventive maintenance compliance, task accuracy, and technician certification levels.
   - **Condition.** Assess the state of lubrication systems and fluids. Track cleanliness targets and oil condition trends, and record any contamination incidents.

- **Response.** Evaluate the efficiency and effectiveness of addressing lubrication-related issues. Measure issue resolution time, completion of root cause analyses, and work order closeout rates.
- **Sustainability.** Focus on environmental and resource efficiency. Monitor lubricant consumption efficiency, waste disposal metrics, and the use of re-refined oils.

2. **Develop Weighted KPIs for Each Category**
   - Assign weights to each KPI based on its importance to the overall lubrication strategy. For instance, PM compliance might be heavily weighted in the execution category.
   - Define performance bands (e.g., red/yellow/green) for each KPI to visualize performance levels easily.

3. **Set Improvement Targets**
   - Establish realistic and measurable improvement targets for each KPI. These targets should align with broader organizational objectives and reflect achievable progress given current resources and constraints.

4. **Implement Internal Benchmarking**
   - Use scorecards to benchmark performance across different departments, sites, or time periods. This helps identify best practices and areas needing attention.
   - Facilitate coaching and recognition programs based on scorecard results to motivate staff and share successful strategies.

5. **Schedule Regular Scorecard Reviews**
   - Integrate the scorecard into quarterly review processes. Use it as a strategic tool to guide discussions on program effectiveness and improvement opportunities.
   - Prioritize budget allocations for tools, training, or system upgrades based on scorecard insights, ensuring that resources are directed toward the most impactful areas.

6. **Use as a Strategic Planning Tool**
   - Leverage scorecard data to guide strategic planning, focusing on areas that will drive the most significant improvements in lubrication practices.
   - Adjust strategies and priorities based on scorecard outcomes to ensure continuous alignment with business goals and performance expectations.

7. **Monitor and Refine**
   - Continuously monitor scorecard effectiveness and relevance. Update KPIs, weights, and targets as necessary to reflect changes in operational priorities or business environments.
   - Involve cross-functional teams in the review and refinement process to ensure comprehensive insights and buy-in.

By following these steps, organizations can effectively utilize lubrication scorecards to drive continuous improvement, enhance program maturity, and achieve long-term operational success.

TELLING THE STORY: TURNING LUBRICATION DATA INTO BUSINESS VALUE

Ultimately, lubrication reporting must tell a compelling story of how fluid management contributes to business success. This involves connecting data to goals that resonate with financial, operational, and safety stakeholders.

Follow these storytelling principles:

- ➤ Start with a problem: downtime event, cost spike, or compliance risk.
- ➤ Show how lubrication practices or insights addressed it.
- ➤ Quantify the impact (avoided cost, extended life, reduced waste).
- ➤ Highlight the people and processes involved.
- ➤ Connect the outcome to organizational goals (reliability, ESG, profitability).

Case studies, infographics, and before-after comparisons make these stories tangible. When lubrication data drives informed decisions and visible improvements, its value becomes irrefutable.

INDUSTRY STORIES

The following are for some examples of stories, industry by industry.

AUTOMOTIVE

Lubricants are essential for engine performance, transmission systems, and other vehicle components. At a major automotive assembly facility specializing in heavy-duty trucks, production suddenly halted when repeated failures emerged within the transmission final drive assembly line. Each unexpected stoppage generated steep operational losses, approaching $50,000 per hour. Beyond the immediate financial hit, the facility faced rising compliance risks linked to warranty claims and emissions certifications, reliant upon consistent component reliability.

The plant's reliability team, equipped with deep expertise in lubrication analysis, quickly focused on lubricant quality and condition as the primary potential issue. Diagnostic tests revealed the lubricant that was initially specified was deteriorating prematurely due to unforeseen high operating temperatures. The resultant lubricant degradation led directly to increased friction, component wear, and subsequent mechanical failures.

To rapidly rectify the problem, the reliability group implemented an intensive lubricant condition monitoring program, using advanced in-line sensors and routine laboratory oil analyses. Insights gained from these procedures allowed the team to confidently identify and transition to a synthetic lubricant specifically engineered for superior thermal and oxidative stability under harsh operating conditions.

Within weeks of deploying the improved lubricant and enhanced monitoring program, the facility recorded remarkable gains in uptime and reliability. Downtime incidents decreased by over 80%, translating into direct operational savings of approximately $750,000 within 6 months. Additionally, component lifespan extended by approximately 40%, significantly reducing maintenance labor and replacement costs. Improved lubricant selection and proactive condition monitoring reduced lubricant consumption and waste disposal by nearly 35%, delivering substantial environmental and sustainability benefits.

Integral to this success was the ongoing collaboration among the reliability engineers, plant technicians, and maintenance supervisors. Training and processes were established to ensure continuous improvement, emphasizing predictive analytics and preventive interventions rather than reactive repairs.

Ultimately, these targeted lubrication practices directly aligned with the organizational objectives of increasing reliability, improving environmental stewardship, and enhancing profitability, firmly establishing lubrication management as a strategic driver of competitive excellence in the automotive manufacturing sector.

MANUFACTURING

Various manufacturing processes require lubricants for machinery, equipment, and tooling to reduce friction and wear. In a large precision manufacturing plant producing critical aerospace components, the production lines experienced persistent, costly downtime due to frequent equipment breakdowns. High-friction tooling and machinery were suffering accelerated wear, resulting in escalating maintenance expenses and production delays that cost the facility approximately $30,000 per hour. In addition, the situation raised alarms around compliance risks tied to quality standards required by aerospace industry certifications.

The reliability team swiftly investigated the issue, immediately identifying inadequate lubrication practices as the primary root cause. Initial assessments revealed inconsistent lubrication schedules and suboptimal lubricant selection that failed under intense operational conditions, leading to excessive friction, heat buildup, and premature tooling degradation.

In response, the reliability engineers implemented a structured lubricant optimization initiative. The team shifted from traditional lubrication scheduling toward predictive maintenance based on real-time lubricant condition monitoring, deploying advanced sensors and periodic laboratory analysis. By selecting high-performance synthetic lubricants with additives tailored specifically for extreme pressure and temperature scenarios, they enhanced equipment durability significantly.

Within just three months, the positive impact was clear. Machinery downtime was reduced by nearly 75%, saving the plant over $600,000 in maintenance and lost production costs. Tool life was extended by approximately 50%, directly reducing procurement and disposal expenses, thus contributing positively to the company's environmental and sustainability metrics. Lubricant consumption also decreased by nearly 30%, significantly cutting waste oil volumes.

Central to this turnaround was the collaborative effort among the reliability engineers, maintenance technicians, and production supervisors. Training workshops and clear standard operating procedures embedded the enhanced lubrication practices deep into the daily operations, transforming the site's culture from reactive to proactive maintenance.

By aligning these improvements with broader organizational goals, the manufacturing plant significantly improved its reliability, enhanced profitability through substantial cost savings, and made measurable strides toward achieving its ESG commitments. The strategic focus on lubrication emerged not merely as maintenance but as a key competitive differentiator in the demanding aerospace manufacturing sector.

AEROSPACE

Lubricants are critical for aircraft engines, landing gear, and other components to ensure safety and reliability. At a prominent aerospace maintenance facility, unexpected downtime disrupted operations, triggered by recurring issues in aircraft landing gear assemblies. Each unscheduled grounding imposed severe financial strain, costing approximately $100,000 per event due to flight delays, expedited repairs, and lost revenue. Equally critical, compliance concerns escalated, as each incident risked violating stringent aviation safety standards and customer commitments.

Upon thorough investigation by the site's reliability and maintenance engineering team, lubricant performance surfaced as a primary factor. The analysis uncovered premature lubricant breakdown under the demanding conditions of landing impacts, extreme temperature swings, and environmental exposure. These breakdowns accelerated component wear and compromised gear integrity, posing both operational and safety risks.

To swiftly mitigate these risks, the reliability team overhauled their lubrication management approach. They selected advanced, aerospace-grade synthetic lubricants specifically engineered for extreme pressure resistance and robust environmental stability. Additionally, they instituted a meticulous lubricant condition-monitoring program involving frequent sampling, laboratory analysis, and data-driven predictive maintenance, enabling proactive interventions before degradation could occur.

Within 4 months, these enhanced practices delivered tangible improvements. Landing gear reliability improved dramatically, decreasing unscheduled downtime events by over 85%, representing a savings of more than $800,000 annually. Component life was significantly extended, and maintenance intervals increased by approximately 40%, substantially reducing overhaul expenses and material waste. Lubricant usage efficiency improved markedly, decreasing lubricant consumption and hazardous waste disposal by nearly 30%, directly benefiting the facility's ESG objectives.

Crucial to these achievements was the integrated collaboration among the reliability engineers, aircraft maintenance crews, and quality assurance teams. Clear, standardized processes were documented, and ongoing technical training ensured that best practices became ingrained into daily maintenance operations.

This targeted approach to lubrication management directly supported the organization's overarching goals: enhancing operational reliability, achieving stringent safety compliance, and boosting profitability. In doing so, the aerospace facility established lubrication excellence not just as essential maintenance but as an integral driver of competitive advantage and sustainable operational performance.

MARINE

Ships and boats utilize lubricants for engine performance, gearboxes, and hydraulic systems. In a bustling marine logistics operation, a fleet of large cargo vessels began experiencing recurring engine and gearbox failures. The increasing frequency of unplanned maintenance caused severe disruptions, with each docking for urgent repairs generating approximately $120,000 in operational losses, port fees, expedited parts procurement, and missed shipping deadlines. Moreover, these repeated incidents heightened compliance risks regarding maritime emissions standards and environmental regulations.

Upon detailed investigation, the marine reliability engineering team pinpointed lubrication deficiencies as the core issue. Laboratory analysis of lubricant samples revealed accelerated contamination and degradation due to seawater ingress, harsh operating conditions, and inadequate lubricant compatibility for maritime environments.

The team swiftly implemented enhanced lubrication practices tailored specifically for marine applications. They selected high-performance marine-grade lubricants formulated to resist corrosion, water contamination, and extreme mechanical stresses common to shipboard machinery. Further, they adopted rigorous lubricant condition monitoring programs involving onboard sensor arrays and scheduled analytical testing, enabling proactive maintenance interventions.

Within 6 months, the improvements were evident. Unplanned downtime events declined by over 70%, directly saving the operation nearly $850,000 in avoided maintenance and disruption-related costs. Component service life extended substantially, increasing by nearly 50%, greatly reducing the overhaul expenditures and environmental waste associated with lubricant disposal. Additionally, improved lubricant handling reduced consumption and waste by approximately 30%, aligning strongly with environmental sustainability goals.

Instrumental in this success were the coordinated efforts among the reliability engineers, vessel maintenance crews, and shoreside logistics managers. Comprehensive training ensured consistent application of new procedures, embedding proactive lubrication management into daily maritime operations.

Ultimately, these refined lubrication practices significantly enhanced operational reliability, reduced the fleet's environmental footprint, and improved profitability. This marine logistics operation demonstrated clearly that strategic lubrication management is not merely a maintenance necessity but a critical driver of competitive advantage, environmental stewardship, and organizational sustainability.

CONSTRUCTION

The heavy machinery and equipment used in construction rely on lubricants for optimal performance and longevity. On a major urban construction project tasked with erecting a large commercial complex, persistent breakdowns in heavy excavation and earth-moving equipment brought progress to a costly halt. Unplanned downtime escalated rapidly, costing the company upward of $20,000 per hour in delays, penalties, equipment repairs, and labor inefficiencies. Adding to the financial strain, compliance issues loomed, threatening the company's contractual obligations to meet stringent deadlines and environmental standards.

The reliability and maintenance team quickly assessed the issue and identified lubricant degradation as the principal cause. Analysis indicated lubricant contamination and premature viscosity loss due to harsh operational conditions, dusty environments, extended heavy loads, and extreme temperature fluctuations. Lubricant-related wear accelerated machinery breakdowns, severely reducing equipment availability.

The team implemented a targeted lubrication optimization initiative, selecting advanced heavy-duty lubricants specifically formulated for severe operating conditions faced by construction equipment. Simultaneously, the team introduced rigorous lubricant condition monitoring practices, deploying field sampling, in-depth laboratory analysis, and real-time monitoring sensors to proactively identify lubricant deterioration.

Within months, the enhanced lubrication strategy yielded impressive results. Equipment downtime dropped by more than 65%, translating into direct savings exceeding $400,000 annually from reduced repair costs, avoided penalties, and improved workforce productivity. Machinery life expectancy increased by nearly 40%, significantly reducing replacement expenditures and the environmental impacts associated with waste disposal. Further, optimized lubricant consumption led to a 25% reduction in waste generation, supporting the company's sustainability targets.

Critical to this transformation was the collaboration among the reliability engineers, equipment operators, and maintenance technicians. Through structured training and clearly documented processes, proactive lubrication management practices became deeply embedded into daily operational routines.

Ultimately, this focused lubrication strategy directly reinforced the organizational goals of improved reliability, profitability, and environmental stewardship, highlighting lubrication as not merely a maintenance requirement but an essential pillar of competitive success in the demanding construction industry.

MINING

Lubricants are vital for the operation of mining equipment and machinery in harsh environments. At a large open-pit mining operation, critical excavation equipment experienced recurring mechanical failures, causing unplanned downtime that dramatically impacted productivity and profitability. Each equipment outage cost the company roughly $50,000 per hour due to halted production, costly repairs, and labor idle time. The severe and abrasive operating environment exacerbated lubricant contamination and accelerated component wear, raising serious compliance and environmental concerns around waste management and emissions.

Facing these urgent challenges, the reliability and maintenance teams launched a detailed assessment of lubricant performance and practices. Early analyses revealed that existing lubricants quickly lost effectiveness under harsh mining conditions, leading to accelerated contamination from dust, moisture, and extreme temperatures. The rapid lubricant deterioration caused significant wear on gears, bearings, and hydraulic components, precipitating frequent breakdowns.

Responding proactively, the reliability group implemented a comprehensive lubrication improvement initiative. They transitioned to specialized, high-performance synthetic lubricants engineered specifically to resist contamination, maintain viscosity stability, and provide exceptional protection in extreme mining environments. Alongside lubricant upgrades, they established an advanced lubricant condition monitoring program, combining regular on-site sampling, real-time sensor data, and off-site laboratory analysis.

Within just 4 months, these improvements markedly transformed operational outcomes. Equipment downtime incidents decreased by nearly 70%, directly saving the mining operation over $1 million annually in avoided downtime costs. Machinery life expectancy improved significantly, with key components demonstrating approximately 50% extended service intervals, substantially lowering replacement expenses and maintenance frequency. Enhanced lubricant selection and management also reduced lubricant waste disposal by roughly 35%, aligning with strict environmental compliance and sustainability objectives.

Integral to achieving these results was the close collaboration among the reliability engineers, equipment operators, maintenance technicians, and environmental compliance specialists. Structured training and documented processes embedded proactive lubrication practices deeply into operational routines, shifting the site's maintenance culture decisively toward preventive and predictive strategies.

Ultimately, by embracing strategic lubrication management, the mining operation significantly enhanced equipment reliability, profitability, and sustainability performance, clearly demonstrating that effective lubrication management is a critical lever for operational excellence in the rigorous mining industry.

AGRICULTURE

Agricultural machinery, such as tractors and harvesters, requires lubricants for effective operation. On a large agricultural operation cultivating vast fields of grain and corn, unexpected breakdowns in critical harvesting and planting machinery began to sharply rise during peak seasons. Each incident of downtime carried substantial financial repercussions, estimated at approximately $10,000 per hour, due to halted field activities, delayed harvests, and subsequent crop losses. Furthermore, these frequent equipment failures posed serious compliance and environmental risks associated with increased lubricant disposal, soil contamination, and fuel consumption inefficiencies.

An in-depth assessment conducted by the farm's reliability and maintenance teams identified lubrication practices as the central issue. Field analyses revealed rapid lubricant contamination from dirt, moisture, and harsh working conditions typical of agricultural environments. Additionally, conventional lubricants used in tractors and harvesters were deteriorating quickly, causing accelerated wear in bearings, transmissions, and hydraulic systems, thereby shortening machinery life and reliability.

In response, the reliability specialists implemented a robust lubrication management program. They transitioned the machinery fleet to advanced agricultural-grade synthetic lubricants specifically engineered to resist environmental contamination, sustain viscosity under high-stress conditions, and deliver consistent performance. To ensure ongoing lubricant effectiveness, they also adopted a comprehensive lubricant monitoring program featuring regular oil sampling, rapid on-site testing, and condition-based sensor analytics.

Within one harvest cycle, these improved practices yielded significant operational enhancements. Downtime events were reduced by more than 60%, saving approximately $300,000 in direct operational costs annually. Equipment lifespan notably improved, with key components extending service intervals by 40%, substantially reducing maintenance expenses and parts replacements. Lubricant consumption dropped by 30%, significantly reducing waste generation and environmental impacts, strongly aligning with the farm's sustainability and ESG commitments.

Central to these improvements was the close collaboration among the reliability engineers, farm mechanics, and machinery operators. Structured training, standardized maintenance protocols, and routine condition monitoring fostered a proactive maintenance culture that reinforced machinery reliability.

Ultimately, the targeted lubrication initiative directly supported the agricultural operation's strategic objectives, ensuring equipment reliability, enhancing profitability through cost-effective operations, and advancing sustainability practices. The experience underscored the premise that effective lubrication management in agriculture is more than routine maintenance; it is an essential foundation for operational efficiency, environmental stewardship, and lasting competitiveness.

FOOD AND BEVERAGE

Food-grade lubricants are essential for machinery in processing and packaging to meet hygiene standards. In a major food and beverage processing facility specializing in dairy products, recurring downtime events plagued the high-speed packaging line, significantly impacting productivity and profitability. Each unexpected stoppage cost the plant approximately $15,000 per hour due to halted production, spoiled inventory, and increased maintenance expenses. Moreover, the frequent mechanical breakdowns heightened compliance risks related to stringent food safety standards and regulatory hygiene requirements.

Upon investigation, the reliability and quality assurance teams quickly pinpointed lubricant selection and handling as the underlying issues. The lubricants in use, although initially classified as food grade, were not robust enough to endure the constant washdowns, high humidity, and temperature fluctuations characteristic of dairy production lines. Lubricant degradation and contamination were accelerating wear and tear on critical conveyor systems, pumps, and filling machinery, compromising equipment reliability and hygienic integrity.

Responding decisively, the team introduced a comprehensive lubrication management program specifically designed for the rigorous demands of food processing. They transitioned to advanced, high-performance NSF H1-certified synthetic lubricants engineered for enhanced resistance to water ingress, microbial growth, and thermal breakdown. To complement this, rigorous lubricant condition monitoring was implemented, employing routine sampling, rapid analytical testing, and real-time sensors to detect early signs of lubricant deterioration or contamination.

Within 3 months, the enhanced lubrication practices delivered remarkable improvements. Unplanned downtime decreased by over 75%, translating into annual savings exceeding $400,000. Equipment lifespan extended by approximately 50%, significantly reducing maintenance costs and downtime associated with replacements and overhauls. Lubricant waste and consumption declined by nearly 30%, positively impacting environmental goals and bolstering ESG commitments.

Critical to the program's success was the close collaboration among the reliability specialists, maintenance technicians, quality assurance personnel, and frontline operators. Detailed training sessions and clearly documented protocols ensured consistent, proactive lubricant management embedded firmly into daily operations.

Ultimately, these improved lubrication practices directly supported the facility's primary organizational objectives, maximizing equipment reliability, ensuring food safety compliance, and improving operational profitability. This initiative underscored the idea that strategic lubrication management is not merely a routine maintenance function but an essential component of competitive advantage and sustainable operational excellence in the food and beverage industry.

PULP AND PAPER

The production of paper products involves lubricants in machinery and equipment for smooth operation. In a high-capacity pulp and paper manufacturing facility, critical papermaking machines faced escalating downtime caused by frequent bearing failures, gear breakdowns, and overheated rollers. Each outage carried substantial financial implications, averaging around $25,000 per hour in lost production, urgent repairs, and delayed orders. Further complicating matters, repeated failures raised serious compliance risks associated with increased waste disposal and environmental contamination and with failing to meet strict regulatory emissions requirements.

An immediate investigation by the reliability engineering team identified inadequate lubrication practices as the primary driver of equipment degradation. Initial analyses revealed that the existing lubricant choices could not withstand the facility's humid environment, extreme pressures, and continuous operation cycles. The lubricants rapidly broke down, became contaminated, and failed to provide the necessary protective film thickness, resulting in accelerated machinery wear.

Addressing this, the team swiftly transitioned to specialized high-performance synthetic lubricants designed explicitly for pulp and paper operations. These lubricants offered superior resistance to moisture, heat, and contamination, significantly reducing friction and extending component life. Concurrently, the reliability group introduced a structured lubricant condition monitoring program involving routine sampling, detailed laboratory analysis, and in-line sensors for real-time lubricant condition assessment.

Within just a few months, the improvements were striking. Equipment downtime dropped by more than 65%, representing approximately $500,000 annually in cost avoidance from minimized disruptions and maintenance expenses. The lifespan of critical components increased notably; bearings and gearboxes showed nearly 40% longer intervals between replacements, considerably lowering operational costs. Enhanced lubricant management also reduced consumption and waste disposal volumes by roughly 30%, positively impacting the facility's environmental goals and aligning with sustainability targets.

The successful outcome hinged on the close collaboration between the reliability specialists, machine operators, and maintenance technicians. Comprehensive training, detailed maintenance protocols, and ongoing lubricant condition monitoring became integral parts of the daily operational culture, ensuring proactive management of lubrication practices.

Ultimately, the enhanced lubrication strategy significantly advanced the pulp and paper plant's key organizational goals: improving operational reliability, profitability, and environmental stewardship. This experience clearly demonstrated that proactive and strategic lubrication management is a cornerstone of sustainable operational excellence in the pulp and paper industry.

CONCLUDING REMARKS

Across diverse industries including automotive, manufacturing, aerospace, marine, construction, mining, agriculture, food and beverage, and pulp and paper, effective lubrication practices consistently emerged as essential for operational reliability, profitability, and environmental sustainability. Each sector initially faced significant challenges characterized by unexpected downtime, escalating maintenance costs, reduced component life, and heightened compliance risks due to improper lubricant selection, management, and contamination.

In each scenario, a structured approach combining advanced lubricant selection, tailored specifically to harsh operational conditions, and comprehensive lubricant condition monitoring practices significantly mitigated these challenges. Real-time sensors, routine oil analyses, and predictive maintenance techniques enabled proactive interventions, resulting in substantial reductions in downtime, often ranging from 60% to over 80%, alongside notable cost savings.

Extended equipment life averaging around 40–50% further amplified financial benefits, substantially reducing the frequency of component replacements and associated labor. Concurrently, improved lubrication strategies consistently reduced lubricant consumption and waste by approximately 30–35%, directly supporting organizational environmental, social, and governance objectives.

Critical to this success was the collaborative synergy of the reliability engineers, maintenance personnel, machine operators, and quality assurance specialists. Structured training and clear operational procedures embedded sustainable lubrication practices deep into daily routines, cultivating a shift from reactive maintenance to predictive and proactive management.

Ultimately, these case studies clearly demonstrate that industrial lubrication is far more than a maintenance necessity. It is a fundamental pillar of operational excellence, competitive advantage, sustainability, and organizational profitability across industries.

INDEX

ABOUT THE AUTHOR

Michael D. Holloway is president of 5th Order Industry (https://5thorderindustry.com/leadership), which provides competency development and training programs for many industries offering standard, customized, and certification preparation courses. He has over 38 years of experience in industry starting with research and development for Olin Chemical and WR Grace, product management for Rohm & Haas, technical marketing and application engineering for GE Plastics, and technical development and management for NCH, ALS, and SGS. He is a subject matter expert in tribology, failure analysis, reliability engineering, and designed experiments for science and engineering. He holds 16 professional certifications, a patent, a Master of Science in Polymer Engineering from the University of Massachusetts, a Bachelor of Science in Chemistry and a Bachelor or Arts in Philosophy from Salve Regina University, Newport RI, and has authored 15 books, contributed to several others, and has been cited in over 1,000 manuscripts and several hundred master's theses and doctoral dissertations.